TOWARDS GLOBAL OPTIMISATION 2

Editors

L.C.W. DIXON

Senior Research Fellow
The Numerical Optimisation Centre
The Hatfield Polytechnic
Hatfield, U.K.

and

G.P. SZEGÖ

Rector
University of Bergamo
Bergamo, Italy

1978

NORTH-HOLLAND PUBLISHING COMPANY
AMSTERDAM · NEW YORK · OXFORD

ISBN 0 444 85171 2

Publishers:

NORTH-HOLLAND PUBLISHING COMPANY
AMSTERDAM • NEW YORK • OXFORD

Sole distributors for the U.S.A. and Canada:
ELSEVIER NORTH-HOLLAND, INC.
52 VANDERBILT AVENUE,
NEW YORK, N.Y. 10017

Library of Congress Cataloging in Publication Data

Main entry under title:

Towards global optimisation 2.

A collection of proceedings from two workshops held in Varenna, Italy, in June 1976 and at the University of Bergamo, Italy, in July 1977, and various other research papers.
Includes index.
1. Mathematical optimization--Congresses.
I. Dixon, Laurence Charles Ward. II. Szegö, G. P.
QA402.5.T72 519.4 78-6493
ISBN 0-444-85171-2

PRINTED IN THE NETHERLANDS

CONTENTS

PREFACE

This volume is devoted to the presentation of recent results obtained in the area of global minimisation routines, both constrained as well as unconstrained and in the related problem of local optimisation routines. The book has practically the same format as the preceding volume "Towards Global Optimisation" published in the same series in 1975. This volume contains among other works, the final results obtained by the joint research programme on Global Optimisation sponsored by C.N.R. at the University of Bergamo, Italy and by S.R.C. at the Numerical Optimisation Centre (N.O.C.), The Hatfield Polytechnic, Hatfield, U.K.

Some of the results which appear in this volume were presented in two workshops held at the Villa Monastero, Varenna, Italy in June 1976 and at the Centro Ricerche per l'ottimizasione Numerica (C.R.O.N.), The University of Bergamo, Italy in July 1977.

It may be interesting for the readers to compare the results and the point of view presented in this book with the one expressed in the preceding volume in order to appraise the actual development in the area which has taken place in the last three years.

The papers in this volume are divided into three parts preceded by an introductory paper. The first part is devoted to unconstrained global optimisation problems and contains twelve papers. The authors of papers describing particular algorithms were required to run a series of tests on a number of standard functions, and to express the results obtained both in terms of number of function evaluations needed as well as of standard times required.

A comparison of the results obtained by the various authors is performed in the introductory paper on "The Global Optimisation Problem: An Introduction" by the editors.

The second part of the book is devoted to local optimisation algorithms and contains seven papers. Many deterministic global minimisation techniques use a local minimisation algorithm as a subroutine; the efficiency of local minimisation routines is therefore of utmost importance for the practical application of global minimisation algorithms.

The third and last part of the volume consists of four papers concerned with some new results on constrained optimisation.

More detailed comments on all papers contained in this volume, as well as the comparison of results are presented in the introductory paper.

January 1978. L.C.W. Dixon & G.P. Szegö.

TOWARDS GLOBAL OPTIMISATION 2
L.C.W. Dixon and G.P. Szegö (eds.)

The Global Optimisation Problem: An Introduction

by

L.C.W. Dixon*

and

G.P. Szegö**

Abstract

In this paper we introduce the papers of the volume paying special emphasis on the philosophy of the standard set of computational tests. A summary of the performance of the various algorithms in these tests is also given.

1. Practical Applications

Many important practical problems can be posed as mathematical programming problems. This has been internationally appreciated since 1944 and has led to major research activities in many countries, in all of which the aim has been to write efficient computer programs to solve subclasses of this problem. An important subclass that has proved very difficult to solve occurs in many practical engineering applications. Let us consider the design of a system that has to meet certain design criterion. The system will include features that may be varied by the designer within certain limits. The values given to these features will be the optimisation variables of the problem. Frequently when the system performance is expressed as a mathematical function of the optimisation variables, this function, which will sometimes be called the objective function, is not convex and possesses more than one local minimum. The problem of writing computer algorithms that distinguish between these local minima and locate the best local minimum is known as the global optimisation problem, and is the subject of this volume.

To emphasize the importance of having an efficient solution to this problem, we note that the different local minima can often correspond to different technologies. Finding the nearest local minimum to the current design is then equivalent to improving current technology and this can frequently only correspond to obtaining an improvement in performance of a few per cent. Moving to a different local minimum, i.e. a different technology, can imply a very significant change in performance.

* The Numerical Optimisation Centre, The Hatfield Polytechnic, England.

** The University Institute of Bergamo, Italy.

The difficulty inherent in such a statement is obvious in that it is difficult to model a new technology before it is implemented and that considerable faith is normally required to attempt to implement an unexpected type of solution to the problem.

One case in which the model of the system could be trusted as it was based on well established physical relations occurred in the design of optical filters. In this problem an examination of the objective function showed that it had many local minima and that the best did not correspond to a local improvement of the current design. In McKeown & Nag (1976) it is reported that when the new design was implemented the predicted improvement in performance occurred.

Other areas where the same phonomenon is well known occur in the design of electrical fitters (Gutteridge (1972)), fitting experimental data (Amadori, Mika, and V. Studnitz (1976)), and system identification (Soderström (1974)).

A much less well known application of these techniques to the problem of the optimal design of a dry cooling tower for a thermal power plant is described in the paper by Archetti & Frontini. In this paper they also discuss some larger optical filter design problems.

Multi extremal problems also arise in the econometric sphere and the application of one of the global optimisation techniques to such a problem is described in the second paper by J. Gomulka. K. Cichocki in a paper presented at the workshop in Bergamo (1977) develops an identification model of the Polish economy which also has multiple minima.

2. A standard test series for comparing results

When the first volume of this series was published in 1975 it became apparent that an acceptable comparison was needed showing the relative performance of the many different methods that had been proposed for solving multiextremal problems. Most authors when presenting their method for publication include the results of tests on a number of standard functions. As these tests are frequently performed on differing machines, with different accuracy and termination requirements and also on a different set of functions, it is very difficult to draw meaningful conclusions on the relative efficiency of the different methods. In the field of unconstrained optimisation it is normal to base efficiency on the number of function (and/or gradient) evaluations required to minimise a set of nearly universally agreed test functions. This approach ignores the relative overhead cost of the different methods completely, on the basis that for small dimensional practical problems these are indeed usually ignorable. In contrast in the field of nonlinear programming Colville (1968) proposed a set of test functions as well as a standard time test, he compared the algorithms both on the basis of the relative numbers of times the correct solution was located and also computed the relationship between the total computer time and the standard time. This method emphasises the relative overhead cost as the cost of a function evaluation of his test functions are often trivial compared to industrial problems.

In designing a test series for multiextremal functions it was decided to combine both criteria. The field of multiextremal problems was restricted to the much narrower field of essentially unconstrained problems in two to six dimensions. A subset of the standard test functions quoted in the literature was then selected. The functions chosen were all relatively cheap to evaluate so that a considerable proportion of the run time could be expected to be spent on algorithmic overheads. A standard computing problem was introduced on which the comparison study would be based. This problem was simply the evaluation of one of the test problems 1000 times.

Each of the contributors to the main section of this volume agreed to present their results on the standard set of function in the proposed format. The inclusion of the standard time test was, however, a late decision and unfortunately this implied that in one case the tests had already been completed and the times had not been recorded, so in that case the time comparison could not be made.

The details of the test series are given in Appendix 1 so that they will be available for the future use of other research workers in the area. The detailed results of each individual test series are given in each authors' contribution; a summary of the results is however given later in this paper.

3. The two basic approaches

Algorithms for the solution of the global optimisation problem divide into two broad classes: deterministic and probabilistic. Even if the basic philosophies behind the two approaches are quite different in most advanced global optimisation algorithms the two basic approaches are combined.

A complete discussion of the relationships between the two classes of algorithms is presented in the first paper by Gomulka.

It must also be pointed out that certain algorithms are designed for the identification of the global minimiser of the function as well as of the absolute minimum function value, others are trying only to identify the absolute minimum function value.

Purely deterministic methods try to guarantee that a neighbourhood of the global minimum will be located. It is now established that this is only possible for restricted classes of function. Well known examples for which deterministic methods can be fruitfully used include convex functions, and one dimensional rational or Lipschitz functions and polynomials. For n-dimensional Lipschitzian functions, whose Lipschitz constant is known, bounds on the possible error of the grid search technique have, of course, been known for a long time.

The grid search technique is an example of a "passive strategy" as defined by Archetti and Betro, i.e. an algorithm where each point of the minimising sequence does not depend on the function values at the previous points. In their paper they show that for the class of Lipschitzian functions two classes of optimal passive strategies exist. They further show that no sequential strategy i.e. no strategy where each point of the sequence depends on the function values at previous points, can guarantee a better value than a particular type of optimal passive strategy with the same number of points, they call this strategy the A-optimal passive strategy. They also prove that no sequential strategy can guarantee any prefixed accuracy with less points than those required by another particular type of optimal passive strategy. It follows from this result that the use of deterministic sequential strategies can only pay when the class of objective functions is more narrowly chosen and the additional properties actually used in the minimisation procedure. Clearly nothing can be said in this respect for functions for which the Lipschitz constant is not known.

A large number of deterministic methods have been investigated in this volume; in her third contribution J. Gomulka describes her experience in implementing two versions of Branin's trajectory method. Though her implementations are quite efficient and often locate the neighbourhood of the global minimum, no subclass of nonconvex functions has been singled out for which success is guaranteed.

In Szegö (1972) Treccani, Trabattoni and Szegö introduce a method for identifying one of the saddle points which must exist on the boundary of the region of attraction of a local minimum by growing a family of ellipses completely around the minimum and completely contained in its region of attraction. This technique requires certain strong assumptions on the objective function, but the convergence of the algorithm is proved. One can use this technique to construct a global minimisation algorithm which allows the identification of the global minimum through the search of all local minima of the function. An implementation of this method for two dimensional problems was described by Corles (1975*) and considerable success was achieved even if the simplifications in the original technique introduced by Corles make a general convergence proof impossible. A closer investigation of the numerical technique used discloses that, though not explicitly stated, major smoothness assumptions on the functions were implied in the implementation of the algorithm. In addition even on functions, on which the method is able to identify the global minimiser and the absolute minimum function value and for which convergence is guaranteed, the difficulties of implementing it in higher dimensions seem to be too great to be worth the effort.

Goldstein & Price (1972) introduced a technique with guaranteed convergence on one dimensional polynomial functions, based upon the idea of descent from local minimum i.e. upon the idea of analising only that area corresponding to function values which are less than a given function value (for example a given local minimum). Szegö (1975) independently proposed a similar technique with the name of global descent method without however giving any numerical results. Treccani extends this approach to more general functions and provides counterexamples of this general idea. Along the same line of thought we should mention the numerical results presented by Levy (1977).

Beale and Forrest describe how the global minimisation problem for functions which satisfy certain assumptions (which are satisfied by all the functions in the comparison tests) can be transformed into a mathematical programming problem. Dependent on the size of the mesh used in the transformation their method is guaranteed to locate a point whose function value is in the neighbourhood of the absolute minimum.

We may conclude by pointing out that for practical purposes purely deterministic methods present considerable difficulties unless some strict assumptions on the objective function can be made and used.

4. Probabilistic Methods

The probabilistic approach to the global optimisation problem relies on the following result.

Assume the area of interest in the global optimisation problem is the finite set S. In the test series this was defined to be a hypercube defined by upper and lower bounds on each variable

$$S = \{\underline{x} : l_i \leq x_i \leq u_i , \quad i = 1, \ldots, n\}.$$

If A is any subset S with measure m and

$$\frac{m(A)}{m(S)} \geq \alpha > 0$$

and if p(A,N) is the probability that at least one point of a sequence of N points, drawn randomly from a uniform distribution over S, falls in A

then $$\lim_{N\to\infty} p(A,N) = 1 .$$

The simplest probabilistic algorithm is the Pure Random Search Algorith - P.R.S.
STEP 1 Evaluate F(x) at N points drawn from a uniform distribution over S. This algorithm has been extensively discussed by Brooks (1958) and Anderssen (1972) who show that if

$$\frac{m(A)}{m(S)} = \alpha$$

then $$p(A_1\ N) = 1 - (1-\alpha)^N.$$

More properties of this method which is also known as the Monte Carlo Method are described in Rubinstein & Weissman (1977).

The number of trial points N required to have a high probability of entering a small region, $\alpha = .0001$ say, round the global minimum, is very high. For this reason most probabilistic methods, combine this type of result with deterministic methods.

A very popular method is the multistart method M.S.

STEP 1 Select $x^{(o)}$ at random

STEP 2 start a local minimisation algorithm from $x^{(o)}$ with stopping criteria ϵg

STEP 3 test whether this is probably a global minimum and if so STOP

STEP 4 return to step 1.

The behaviour of local minimisation algorithms used in this way and the conditions that must be built into them for theoretical results to be possible are discussed in detail in Dixon et al (1976). In Dixon (1977) it is also shown that modifications typified by Hartman's S2 algorithm (1972) can lessen the probability of success on particular types of function.

Hartman's S2 algorithm consists of five steps

STEP 1 $v^+ = +\infty$

STEP 2 select $x^{(o)}$ at random

STEP 3 if $F(x^{(o)}) > v^+$ return to 2

STEP 4 perform a local minimisation from $x^{(o)}$ to M_j and set $v^+ = F(M_j)$

STEP 5 return to 2.

In this algorithm an additional local search is only initiated if a point is found randomly that is better than the current local minimum. On many functions, including some in the test series, this is very unlikely. An alternative strategy to this is to locate a local minimum and to then seek an improved point by performing a number of line searches out from this point. These line searches can be performed with varying degrees of accuracy. One single completely random step is equivalent to Hartman's S2 method. Other versions that were tested included that suggested by Bremmermann (1972) of fitting a quartic through 5 equispaced points along the line, where the step used is random, and also the powerful one dimensional implementation due to Zilinskas of the Bayesian method described in Motskus et al. This method was found in tests at Hatfield to be an exceedingly reliable method for finding the global minimum along the line. However, on some of the test functions the relative size of the angle subtended on a hypersphere round a local minimum by its level set was so small, that no improved point was found in 2000 linear searches. For this reason probabilistic methods of improving on a local minimum point also seem doomed to be very inefficient on many functions.

The same test series also indicated that methods which undertake local line searches from random points, and accept the best point found along the line as the start for the next line search, frequently locate local minima very accurately. They then effectively revert to the algorithms just discussed above and have the same inherent disadvantage.

It was for these reasons amongst others that the multistart method has remained so popular. It has the further advantage that when features of the objective function are known these can be used to guide the choise of the initial selection which is rarely taken purely at random in practice. When few features of the objective function are known and random points must be taken,it can however be very expensive on function evaluations,as indicated by the results obtained by this method by Mockus et al and reported in their paper.

To overcome the high cost of the M.S. method, Becker & Lago (1970) introduced the concept of clustering into global optimisation. The basic assumption in clustering is that the points with low function values will be clustered in the neighbourhood of local minima. As further searches from points in the same neighbourhood should all lead to the same local minimum, a saving in cost can be obtained by continuing the search from a reduced number of points. Törn has devised a successful implementation of this approach based on a random search algorithm, while Price reports on an implementation based on the Nelder & Mead Simplex approach. Gomulka also reports on her experience using Törn's algorithm and on her own modification using a variable metric algorithm for the basic searches. All these algorithms are consistantly successful on the functions in the standard test series, but Gomulka found that the variable metric based algorithm could succeed on more difficult econometric functions on which the random search based algorithm failed. When running these algorithms considerable additional information about the function can be obtained which can frequently be very valuable.

Another approach to the problem of combining stochastic and deterministic methods is outlined in the paper by Fagiuoli et al. In this method the region of interest is covered by predetermined set of cells and a stochastic automaton set up to govern the choice of cell. The authors claim their method of constructing stochastic automata is more efficient than the related method by McMurty & Fu.

Essentially after a set of initial samples have been taken, a point is selected in the cell that is most likely to contain an unfound global optima. A local constrained minimum in that cell is obtained by recursive quadratic programming (described in the paper by Bartholomew-Biggs), the stochastic automata is then updated and the method repeated till the probability of further improvements is significantly reduced.

While the overheads involved in the clustering algorithms and the stochastic automata approach are obviously higher than most of the simpler strategies discussed earlier they are still insignificant when compared to even small industrial problems. As industrial problems become larger so it becomes easier to justify even higher overheads if these can reduce the number of function evaluations still further. The Bayesian approach described by Mockus et al is designed to solve problems where the cost of the function evaluation is very large, and therefore large overheads can also be ignored. As will be seen later it is very successful in this aim.

Both the stochastic automata approach and the Bayesian approach have in built statistical assumptions from which the probability that the algorithm has already located a neighbourhood of the global optimum can be obtained. These probabilities are then used as terminating criteria. In the first volume of this series Archetti proposed a sampling technique based upon an idea of Chichinadze (1967). This has been developed into an algorithm which is described in the paper by De Biase & Frontini; in this algorithm an approximation is constructed to the probability distribution of the objective function, i.e. $P(v)$ is the probability that $F(x) < v$. An estimate is then made of the root $P(v) = 0$ and considerable theoretical results are established to guarantee the uniform convergence of the approximation to $P(v)$ and hence of the ultimate accuracy of the root. An efficient local minimisation algorithm is incorporated to determine the position of the global optimum accurately. The algorithm seems quite efficient at locating the global optimum, but unfortunately the theoretical convergence of the root to the global minimum value does not seem to occur from the results reported in their paper.

5. Numerical results

The detailed numerical results of each contributor on the standard test function will, of course, be found in their papers in this volume. It did seem sensible to gather this data together for comparison purposes, and to include some data on the random line search methods that would not otherwise be included in this volume though they are reported elsewhere, Dixon (1977).

The least successful algorithms tested on the standard functions were these random line search routines, as they frequently became trapped by local minima. Results are included in the table for three variants, Bremmermann's original implementation, a slight modification that accepted the best point found along the line even if it was not the predicted minimum and a similar approach when the line search was replaced by a Bayesian method due to Zilinskas. Considerable testing showed that this was highly reliable in finding the global minimum along the line requiring about 30 function evaluations. But that from the local minimum of SQRN5 the window of the global minimum was so small that 2000 random directions did not penetrate it.

Two tables of results are given, in the first the number of function evaluations required to find the global minimum or terminate is recorded. In the second the time required, divided by the standard time of that machine, is recorded. An L indicates a local minimum was accepted. The standard time was obtained by calling the subroutine for evaluating SQRN5 1000 times at the point (4, 4, 4, 4).

Table 1 — Function Evaluations

Line search	SQRN5	SQRN7	SQRN10	HARTMAN 3	HARTMAN 6	GP	RCOS
Bremmermann	340L	1700L	2500L	505L	L	210L	250
Mod Bremmermann	375L	405L	336L	420L	515	300	160
Zilinskas	L	12121L	8892L	8641			5129
Trajectory							
Gomulka/Branin	5500	5020	4860				
Clustering							
Törn	3679	3606	3874	2584	3447	2499	1558
Gomulka/Törn	6654	6084	6144				
Gomulka/V.M.	7085	6684	7352	6766	11125	1495	1318
Price	3800	4900	4400	2400	7600	2500	1800
Sampling							
Fagiuoli	2514	2519	2518	513	2916	158	1600
De Biase/Frontini	620	788	1160	732	807	378	597
Mockus	1174	1279	1209	513	1232	362	189

The Beale-Forrest approach can not be included in this table as it does not evaluate the function.

Table 2 — Standard Time

Line search	SQRN5	SQRN7	SQRN10	HARTMAN 3	HARTMAN 6	GP	RCOS
Bremmermann	1 L	8 L	17 L	2 L	L	0.5L	1 L
Mod Bremmermann	1.5L	1.5 L	2 L	2 L	3	0.7	0.5
Zilinskas	L	282 L	214 L	175			80
Trajectory							
Gomulka/Branin	9	8.5	9.5				
Clustering							
Törn	10	13	15	8	16	4	4
Gomulka/Törn	17	15	20				
Gomulka/V.M.	19	23	23	17	48	2	3
Price	14	20	20	8	.46	3	4
Sampling							
Fagiuoli	7	9	13	5	100	0.7	5
De Biase/Frontini	23	20	30	16	21	15	14
M.P. Package							
Beale/Forrest	96	258	1059	117	255	1407	1.5

It should be stressed at this point that the Beale-Forrest method is not designed to solve this type of problem. These are all small highly nonlinear functions while it is meant for large mainly linear problems. It did, however, successfully solve all the problems, as did the clustering methods and the sampling techniques.

6. Local optimisation algorithms

It will have been noticed that most of the efficient global optimisation algorithms discussed in section 4 were based upon well known local routines. Random search routines, Nelder & Mead simplex and the variable metric method were amongst those mentioned. Improvements in the understanding of the behaviour of unconstrained algorithms and in their performance are both important to the future development of algorithms for the global optimisation problems. The second part of this volume therefore again contains selected papers on this problem.

Spedicato presents two papers, in the first he reviews the present state of quasi-Newton variable metric methods, which are generally accepted to be the most efficient class of general purpose unconstrained optimisation algorithms. In his second paper he gives details of his computational experience using these methods on moderately sized problems, thus indicating the probable choice of algorithm for using in combination with, say, clustering on problems larger than those included in our test series. Crouch's note introduces another modification to the quasi-

Newton method which maintains conjugate directions without exact line searches being undertaken. Two previous ways of achieving this have been suggested by Dixon (1973) and Nazareth (1977) and the properties of such modifications are the subject of continuing investigation. The paper by Gaviano investigates the necessary and sufficient conditions for gradient related algorithms to converge to the minima of convex functions or stationary points of nonconvex functions, whilst that by Resta & Sutti discusses this problem for more general algorithms. Both papers mark a considerable advance on the previous position in these areas. A new algorithm for seeking the minimum of a differentiable function without estimating the gradient had been presented by Carla Sutti in the first volume of this series. The convergence of this algorithm can be established by this method. Numerical results obtained with her algorithms are contained in her own paper in this volume and compare favourably with those from other related methods for which such convergence properties have not been obtained.

The final paper in part two approaches the problem of stochastic optimisation, i.e. the problem where each evaluation of the objective function is subject to random error. Few implementable algorithms have been suggested that can theoretically solve this problem and the one presented here is a major advance as the authors, Betro' & De Biase, show that it is unnecessary to estimate the gradient and Hessian independently. This had been a necessary feature of previous proofs of convergence and had made the implementations very inefficient. The prospect of being able to incorporate such an algorithm with a clustering algorithm to give a global stochastic optimisation routine is quite exciting.

7. Constrained Optimisation

There have as yet been few attempts to investigate the global constrained optimisation problem except for the introduction of lower and upper bounds on the variables. When more general equality or inequality constraints are included, modification of efficient nonlinear programming algorithms will be required. The performance of two of the most efficient recent methods of nonlinear programming are compared in the paper by Bartholomew-Biggs. The two chosen methods are the ideal penalty function algorithm developed by Fletcher (1972) and his own recursive equality quadratic programming algorithm. An analysis is given which indicates in which circumstances each algorithm should be preferred.

Another recent development in this area is introduced in McKeown's paper. The method which he terms functional linear programming was originally devised to solve a problem in structural engineering which contained many features which often prove difficult or insurmountable for more orthodox algorithms. These include variables that can only take integer values, regions in which the objective function does not exist, multiple finite minima. The problem also had the unusual feature that the total number of variables was itself a variable of the problem. The functional linear programming approach overcomes all these difficulties on the particular structural problem and on any other problem that can be posed in this form. Unexpectedly it can also be shown that the dual of any convex programming problem can be posed as a functional linear programming problem. In the paper by Resta, Sideri & Treccani a mathematical treatment is given of a convex programming method based on the approach used in functional linear programming. They have unfortunately not yet completed the convergence proof for problems in more than three dimensions. In the final paper of the volume Resta presents his numerical experience of solving convex programming problems by the functional linear programming approach. He has used two of the problems tested by Barthomolew-Biggs and the results in the two papers can be compared.

Appendix 1 Test Functions for Global Optimisation

1. Shekel's family (SQRIN)

$$f(x) = -\sum_{i=1}^{m} \frac{1}{(x - a_i)^T(x - a_i) + c_i} \qquad x = (x_1, \dots, x_n)^T$$

$$a_i = (x_{i1}, \dots, x_{in})^T$$

$$c_i > 0$$

Region of interest: $0 \leq x_j \leq 10 \quad j = 1, \dots, n.$

Consider 3 cases from the table below with

Data: $n=4$; $m=5, 7, 10$

i	a_i				c_i
1	4.	4.	4.	4.	.1
2	1.	1.	1.	1.	.2
3	8.	8.	8.	8.	.2
4	6.	6.	6.	6.	.4
5	3.	7.	3.	7.	.4
6	2.	9.	2.	9.	.6
7	5.	5.	3.	3.	.3
8	8.	1.	8.	1.	.7
9	6.	2.	6.	2.	.5
10	7.	3.6	7.	3.6	.5

2. Hartman's family

$$f(x) = - \sum_{i=1}^{m} c_i \exp\left(- \sum_{j=1}^{m} a_{ij}(x_j - p_{ij})^2\right)$$

where $x = (x_1, \ldots, x_n)$, $p_i = (p_{i1}, \ldots, p_{in})$, $a_i = (a_{i1}, \ldots, a_{in})$.
p_i is an approximate location of i'th local minimum, a_i is proportional to eigenvalues of Hessian at i'th local minimum, $c_i > 0$ is the height (depth?) of i'th local minimum (assuming that the interference of different local minima is not too strong).

Data $0 \leq x_i \leq 1$. (1) m = 4, n = 3

i	a_{ij}			c_i	p_{ij}		
1	3.	10.	30.	1.	0.3689	0.1170	0.2673
2	.1	10.	35.	1.2	0.4699	0.4387	0.7470
3	3.	10.	30.	3.	0.1091	0.8732	0.5547
4	.1	10.	35.	3.2	0.03815	0.5743	0.8828

(2) m = 4, n = 6

i	a_{ij}						c_i
1	10.	3.	17.	3.5	1.7	8.	1.
2	.05	10.	17.	.1	8.	14	1.2
3	3.	3.5	1.7	10.	17.	8.	3.
4	17.	8.	.05	10.	.1	14.	3.2

	p_{ij}					
i=1	0.1312	0.1696	0.5569	0.0124	0.8283	0.5886
i=2	0.2329	0.4135	0.8307	0.3736	0.1004	0.9991
i=3	0.2348	0.1451	0.3522	0.2883	0.3047	0.6650
i=4	0.4047	0.8828	0.8732	0.5743	0.1091	0.0381

3. Branin (RCOS)

$$f(x_1, x_2) = a\,(x_2 - b\,x_1^2 + c\,x_i - d)^2 + e\,(1-f)\cos x_1 + e$$

$$a = 1,\quad b = 5.1(4\pi^2),\quad c = 5/\pi,\quad d = 6,\quad e = 10,\quad f = 1/(8\pi)$$

Region of interest = $-5 \leq x_1 \leq 10$, $0 \leq x_2 \leq 15$

There are three minima, all global, in this region.

4. Goldstein & Price (GOLDPR)

$$f(x_1,x_2) = \left[1 + (x_1+x_2+1)^2 \; (19-14x_1+3x_1^2-14x_2+6x_1x_2+3x_2^2)\right].$$
$$\left[30 + (2x_1-3x_2)^2 \; (18-32x_1+12x_1^2+48x_2-36x_1x_2+27x_2^2)\right].$$

Region of Interest = $-2 \leq x_{1,2} \leq 2$

Four local minima. Global minima at (0, -1) with the value f=3.

5. Standard Time

Evaluate Shekel (SQRIN) n = 4, m = 5 as a subroutine. Calling the subroutine 1000 times with $x_j = 4$, on the same machine/compiler/language etc. as that used in the test.

References

References in the list below marked * refer to papers in the first volume, "Towards Global Optimisation", North-Holland (1975).

1. Amadori R., Mika K., Studnitz I.V. (1976). A comparison of the performance of unconstrained Minimisation algorithms for the solution of special nonlinear least squares problems. Kernforschungsanlage Jülich Rept 1277.
2. Anderssen R.S., (1972). In "Optimisation". Editors Anderssen R.S., Jennings L.S., & Ryan D.M., University of Queensland Press.
3. Archetti F., (1975)*. A sampling technique for global optimisation. Vol. 1 pp 158-165.
4. Becker R.W. & Lago G.V. (1970). In "Proceedings of the 8th Allerton Conference on Circuits & System Theory".
5. Bremmermann H. (1972). Mathematical Biosciences, Vol. 9., pp 1 - 15.
6. Brookes S.H. (1958). Op. Res. 6 pp 244 -251.
7. Chichinadze V.K. (1967). Eng. Cyb. 1 pp 115 - 123.
8. Cichocki K. (1977). Paper presented at the Summer School on Optimisation held at the University of Bergamo.
9. Colville A.R. (1968). A comparative study on Nonlinear Programming Codes. I.B.M. TR 320-2949.
10. Corles C.R. (1975)*. The use of regions of attraction to identify global minima. Volume 1 pp 55 - 96.
11. Cutteridge O.P.D. (1974). Powerful 2-part program for solution of nonlinear simultaneous equations. Electronic Letters Vol. 11, No. 10, pp 182 - 184.

12. Dixon L.C.W. (1973). J. Inst. Maths. Applics. Vol. 11, pp 317 - 328.
13. Dixon L.C.W. (1977). Global optimisation without convexity. Paper presented at the International Summer School on "Design and implementation of optimisation software", Urbino, Italy.
14. Fletcher R. (1973). Math. Prog. Vol. 5, No. 2, pp 129.
15. Goldstein A.A. & Price J.F. (1971). Maths of Computation, Vol. 25, pp 115.
16. Hartman J.K. (1972). Naval postgraduate School, Monteray, Rep. NP5 55HH72051A
17. Levy A.V. (1977). Paper presented at the conference on "State of the art of numerical analysis", Dundee, Scotland.
18. McMurtry G.F. & Fu K.S. (1966). I.E.E.E. Auto. Control AC-11 No. 3 pp 379-87.
19. McKeown J.J. & Nag A. (1976). An application of optimisation techniques to the design of an optical filter. In "Optimisation in Action", Editor Dixon L.C.W., Academic Press.
20. Nazareth L. (1977). A conjugate direction algorithm with line searches. J.O.T.A. (to appear).
21. Rubinstein Y & Weissman I. (1977). The Monte Carlo method for global optimisation. Op. Res., Stats & Economics - Mimeograph Series 187.
22. Soderström T. (1974). Automatica Vol. 10 pp 617 - 626.
23. Sutti C. (1975)*. A new method for unconstrained minimisation without derivatives. Vol. 1 pp 277 - 289.
24. Szegö G.P. (1972). Minimisation algorithms, Academic Press.
25. Szegö G.P. (1975). Numerical methods for global minimisation. Proc. I.F.A.C. Conference, Boston.
26. Zilinskas A. (1976). Optimisation of one dimensional multimodal functions. (Private communication).

PART ONE

GLOBAL OPTIMISATION

TOWARDS GLOBAL OPTIMISATION 2
L.C.W. Dixon and G.P. Szegö (eds.)

DETERMINISTIC VS PROBABILISTIC APPROACHES TO GLOBAL OPTIMISATION

Dr. J. Gomulka
The Numerical Optimisation Centre
The Hatfield Polytechnic
Hertfordshire, England

This paper discusses the experience gained during a three year investigation of the global optimisation problem. The main theme of this report is the considerable advantage held by certain probabilistic approaches over the deterministic approach to this problem.

1. INTRODUCTION

The aim of the project was to study possible ways of devising an algorithm for finding the global as opposed to a local minimum of fairly general functions. The demand was that an algorithm be useful in at least some practical situations, where there is little or no "a priori" information on the possible number and location of local minima.

From the beginning the main effort was concentrated on a search for a deterministic algorithm.

Leaving aside methods devised for separable and other special classes of functions, mainly in the context of mathematical programming, there were two kinds of algorithms proposed for essentially unconstrained global minimisation under mild assumptions, namely grid methods and trajectory methods.

2. GRID METHODS

Grid methods were explored in Evtushenko (1970) and Shubert (1972) for functions satisfying a Lipschitz condition (or a similar condition limiting local rate of change of function values). The algorithms proposed as the result of this approach have two major drawbacks. Firstly, the number of grid points, which is equal to number of function evaluations, grows very rapidly with the dimension of the problem making the algorithms impractical for functions of 3 or perhaps even 2 variables. Secondly, the Lipschitz constant of a minimised function has to be known or estimated before starting the minimisation. The number of grid points is, roughly, proportional to L^N, where L is the Lipschitz constant of the function and N the number of independent variables. Hence overestimating L raises the cost considerably. On the other hand obtaining a close estimate of L could in most practical cases pose a problem comparable in difficulty with the original one.

It was not apparent how any of these obstacles could be overcome either by improvements in the original algorithms or by substituting a different class of functions, more restrictive than Lipschitzian but still broad enough for applications. If the demand is that the value and position of global minima be determined up to a given accuracy, with certainty, then it seems unavoidable that the distance between consecutive grid points be related to the accuracy and to the maximal possible local rate of change of a minimised function. Estimates of a bound on the rate of change may vary according to a class of function and other factors. But invariably they contain some constants depending on a particular function and region, like the Lipschitz constant of a function itself or its derivatives. These constants can sometimes be calculated analytically, for functions given by simple analytic expressions. In most practical problems one would either have to rely on "physical intuition" or estimate relevant constants by statistical techniques. But then one could hardly claim that the algorithm as a whole is deterministic, i.e. is certain to give the correct solution. It might be easier to estimate the global minimum itself by statistical techniques.

3. TRAJECTORY METHODS

Another attempt at a deterministic method was made by Branin (1971) - (1972). It looked more hopeful. In fact most of the time and effort, especially during the first two years of this project, went into developing the approach.

The method is based on the properties of the following equation:

$$g(x) = k\, g_o, \tag{1}$$

where g is the gradient of an objective function f and k is a scalar. To be more precise, let x_o be a non-stationary point in the domain D of f and let $g_o = g(x_o)$. Points in D at which gradient is parallel to g_o satisfy (1), for different values of k. This set of points will be called here a curve of constant gradient direction and denoted by C_{go} or C_{xo}. Whatever g_o, every stationary point x* belongs to C_{go}, since it satisfies (1) with k = 0. Moreover, if the hessian at x* is nonsingular, then C_{go} actually passes through x* in the sense that the part of it in the neighbourhood of x* is an arc containing x*. In fact C_{go} is, at each of its nonsingular points x, tangent to the Newton-Raphson vector -Hg, where H is the hessian matrix at x. The idea of a trajectory method is to start at an arbitrary point x_o in D and follow an arc of C_{go} in the hope of discovering stationary points, which all belong to C_{go}. It can be shown that this continuation process can lead either to the boundary of D or return to x_o along a closed curve. A detailed discussion of this and other properties of equation (1) is given in Gomulka (1977).

In that paper two algorithms are described for tracing an arc of C_{go} and marking stationary points along it, and results obtained with these algorithms on several test functions are presented.

One of the algorithms is based on integrating an ordinary differential equation

$$\dot{x} = \pm H^{-1} g$$

(H^{-1} denotes here the inverse hessian of f), whose trajectories coincide with the curves of constant gradient direction. The other uses an extrapolation technique to generate the same solutions. Both contain a correcting loop based on the relation

$$\Delta x = H^{-1} \Delta g,$$

which ensures that the direction of the gradient remains constant, thus avoiding an accumulation of error and a gradual deviation from the original trajectory, which is a curse of most integration processes. The second algorithm seems to be quite efficient and although there is still much scope for improvements, enough has been done to show that the problem of finding all stationary points lying on any connected arc of C_{go} can be solved.

However, in general C_{go} consists of several disconnected branches, each of which is either an arc with its ends on the boundary of D or a closed curve contained in D. To be sure of finding the global minimum one would have to search all such branches. Investigating one of them amounts in fact to a local search, more extensive than a simple local minimisation, because several - and occasionally even all - local minima can be found, but this still gives no guarantee of locating the global minimum.

A trajectory algorithm can be extended to include so called "secondary searches" Hardy (1975). As mentioned before, at each stationary point of f all curves C_g (with different g) intersect. It is therefore possible to take any stationary point x* found on C_{go} as an initial point of an arc of C_{g1}, $g_1 \neq g_o$, and with luck, find some new stationary point on it. Different criteria could be used for choosing x* and g_1. For example, x* could be a saddle point and C_{g1} a trajectory going in the direction orthogonal to C_{go} and giving maximal (local) decrease of f compatible with the orthogonality constraints.

In actual tests on functions of 2 variables he undertook secondary searches from all stationary points, in the direction orthogonal to the original one. The search was ended when no new stationary points were discovered on any additional trajectory.

Perhaps predictably, the tests showed that by incorporating secondary searches the chance of finding all local minima was greatly increased. However, the stopping criterion mentioned above is not satisfactory. It is not difficult to construct examples in which, according to this criterion, the search should be terminated although the global minimum had not yet been found. In his reported tests the global optimum of one function was never located.

In the context of the trajectory method the question of a conclusive stopping criterion is really most important. Trajectory algorithms are rather expensive. Moreover, at present they rely on the availability of analytic second derivatives which is a serious drawback in practical applications. It is true that they are in a relatively early stage of development and probably can be very much improved. The effect of numerical derivatives could be tried almost immediately. Also, as in the variable metric local optimisation algorithms, the hessian could be updated instead of computed from scratch at every iteration.

However, some degree of inefficiency is built into the basic idea of the trajectory method. We have to follow an arc of the curve C_{go} of constant gradient direction with enough accuracy to prevent the search missing a stationary point lying on it or, worse, jumping inadvertently onto another branch of C_{go} and thus missing a whole part of the initial arc. These accuracy demands themselves necessitate the calculation of a certain number of points along a trajectory, whatever the algorithm used. This minimal necessary number of function evaluations is influenced by the vagaries of a particular trajectory, which could not be predicted when the choice of the initial point was made. Thus we may be forced to compute many points with high function values and no direct interest to the minimisation problem only because the trajectory happens to take a long roundabout route or because the necessary accuracy is difficult to maintain. This may be caused by many reasons including a badly conditioned hessian. In local minimisation the objective of each iteration is to find a point with a smaller function value, so that each point is "better" than the previous one. A trajectory algorithm is not guided by the function value and no such systematic improvement occurs. If for any reason it has to be stopped in the middle, then the information obtained until that moment may well be quite useless.

All these factors seem to indicate that whatever the merits of a particular implementation, a trajectory algorithm with secondary searches will on average demand more function evaluations than a few runs of an efficient local optimiser from several points in the region. This simple approach is currently the most popular form of "global" optimisation. Therefore the usefulness, if any, of a trajectory method must lie in its ability to provide a definitive answer, an assurance that the best point obtained is indeed the global minimum.

4. APPLICATION OF MORSE'S THEORY

In this direction only a minor success has been achieved. Namely, it has been noticed that one of the Morse theorems on critical points (Morse-C) can serve as a necessary condition for a global minimum, in connection with a trajectory algorithm or any other method which aims to find all stationary points in a given region D. Let us recall that the index of a stationary point is the number of negative eigenvalues of the hessian matrix and let m_k be the number of stationary points with index k present in D (e.g. m_o is the number of local minima in D). The theorem in question sets out conditions which have to be satisfied by the number m_k, $k = 0, 1, \ldots$.

For each stationary point found on the initial and all secondary trajectories the index can be calculated and then the Morse conditions can easily be checked. If they are not satisfied, then clearly all the stationary points in D have not been discovered and a further search is needed. However, the fulfillment of the Morse conditions does not mean that no undiscovered stationary points exist. We have here only a necessary condition for stopping and not a sufficient one.

The theorem can be applied in the simple way described above only if everywhere on the boundary of D the gradient points away from D. This restriction can be removed and an almost arbitrary behaviour of the function on the boundary of D can be accommodated, though at a rather considerable cost.

The main weakness of the stopping criterion derived from the Morse theory, namely that it constitutes only a necessary condition and not a sufficient one, is of a more fundamental character and cannot be overcome without reformulating the whole problem. The reason for that lies in a simple and well known fact, that for any differentiable function f a new local minimum of an arbitrary depth can be created by only changing f on a given arbitrary subregion N. A new modified function can enjoy the same differentiability properties which f had. If a new minimum is deep enough, it will be a global minimum of the modified function. On the other hand the difference between the new and old function is absolutely undetectable from outside N. This shows that a global minimum can be completely invisible from outside a certain, perhaps very small, subregion N.

One can try to resist these pessimistic conclusions by arguing that practical functions do not jump suddenly into deep narrow holes and even if they sometimes did, then a minimum of this type, being extremely sensitive to small changes in variables, would be of no practical significance. It is therefore quite irrelevant for application purposes whether an algorithm can discover such minimum or not.

This is a valid point, although it does not by itself resolve the difficulty. But it makes clear the fact, that a problem of numerical global optimisation not only cannot be solved (in a deterministic sense), but cannot even be properly stated without some problem-dependent constants. To make use of practical remedies suggested above one would need a more exact description of those insignificant local minima, which are to be excluded from consideration. Does "insignificant" in this context mean "with large eigenvalues of hessian"? Or "with a small region of attraction"? Is a region of attraction to be small in volume, or in diameter, or in some other sense? Not only must these general questions be answered, but perhaps more importantly, for any considered objective function all "large" and "small" constants have to be given numerical values.

Let us suppose for the sake of argument that this can be done, so that the problem "to find a global minimum which is not insignificant" (or alternatively, "to find a global minimum assuming that an objective function has no insignificant local minima") can be well defined. To ensure that the solution of this problem is obtained without imposing further assumptions, the network of trajectories would have to be so dense that no subregion of "significant" size could be squeezed in between them. In other words, every significant subregion would have to contain at least one calculated point of a trajectory. To obtain a less stringent condition, restrictions on the changeability of an objective function would have to be made (Lipschitz condition, bounds on derivations or their combinations etc.).

It certainly follows that the function-dependent constants, which were among the strongest reasons for abandoning grid algorithms, have caught up with trajectory methods. In fact when trajectory methods are treated as strictly deterministic, they exhibit remarkable similarities to the grid methods. In both cases the stopping criterion involves a sufficiently dense covering of the whole region by points of function (or gradient) evaluation. Moreover, the density that is sufficient, is determined by two factors: namely the accuracy demanded and the constants characterising an objective function. In the case of grid methods accuracy is expressed explicitly as the maximal acceptable error in the computed value of global minimum. In the trajectory approach it is implicit in the definition of an "insignificant minimum", which, as was argued before, is in this framework a necessary part of an optimisation problem.

5. GENERAL COMMENTS ON DETERMINISTIC METHODS

Grid methods and trajectory methods, which at first glance look like completely different approaches to global optimisation, after more careful analysis appear to have a common underlying pattern. Looking closer at the argument used to establish this pattern for trajectory methods, at its relative generality and independence of special features of the trajectory approach, one is strongly inclined to believe, that it could be applied with only minor changes to any other deterministic method of global optimisation.

If this is true, then the case against deterministic global optimisers in general is in fact as strong as the case against grid methods: the same reasons apply.

To sum up, our investigation of the trajectory approach led to the following conclusions. A trajectory algorithm was meant to stop after finding all local minima. But in the case of an arbitrary differentiable function that is impossible, since the region of attraction of a local minimum can be arbitrarily small. Therefore an accuracy limit has to be set defining local minima which can be excluded from consideration. The task of finding all remaining minima, although possible in principle, would in all probability in most cases demand much more computing time than can be allowed in practice. Reduction in the volume of computation could be achieved by imposing bounds on the rate of change of the objective function. Disadvantages of such bounds, which were discussed earlier in connection with grid methods, are in our view serious enough to make the whole method impractical. For this reason further work in this direction is unlikely to yield a deterministic global optimisation algorithm of general use.

This does not preclude the usefulness of trajectory algorithms, existing or improved, in some special circumstances. For example, if one local minimum attracts most local searches started from random points, but it is known or suspected that another minimum exists and is of interest to the problem, then a trajectory which departs from the dominating minimum may offer the best chance of finding the other one. However, a general purpose global optimisation algorithm is better sought among probabilistic methods.

Although in the course of this project probabilistic methods were given attention and some of them were examined in detail, no systematic study of the probabilistic approach to global optimisation was undertaken. The choice of algorithms which have been considered and which are reported below, was to a large degree governed by chance and should not be regarded as the result of a comparative assessment.

6. BREMMERMAN'S METHOD

The first probabilistic method put to test was a relatively well known algorithm due to Bremmerman (1972), recommended by its author as an efficient method of finding global minima of functions of up to 100 variables. At the start of the k'th iteration x_k denotes the best point until now. A random vector v is sampled, the probability distribution of v being n-dimension normal. The objective function is evaluated at four points along the line $x_k + t\,v$, namely at $x_k \pm h\,v$, $x_k \pm 2h\,v$, where h is a fixed scalar parameter, and a fourth degree interpolation polynomial P(t) is fitted to these data (x_k itself is the fifth point). The value $t = t*$, for which P(t) attains its absolute minimum, is found by solving algebraically the cubic equation $P'(t) = 0$. Then $f(x_k + t^*v)$ is computed and if it is better than $f(x_k)$, then $x_{k+1} = x_k + t^*\,v$; otherwise $x_{k+1}=x_k$. The algorithm stops after a fixed number of iterations.

Bremmerman's algorithm was tested on 10 functions of 2 to 4 variables. First it must be noticed, that the value of h cannot be too small, since then the algorithm easily gets stuck in a local minimum. On the other hand, too large a value for h often blocks any progress at all. The intermediate, one might call the "efficient" values of h do not in general ensure a result with high accuracy. In our examples after attaining the minimum function value with 4 digits and coordinates of the position with 2 - 4 digits further progress was very slow and insignificant. With this accuracy the absolute minimum was found in 10 - 30 iterations (i.e. 50 - 150 function evaluations) for 5 out of our 10 functions, all of them of 2 variables. On a further two functions of 2 variables the result was achieved in 200 - 350 iterations. On 2 functions of 3 variables the result was not reached in 250 iterations and there was practically no progress. Finally, on one function of four variables there was a very slow quasi-steady progress, but after 800 iterations the answer was still very inaccurate. The effect of wrong choice of h was illustrated on an "easy" two-dimensional function, which for h = .1 was minimised in 15 iterations, for h = .05 in 150 iterations and for h = .005 not at all.

We considered these results discouraging, mainly because of their erratic, unpredictable character. This seemed particularly dangerous, since the algorithm itself does not offer any checks or controls. Consequently, no further work was done in this direction.

Two other algorithms with probabilistic background, which had been proposed recently were considered. One was a clustering method due to A. Törn (1975), and the other a method developed by a group of Italian mathematicians, led by F. Archetti (1975).

7. TORN CLUSTERING ALGORITHM

Törn kindly supplied the N.O.C. with a listing of his algorithm and considerable experience has been obtained both with this original listing and with a modified routine. This experience is given in detail in Gomulka (1977 b).

Our conclusions were that the Törn algorithm is a valuable tool for dealing with global optimisation problems, especially if some alternative local minima are also of interest. With a proper choice of parameters it gives, in general, better results than repeated local optimisation either by increasing the confidence or by a saving in computer time.

In many cases its efficiency was considerably enhanced by operating it interactively.

8. THE ARCHETTI APPROACH

The theoretical foundations of this method were first formulated by F. Archetti (1975). This has recently been improved Archetti & Betró (1975) and Betrō & De Biase (1976) who showed that an algorithm with uniform convergence could be obtained if certain modifications are incorporated.

Numerical results with the improved method are reported by De Biase & Frontini (1977) on a number of standard test functions.

A few preliminary experiments were performed on an early listing at the N.O.C. In each case the global minimum point was found. Again in each case this was located by the local minimisation method incorporated in the algorithm, and at this point the predicted value and the located value of the objective function at the global minimum were <u>not</u> in close agreement. This lack of agreement is also confirmed in the detailed results reported in De Biase & Frontini. The success of the method does not therefore appear to be related to the theoretical result by which these two values should converge uniformily.

During our preliminary investigations the method was applied to the wellknown Rosenbrock function

$$f(x) = 100\,(x_1^2 - x_2)^2 + (1 - x_1)^2$$

$$-1 \leq x_i \leq 5\ .$$

The profile of their function $\psi(\xi)$ is shown in Fig 1 . This function is approximately the probability that $f(x) < \xi$ and the minimum should correspond to the root of $\psi(\xi) = 0$. At the stage plotted the predicted value was -322 not 0. This is due to the great domination of the region with $f(x) > 1000$ in this box.

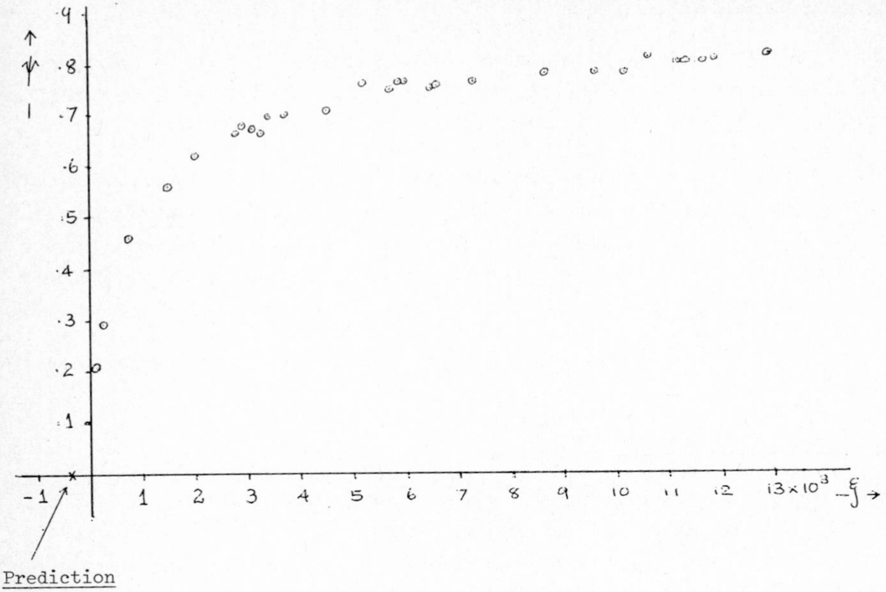

FREQUENCY CURVE $\psi(\xi)$
FOR ROSENBROCK'S FUNCTION

It is also possible to construct situations in which for a limited sample of points the predicted and located function values could agree at a local minimum, causing premature termination. It is possibly only fair to state that this has never occurred in practice and their results are very impressive.

9. CONCLUSIONS

Theoretical reasons have been advanced indicating that it is unlikely that an efficient deterministic algorithm for global optimisation could be devised for general functions. It is therefore expected that general purpose global optimisation packages will be probabilistic in nature. Two possible approaches have been briefly discussed.

ACKNOWLEDGEMENT

The author akcnowledges the financial support of the S.R.C. while working on this project.

REFERENCES

1. Archetti, F. (1975*). A sampling technique for global optimisation - In "Towards Global Optimisation" Eds. L.C.W. Dixon & G.P. Szegö, North-Holland, 1975.
2. Archetti, F. & Betró, B. (1975). University of Pisa Series A. No. 21.
3. Betró, B. & De Biase, L. (1976). University of Pisa Series A. No. 31.
4. Branin, F.H. (1972). Widely Convergent Method for Finding Multiple Solutions of Simultaneous Nonlinear Equations. I.B.M.J. R & D. Sept. 1972.
5. Branin, F.H. (1971). Solution of Nonlinear DC Network Problem via Differential Equations. Mem. Mexico 1971, International I.E.E.E. Conf. on Systems Networks & Computers, Oaxtepex, Mexico.
6. Bremmerman, H. (1970). Mathematical Biosciences 9 pp 1-15.
7. De Biase, L. & Frontini, J. (1977**). A stochastic method for global optimisation: its structure & numerical performance. In ref. 8.
8. "Towards Global Optimisation 2" eds. L.C.W. Dixon & G.P. Szegö, North-Holland.
9. Etvushenko, Y.G. (1971). Zh. Vychisl. Mat. mat FIZ. 11. 6, pp 1390-1403.
10. Gomulka, J. (1977**). "Two implementations of Branin's method; numerical experience". In ref. 8.
11. Gomulka, J. (1977**). "A user's experience with Törn's clustering algorithm". In ref. 8.
12. Hardy, J. (1975*). An implemented extension of Branin's method. In "Towards Global Optimisation". Eds. L.C.W. Dixon & G.P. Szegö, North-Holland 1975.
13. Morse, M. & Cairns, S. (1969). "Critical point theory in global analysis and differential topology". Academic Press 1969.
14. Shubert, B.O. (1972). Siam J. of Num. An. 9. 3. pp 379-388.
15. Törn, A. (1974). Global optimisation as a combination of global and local search: Abo Akademi Press.

TOWARDS GLOBAL OPTIMISATION 2
L.C.W. Dixon and G.P. Szegö (eds.)

A PRIORI ANALYSIS OF DETERMINISTIC STRATEGIES FOR GLOBAL OPTIMIZATION PROBLEMS (*)

F.ARCHETTI, B.BETRO'
ISTITUTO DI MATEMATICA
UNIVERSITA'DEGLI STUDI DI MILANO
MILANO, ITALIA

In this paper the global optimization problem is considered for functions $f \in L_{\ell,\rho}(K) \equiv \{f: |f(x)-f(y)| \leq \ell\rho(x,y);\ x,y \in K\}$ where ρ is some distance function and $K \subset R^N$ is a compact set. A strategy is a vector function defined over $L_{\ell,\rho}(K)$ whose values are finite subsets of K. Two optimality criteria for strategies are introduced: the existence as well as some properties of optimal strategies are proved. Some methods are given for their computation.

INTRODUCTION

In this paper we are concerned with the problem

P1) find x^*, f^* such that $f^*=f(x^*)=\min_{x \in K} f(x)$

where K is a compact set in R^N and $f:R^N \to R$ is continuous on K. The set $X^* \equiv \{x^* \in K: f^*=f(x^*)=\min_{x \in K} f(x)\}$ will be called the "solution set" for P1).

The problem P1) has been solved for some classes of functions (e.g. convex functions), for which X* can be characterized analytically, and many algorithms have been shown to converge.

When the objective function f has more local minima the usual optimization techniques are, in general, likely to fail to find the global minimum and converge to any of the local extrema.

(*) This work has been partially supported by "Gruppo Nazionale per l'Informatica Matematica" of the Italian National Research Council.

The global optimization problem, i.e. the problem of finding the global extrema of functions with many local minima has been gaining increasing attention in the last years and many algorithms, of very different nature, have been proposed in the literature (see Dixon et al.(1974), Dixon et al. (1975), Dixon and Szegö (1975)).
The first steps towards a theoretical analysis of numerical methods have been made by Sukharev (Sukharev (1971) and Ivanov (Ivanov (1972)).
In this paper a new optimality criterion is introduced; its relevance as well as its relation to different optimality criteria is stressed proving some new results about the existence and the properties of optimal strategies.

First one has to remark that P1) is not well posed in the classica sense; even if X^* contains a single element x^*, the continuous dependence of X^* on the data of the problem cannot in general be ensured. For example, let $f(x)=f_\delta(x)=\cos(\pi x)-\delta x$, $K\equiv[-2,2]$.
If $\delta>0$, then $x^*\simeq 1$, while for $\delta<0$ $x^*\simeq -1$.

A less ambitious aim would be to solve the problem

P2) find $f^*=\min_{x\in K} f(x)$.

Provided that $f\in C(K)$, K compact in R^N, this problem is well posed as shown by the relation $|f^*-g^*|\leq\|f-g\|$, where $\|\cdot\|$ denotes the supremum in K.

If $f\in L_{\ell,\rho}(K)\equiv\{f:|f(x)-f(y)|\leq\ell\rho(x,y);\ x,y\in K\}$ where ρ is some continuous distance function, then problem P2) can be solved, within a prefixed accuracy ε, once a value of ℓ is known, and some "space covering" algorithms have been proposed in the literature for this purpose.

Rather than deal with specific algorithms, in this paper the authors are concerned with solving some theoretical problems underlying the design of deterministic algorithms and with evaluating the computational complexity of P2) using "space covering" techniques.

In §1 the relevance of some "quantitative" conditions on f is pointed out; "passive" strategies for global optimization are introduced in §2 with the related concepts of accuracy in $L_{\ell,\rho}(K)$.

Two optimality criteria for strategies are next introduced and in §3 optimal strategies are shown to exist and characterized by some properties.
In §4 the equivalence between global optimization and covering problems is analyzed, connecting optimal strategies for P2) in $L_{\ell,\rho}(K)$ and optimal coverings of K.
"Sequential" strategies are introduced in §5 and shown to be equivalent, in terms of guaranteed accuracy in $L_{\ell,\rho}(K)$, to passive ones.
Finally, in §6 and §7, some methods to obtain optimal strategies are considered.

1. A PRIORI QUANTITATIVE CONDITIONS

Even if the global optimization problem $\min_{x\in K} f(x)$, where K is a compact set in R^N, is well posed in the dependent variable, as shown in the introduction, it is impossible to guarantee, with merely qualitative assumptions, an approximation $\hat{f}$ to f^*, of prefixed accuracy ε, in some finite number of function evaluations.
Let, for example, $f \in C^\infty([0,1])$; then, for any prefixed $\varepsilon>0$, there is a neighborhood U_ε of a point $x^*\in X^*$ such that, for $x\in U_\varepsilon \cap [0,1]$ we have $f(x)-f^*<\varepsilon$.
However, if only "qualitative" assumptions are made about f, we cannot know how large U_ε is and whether, whichever the points x_j, for a fixed n, $\min_{j=1,\dots,n} f(x_j)-f^*<\varepsilon$; indeed it is always possible to define a function $g(x) \in C^\infty([0,1])$ such that $g(x_j)=f(x_j)$, $j=1,\dots,n$ and $\min_{j=1,\dots,n} g(x_j)-f^*>\varepsilon$.
The introduction of "quantitative" assumptions about f enables to overcome the shortcoming as shown by the following argument.
Let $f(x) \in C([a,b])$ and $W(f,\delta)$ be its modulus of continuity:

$$W(f,\delta) = \sup_{\substack{|x-y|\leq\delta \\ x,y\in[a,b]}} |f(x)-f(y)|$$

A function $W(\delta)$ is assumed such that:

(1) $\lim_{\delta\to 0^+} W(\delta)=0$ and $W(f,\delta)\leq W(\delta)$, $\quad \delta\geq 0$.

For any prefixed $\varepsilon>0$ we can choose $\delta>0$ such that $W(\delta)\leq\varepsilon$. Next f is evaluated at the points $x_j\in(a,b)$, $j=1,\dots,n$ such that

(2) $\max_{x\in[a,b]} \min_{j=1,\ldots,n} |x-x_j| \leq \delta$.

A partition of [a,b], satisfying this relation, can be accomplished, for instance, taking

$$n=\left\lceil \frac{b-a}{2\delta} \right\rceil' \text{ (*) and } x_j=a+(2j-1)\delta \quad j=1,\ldots,n.$$

If $\hat{f}_n=\min_{j=1,\ldots,n} f(x_j)$ we have $\hat{f}_n-f^* \leq f(\bar{x}_n)-f^*$, where $|x^*-\bar{x}_n| = \min_{j=1,\ldots,n} |x^*-x_j|$, for some $x^*\varepsilon X^*$.
By (1) and (2) we may derive:

$$0\leq \hat{f}_n-f^*\leq W(f,\delta)\leq W(\delta)\leq\varepsilon \quad .$$

If $f\varepsilon C^1([a,b])$ and $|f'(x)|\leq M$ we may assume $W(\delta)=M\delta$. The number of function evaluations required to approximate f* within the prescribed accuracy is about

$$\frac{b-a}{2\varepsilon} M.$$

If $|f^{(r)}(x)|\leq M$, for $r>1$, it may be shown (see Brent (1973)) that the maximum number of function evaluations required to ensure an accuracy ε is about

$$\left(\frac{M}{2\varepsilon}\right)^{1/r} (b-a).$$

In the sequel we shall be concerned with the wider class of functions for which a Lipschitz condition holds with constant ℓ for some distance ρ , over a compact $K\subset R^N$, i.e.

$$f\varepsilon L_{\ell,\rho}(K)\equiv \{ f:K\to R, \quad |f(x)-f(y)| \leq \ell\rho(x,y) \text{ for } x,y \,\varepsilon\, K \}$$

where $\rho(x,y)$ satisfies the following conditions for any $x,y,z\,\varepsilon\,K$.

1) $\rho(x,y) = \rho(y,x)$ 2) $\rho(x,y) \geq 0$

3) $\rho(x,y) \leq\rho(x,z)+ \rho(y,z)$ 4) $\rho\varepsilon C(K\times K)$.

Condition 4) implies that if $f\varepsilon L_{\ell,\rho}(K)$ then f is continuous on K.

(*) $[\cdot]'$ means the least integer not less than $\cdot$; $[\cdot]$ the greatest integer not greater than $\cdot$.

2. PASSIVE STRATEGIES: DEFINITIONS AND BASIC PROPERTIES

For any n, a strategy is any vector valued function with domain $L_{\ell,\rho}(K)$ and range K^n (*).

If the strategy is constant in $L_{\ell,\rho}(K)$, that is it maps $L_{\ell,\rho}(K)$ into a single point of K^n, then the strategy is termed passive.
Passive strategies mapping $L_{\ell,\rho}(K)$ into $(x_1,x_2,\ldots,x_n)$, $x_i \in K$, will be indicated in the following by S_n; also the set of points $(x_1,x_2,\ldots,x_n)$ will be called a passive strategy, or briefly a strategy, writing $S_n \equiv (x_1,x_2,\ldots,x_n)$.

When the strategy is allowed to depend on $f \in L_{\ell,\rho}(K)$ so that its values $(\hat{x}_1,\hat{x}_2,\ldots,\hat{x}_n)$, $\hat{x}_j \in K$ are

$$\hat{x}_2 = \chi(\hat{x}_1, f(\hat{x}_1))$$

$$\cdots\cdots\cdots\cdots$$

$$\cdots\cdots\cdots\cdots$$

$$\hat{x}_n = \chi(\hat{x}_1,\ldots,\hat{x}_{n-1}, f(\hat{x}_1),\ldots,f(\hat{x}_{n-1}))$$

where χ is some selection rule, then the strategy is termed sequential and indicated by $\hat{S}_n$. We remark that passive strategies can be viewed as particular sequential ones.
The set of all sequential strategies ranging in K^n will be indicated by $\hat{K}_n$.
§'s 2-4 are concerned with the properties of passive strategies. Sequential strategies will be considered again in §5.

Let $S_n \equiv (x_1,x_2,\ldots,x_n)$ be a passive strategy. We measure its effectiveness for the function f by the accuracy

$$A(f,S_n) = \min_{i=1,\ldots,n} f(x_i) - f^* \quad .$$

A more significant index associated to S_n is its performance with respect to the whole class of functions under consideration:

$$A(S_n) = \operatorname{Sup}_{f \in L_{\ell,\rho}(K)} A(f,S_n),$$

which is called the guaranteed accuracy of S_n in $L_{\ell,\rho}(K)$.

(*) K^n means the cartesian product KxKx...xK n times.

We may introduce the following definitions of optimality for strategies

Definition 1: S^*_n is said to be A-optimal (A is for "respect to accuracy") when

$$A(S^*_n)=\inf_{S_n \varepsilon K^n} A(S_n) .$$

Definition 2: Given $\delta>0$, if there exists an n^* and a strategy S_{n^*} such that $A(S_{n^*})\leq\delta$, while no strategy S_n exists, for $n<n^*$, such that $A(S_n)\leq\delta$, then S_{n^*} is said to be n-optimal (n is for "respect to the number of points").

In order to prove the existence of both A- and n-optimal strategies the following lemma is needed:

Lemma 1. For any strategy $S_n\equiv(x_1,x_2,\dots,x_n)$

(3) $$A(S_n)=\mathrm{Sup}_{a\varepsilon K} \min_{i=1,\dots,n} \ell\rho(x_i,a).$$

Proof. Let $h_a(x)=\ell\rho(x,a)$, $a,x \varepsilon K$, $f(x^*)=f^*$, $x^*\varepsilon K$.

By the properties of ρ we have, for $a,x,y \varepsilon K$

$$\ell\rho(x,y)\geq\ell|\rho(x,a)-\rho(y,a)| = |h_a(x)-h_a(y)| .$$

For the strategy S_n

$$A(h_a,S_n)=\min_{i=1,\dots,n} h_a(x_i)-\min_{x \varepsilon K} h_a(x)= = \min_{i=1,\dots,n} \ell\rho(x_i,a).$$

If $\bar{i}$ is such that $\rho(x_{\bar{i}},x^*)=\min_{i=1,\dots,n} \rho(x_i,x^*)$ then

$$A(h_{x^*},S_n)= \min_{i=1,\dots,n} \ell\rho(x_i,x^*)=\ell\rho(x_{\bar{i}},x^*)\geq f(x_{\bar{i}})-f(x^*) \geq \geq\min_{i=1,\dots,n} f(x_i)-f(x^*)=A(f,S_n)$$

and

$$\mathrm{Sup}_{a\varepsilon K} A(h_a,S_n)\geq A(f,S_n).$$

As this inequality holds for every $f \varepsilon L_{\ell,\rho}(K)$, we get:

$$\mathrm{Sup}_{a\varepsilon K} A(h_a,S_n)\geq \mathrm{Sup}_{f\varepsilon L_{\ell,\rho}(K)} A(f,S_n)=A(S_n)$$

but h_a too belongs to $L_{\ell,\rho}(K)$ and thus

$$\mathrm{Sup}_{f\varepsilon L_{\ell,\rho}(K)} A(f,S_n)\geq \mathrm{Sup}_{a\varepsilon K} A(h_a,S_n) ,$$

whence

$$A(S_n)=\mathrm{Sup}_{f\varepsilon L_{\ell,\rho}(K)} A(f,S_n)=\mathrm{Sup}_{a\,\varepsilon K} A(h_a,S_n) = \\ =\ell\,\mathrm{Sup}_{a\varepsilon K}\, \min_{i=1,\dots,n}\rho(x_i,a) \;;$$

this proves the lemma.

3. THE EXISTENCE OF OPTIMAL STRATEGIES

We state the following theorem:

Theorem 1. Let K be a compact set in R^N. Then

a) for any n an A-optimal strategy exists;

b) for any δ an n-optimal strategy exists.

Proof.a) We show that $A(S_n)$, $S_n\equiv(x_1,x_2,\dots,x_n)$, is continuous with respect to $x_1,x_2,\dots,x_n$.
Let $S_n'\equiv(x_1',x_2',\dots,x_n')$, $x_i'\varepsilon K$, $S_n'\neq S_n$.
By the triangle inequality we have:

$$\min_{i=1,\dots,n}\{\rho(x_i',a)-\rho(x_i',x_i)\}\leqslant \min_{i=1,\dots,n}\rho(x_i,a)\leqslant \\ \leqslant\min_{i=1,\dots,n}\{\rho(x_i',a)+\rho(x_i',x_i)\}$$

and thus

$$\min_{i=1,\dots,n}\rho(x_i',a)-\max_{i=1,\dots,n}\rho(x_i',x_i)\leqslant \\ \leqslant\min_{i=1,\dots,n}\rho(x_i,a)\leqslant\min_{i=1,\dots,n}\rho(x_i',a)+ \\ +\max_{i=1,\dots,n}\rho(x_i',x_i);$$

this implies

$$\mathrm{Sup}_{a\varepsilon K}\min_{i=1,\dots,n}\rho(x_i',a)-\max_{i=1,\dots,n}\rho(x_i',x_i)\leqslant \\ \leqslant\mathrm{Sup}_{a\varepsilon K}\min_{i=1,\dots,n}\rho(x_i,a)\leqslant \\ \leqslant\mathrm{Sup}_{a\varepsilon K}\min_{i=1,\dots,n}\rho(x_i',a)+\max_{i=1,\dots,n}\rho(x_i',x_i).$$

Finally

$$-\max_{i=1,\dots,n}\rho(x_i,x_i')\leqslant\mathrm{Sup}_{a\varepsilon K}\min_{i=1,\dots,n}\rho(x_i,a)+ \\ -\mathrm{Sup}_{a\varepsilon K}\min_{i=1,\dots,n}\rho(x_i',a)\leqslant\max_{i=1,\dots,n}\rho(x_i',x_i).$$

By the continuity of ρ, as $x_i'\to x_i$, $\rho(x_i',x_i)\to 0$ and hence

$$\mathrm{Sup}_{a\varepsilon K}\min_{i=1,\dots,n}\rho(x_i',a)\to\mathrm{Sup}_{a\varepsilon K}\min_{i=1,\dots,n}\rho(x_i,a)\;,$$

which proves, by (3), the continututy of $A(S_n)$ with respect to $x_1,\dots,x_n$.
As K is a compact set, such continuity ensures that the infimum is

attained and an A-optimal strategy S_n^* exists.

b) We have to show that, for any $\delta>0$, the set

$$I_\delta \equiv \{n \text{ for which a strategy } S_n \text{ exists such that } A(S_n)\leq\delta\}$$

is non empty.

Let's consider the sets $D(x)\equiv\{y:\rho(x,y)<\delta/\ell\}$: they are open by the continuity of ρ, and clearly $K\subset\bigcup_{x\in K} D(x)$.

Therefore, by the Heine-Borel theorem, we can single out a finite number of sets $D(x_i)$, $i=1,\ldots,n$, covering K.

A fortiori this result is true also for the sets $\bar{D}(x_i)\equiv\{y:\rho(x_i,y)\leq \leq\delta/\ell\}$. By (3) it turns out $A(S_n)\leq\delta$, where $S_n\equiv(x_1,x_2,\ldots,x_n)$; hence I_δ is non empty.

From the definitions and the existence of both A-optimal and n-optimal strategies some interesting properties easily follow.

We list here three of them:

1. The sequence $\{A(S_n^*)\}$ is not increasing. Indeed, if we consider the strategy $S_n^*\equiv(x_1^*,x_2^*,\ldots,x_n^*)$ and the strategy $\bar{S}_{n+1}\equiv(x_1^*,x_2^*,\ldots,x_n^*,x_n)$, $x_n \in K$ then

$$A(S_n^*)\geq A(\bar{S}_{n+1})\geq A(S_{n+1}^*).$$

2. $A(S_n^*)\to 0$ as $n\to\infty$. Indeed, for any $\delta>0$, by the existence of n-optimal strategies we can find n^* and S_{n^*} such that $A(S_{n^*})\leq\delta$; thus $A(S_{n^*}^*)\leq\delta$ and, by property 1., $A(S_n^*)\leq\delta$ for any $n\geq n^*$.

3. n-optimal strategies can be constructed from A-optimal ones. Indeed let $\delta>0$ and, for any n, S_n^* be an A-optimal strategy an n^* be such that

$$n^*=\min\{n:A(S_n^*)\leq\delta\} .$$

The number n^* exists by property 2. and $A(S_{n^*}^*)\leq\delta$; as $A(S_n^*)>\delta$ for $n<n^*$ also $A(S_n)>\delta$ for any $n<n^*$ and for any strategy S_n; this implies that $S_{n^*}^*$ is n-optimal.

In the proof of theorem 1.b, the connection between strategies and coverings was introduced. In the next section we are going to better clarify and further develop this connection.

4. PASSIVE STRATEGIES AND COVERINGS

Let K be a compact set in R^N. We recall that a covering of K is a family of sets D_α, $\alpha\varepsilon A$, such that $\cup_{\alpha\varepsilon A} D_\alpha \supset K$.
In what follows, we shall consider coverings $C_{\delta,n} \equiv \{D^\delta(x_i)\}$, $i=1,\ldots,n$ where $D^\delta(x_i) \equiv \{x: \rho(x_i,x) \leq \delta\}, x_i \varepsilon K$.
We introduce the following definition:

Definition 3: δ is the radius of $C_{\delta,n}$ and x_i, $i=1,\ldots,n$ are its centres.

Two optimality criteria can be stated for these coverings:

Definition 4: A covering $C_{\delta^*,n}$ of K is said to be r-optimal (r is for "respect to radius") if no covering $C_{\delta,n}$ of K exists with $\delta<\delta^*$.

Definition 5: A covering C_{δ,n^*} of K is said to be n-optimal if no covering $C_{\delta,n}$ of K exists with $n<n^*$.

As ρ is bounded on K, for each strategy $S_n \equiv (x_1,x_2,\ldots,x_n)$ a set Δ of positive numbers exists, such that, for $\delta\varepsilon\Delta$, it is $\cup_{i=1}^n D^\delta(x_i) \supset K$.
Let $\bar{\delta}=\inf \Delta$: we call $C_{\bar{\delta},n}$ the characteristic covering for S_n.
Thus relation (3) says that, for any strategy S_n, $A(S_n)$ is just ℓ times the radius of the characteristic covering for S_n. On the other hand to any covering we may link the strategy whose points are the centres of the covering.

We can easily prove the following theorem:

Theorem 2. Let S^*_n be an A-optimal strategy: its characteristic covering $C_{\delta^*,n}$ is r-optimal with $\delta^*=A(S^*_n)/\ell$.
Let $C_{\delta^*,n}$ be an r-optimal covering: the strategy $S^*_n \equiv (x^*_1,x^*_2,\ldots,x^*_n)$, where x^*_i, $i=1,\ldots,n$, are the centres of $C_{\delta^*,n}$ is A-optimal and $A(S^*_n)= =\ell\delta^*$.
Let S_{n^*} be an n-optimal strategy for $\delta>0$: its characteristic covering $C_{\delta/\ell,n^*}$ is n-optimal.
Let C_{δ,n^*} be an n-optimal covering: $S_{n^*} \equiv (x_1,x_2,\ldots,x_{n^*})$, where x_i, $i=1,\ldots,n^*$ are the centres of C_{δ,n^*} is n-optimal with $A(S_{n^*}) \leq \ell\delta$.

Proof. If a strategy S^*_n is A-optimal with guaranteed accuracy δ then the characteristic covering for S^*_n is r-optimal with radius δ/ℓ, by relation (3). If a covering is r-optimal with radius δ^*, then it coincides with the characteristic covering for the strategy S_n^* with points at its centres, and hence, by (3), the strategy S_n^* is A-optimal with guaranteed accuracy $A(S_n^*)=\ell\delta^*$.

If a strategy S_{n^*} is n-optimal for a fixed $\delta>0$, then, by (3), it is

$$\mathrm{Sup}_{a\varepsilon K}\ \min_{i=1,\dots,n}\ \rho(x_i,a)\leq\delta/\ell,\quad S_{n^*}\equiv(x_1,x_2,\dots,x_{n^*})$$

while, for $n<n^*$, for any n-tuple $(x_1,x_2,\dots,x_n)\ \varepsilon\ K^n$,

$$\mathrm{Sup}_{a\varepsilon K}\min_{i=1,\dots,n}\ \rho(x_i,a)>\delta/\ell.$$

This prevents the existence of any covering $C_{\delta/\ell,n}$ of K for $n<n^*$ and thus the covering $C_{\delta/\ell,n^*}$ with centres at the points of S_{n^*} is n-optimal. Conversely, if $C_{\delta/\ell,n^*}$ is an n-optimal covering, then, if $\bar{x}_1, \bar{x}_2,\dots,\bar{x}_{n^*}$ are its centres,

$$\mathrm{Sup}_{a\varepsilon K}\ \min_{i=1,\dots,n^*}\ \rho(\bar{x}_i,a)\leq\delta/\ell$$

while, for $n<n^*$, for any n-tuple $(\bar{x}_1,\bar{x}_2,\dots,\bar{x}_n)\ \varepsilon\ K^n$,

$$\mathrm{Sup}_{a\varepsilon K}\ \min_{i=1,\dots,n}\ \rho(x_i,a)>\delta/\ell.$$

Hence, considering the strategy $S_{n^*}\equiv(\bar{x}_1,\bar{x}_2,\dots,\bar{x}_{n^*})$, its characteristic covering has radius not greater than δ/ℓ, while any other strategy S_n for $n<n^*$ has a characteristic covering with radius greater than δ/ℓ.

This fact and (3) prove also the second part of the theorem.

We remark that, by theorem 2, optimal strategies do not depend on ℓ, but only on ρ and K.

5. PASSIVE AND SEQUENTIAL STRATEGIES

In this section the effectiveness of sequential strategies against passive ones is discussed.

The concepts of accuracy and guaranteed accuracy, as well as the optimality criteria introduced in §2, can be easily extended to sequential strategies:

$$A(f,\hat{S}_n)=\min_{i=1,\dots,n}\ f(\hat{x}_i)-f^*,\quad \hat{S}_n(f)\equiv(\hat{x}_1,\hat{x}_2,\dots,\hat{x}_n)$$

is the accuracy given by $\hat{S}_n$ for that particular function f;

$$A(\hat{S}_n)=\mathrm{Sup}_{f\varepsilon L_{\ell,\rho}(K)}\ A(f,\hat{S}_n)$$

is the accuracy guaranteed by $\hat{S}_n$ in $L_{\ell,\rho}(K)$.
A strategy $\hat{S}^*_n$ is said to be an A-optimal sequential strategy when

$$A(\hat{S}_n^*)=\inf_{\hat{S}_n\varepsilon \hat{K}_n} A(\hat{S}_n).$$

Given $\delta>0$, if there exists an n^* and a sequential strategy $\hat{S}_{n^*}$ such that $A(\hat{S}_{n^*})\leq\delta$, while no strategy $\hat{S}_n$ exists for $n<n^*$ such that $A(\hat{S}_n)\leq\delta$, then $\hat{S}_{n^*}$ is said to be n-optimal.

One could think that sequential strategies improve the performance of passive ones, in terms of guaranteed accuracy: this is not the case, as shown by theorems 3 and 4.
Theorem 3 states that no sequential strategy can guarantee a better accuracy than an A-optimal passive one with the same number of points; theorem 4 states that no sequential strategy can guarantee a prefixed accuracy $\delta>0$, with less points than those required by an n-optimal passive strategy.

<u>Theorem 3</u>. Let $m_1=\inf_{\hat{S}_n\varepsilon \hat{K}_n} A(\hat{S}_n)$ and $m_2=\inf_{S_n\varepsilon K^n} A(S_n)$; then $m_1=m_2$.
For the proof we refer to Sukharev (1971).

<u>Theorem 4</u>. Let $n_1=\min$ {n for which a passive strategy S_n exists such that $A(S_n)\leq\delta$}
and $n_2=\min$ {n for which a sequential strategy $\hat{S}_n$ exists such that $A(\hat{S}_n)\leq\delta$};
then $n_1=n_2$.

<u>Proof</u>: as any passive strategy is also a sequential one, clearly $n_2\leq n_1$. By way of contradiction, let $n_2<n_1$. It turns out

$$A(S_{n_2})>\delta \quad \text{for any } S_{n_2}\varepsilon K^{n_2};$$

as K is compact, theorem 1 implies that

$$\inf_{S_{n_2}\varepsilon K^{n_2}} A(S_{n_2})=\min_{S_{n_2}\varepsilon K^{n_2}} A(S_{n_2})$$

and hence

$$\inf_{S_{n_2}\varepsilon K^{n_2}} A(S_{n_2})>\delta. \tag{4}$$

By theorem 3 we have

$$\inf_{S_{n_2} \varepsilon K^{n_2}} A(S_{n_2}) = \inf_{\hat{S}_{n_2} \varepsilon \hat{K}_{n_2}} A(\hat{S}_{n_2});$$

as $\inf_{\hat{S}_{n_2} \varepsilon \hat{K}_{n_2}} A(\hat{S}_{n_2}) \leq \delta$, the above inequality and (4) lead to the desired contradiction.

The above results are concerned only with the guaranteed accuracy in $L_{\ell,\rho}(K)$: thus for particular functions in $L_{\ell,\rho}(K)$, sequential strategies may perform better than passive ones:for this reason some research has been devoted to develop sequential algorithms (see Shubert (1972), Strongin (1973), Evtushenko (1971)).

6. CONSTRUCTION OF OPTIMAL STRATEGIES

First we consider the A-optimality: let $K \equiv [0,1]^N$ and

(5) $\rho(x,y) = \max_{i=1,\ldots,N} |x^i - y^i|$, $x \equiv (x^1,\ldots,x^N)$, $y \equiv (y^1,\ldots,y^N)$.

The construction of an A-optimal strategy is connected, by theorem 2, with the construction of an r-optimal covering of K with n centres. In Sukharev (1971) the r-optimal covering of K with n hypercubes $D(x_i,r) \equiv \{x: \rho(x,x_i) \leq r\}$ is given for any n; the result can be stated as follows:

Theorem 5. a) If $n=\nu^N$ for some natural number ν then the strategy $S_n^* \equiv (x_1^*, x_2^*, \ldots, x_n^*)$, where x_i^*, $i=1,\ldots,n$, are the centres of the hypercubes with edges of length $1/\nu$ partitioning K, is A-optimal.
b) If $\nu^N < n < (\nu+1)^N$, then the strategy $S_n^* \equiv (x_1^*, x_2^*, \ldots, x_n^*)$, where x_i^*, $i=1,\ldots,\nu^N$ are the same as in a) and x_i^*, $i=\nu^N+1,\ldots,n$ are selected arbitrarily, is A-optimal.
In both case the guaranteed accuracy is

$$A(S_n^*) = 1/2\,[\sqrt[N]{n}].$$

The method in which the function to be optimized is evaluated at the points of the strategy S_n^* for $n=\nu^N$, is usually termed the "grid search" method.
With the above assumptions about K and ρ, it is easy to derive, by property 3 of §3, n-optimal strategies.
For any $\delta>0$, let $n^* = \min\{ n: A(S_n^*) \leq \delta\}$; it turns out

$$n^* = ([1/2\delta]')^N.$$

Thus the strategy $S^*_{n^*}$ is n-optimal.
For a general set K, the problem of finding an r-optimal covering of K, and hence an A-optimal strategy, is not easy to solve; it leads to the problem of optimizing the nxN variables function

$$\max_{a\in K} \min_{i=1,\dots,n} \rho(x_i,a) \ .$$

In the next section we are going to outline a technique by which n-optimal coverings C_{δ,n^*} of K, and therefore n-optimal strategies with guaranteed accuracy not larger than δ/ℓ, can be approximated.

7.n-OPTIMAL STRATEGIES AND INTEGER PROGRAMMING PROBLEMS

Let C_{δ,n^*} be an optimal covering of K of radius δ. We assume that ρ is given by (5), although the following results can be derived also for different distance functions. Moreover, for sake of simplicity we consider N=2.

Let $\mathbf{H}$ be a regular lattice of size $h=\delta/m$, m integer, and H be the set of lattice points (knots) of $\mathbf{H}$ lying within K.
In the sequel K is assumed to be such that, for sufficiently large m, for any point $c\in K$, a knot $p\in H$ exists such that $\rho(c,p)<h$.

Let Q be the set of those squares of side h, carved by $\mathbf{H}$ in R^2, whose intersection with K is non empty. Now we are going to consider separately the problem of covering H and Q; they will be shown to be equivalent to two "set covering" problems (in the sense of Opérations Research) and will therefore be reduced to integer linear programming (ILP) problems.

1) The n-optimal covering of H.

Let $D^\delta(\bar{x})\equiv\{x:\ \rho(x,\bar{x})\leq\delta\}$: with the above distance function, $D^\delta(\bar{x})$ is a square with centre $\bar{x}$ and side 2δ.

We shall consider coverings of H given by any family of sets $D^\delta(a_i)$, $i=1,\dots,n$, $a_i\in K$, such that $H\subset\bigcup_{i=1}^n D^\delta(a_i)$.

First we prove that for any such covering of H, n points x_i can be selected in H such that $H\subset\bigcup_{i=1}^n D^\delta(x_i)$.

<u>Theorem 6</u>. Let a and c be any two points in K such that $\rho(a,c)\leq\delta$.

If m is such that $p \in H$ exists for which $\rho(c,p)<h$, then also $\rho(a,p) \leq \delta$.

Proof. Let $a \equiv (a_1, a_2)$, $c \equiv (c_1, c_2)$ and $p \equiv (p_1, p_2)$.
The relation $\rho(a,c) \leq \delta$ implies that

$$(6) \qquad \begin{aligned} a_1 - mh \leq c_1 \leq a_1 + mh \ , \\ a_2 - mh \leq c_2 \leq a_2 + mh \ . \end{aligned}$$

Without loss of generality we can assume that the points of H have coordinates which are multiples of h. Therefore, for some integers i,j, $a_1 = ih$ and $p_1 = jh$; thus the inequality $\rho(c,p)<h$ implies

$$j-1 \leq \left[\frac{c_1}{h}\right] \leq j$$

$$j \leq \left[\frac{c_1}{h}\right]' \leq j+1$$

or equivalently

$$(7) \qquad p_1 - h \leq \left[\frac{c_1}{h}\right] h \leq p_1$$

$$(8) \qquad p_1 \leq \left[\frac{c_1}{h}\right]' h \leq p_1 + h \ .$$

By (6) it follows:

$$a_1 - mh \leq \left[\frac{c_1}{h}\right] h \leq a_1 + mh$$

$$a_1 - mh \leq \left[\frac{c_1}{h}\right]' h \leq a_1 + mh \ .$$

These inequalities, together with (7) and (8) imply

$$a_1 - mh \leq p_1 \leq a_1 + mh \ .$$

The same argument, applied to a_2, c_2 and p_2 leads to the required inequality $\rho(a,p) \leq \delta$.

As H is a finite set, an n-optimal covering of H by a family of sets $D^{\delta}(x_i)$, $x_i \in H$, $i=1,\ldots,n(h)$ obviously exists; let it be $C^{\ell}_{\delta,n(h)}$.
No covering of H with $n<n(h)$ centres can exist even if the centres x_i, $i=1,\ldots,n$ are allowed to be in K; if so, by theorem 6, the n-optimality of $C^{\ell}_{\delta,n(h)}$ would be violated. Therefore any n-optimal covering of H can be associated to $C^{\ell}_{\delta,n(h)}$.
Moreover, as $K \supset H$, we have $n(h) \leq n^*$.

The problem of computing $C^{\ell}_{\delta,n(h)}$ can be handled in the following way: to any point $p_i \varepsilon H$, $i=1,\dots,r$ a variable z_i is associated, which can assume only the values 0 and 1.

Let $I_i \equiv \{j: \rho(p_j,p_i) \leqslant \delta\}$: the points p_j, $j \varepsilon I_i$, are the only ones which can be centres of a square of side 2δ containing p_i.

Therefore, the covering condition for p_i requires that at least one point p_k, $k \varepsilon I_i$, is assumed as a centre of the covering.

Let $z^* \equiv (z_1^*, z_2^*, \dots, z_r^*)$ be a solution of the following ILP problem:

$$\min \sum_{i=1}^{r} z_i$$

subject to the constraints:

$$\sum_{j \varepsilon I_i} z_j \geqslant 1 \qquad i=1,\dots,r;$$

then we can assume

$$D^{\delta}(p_i) \,\varepsilon\, C^{\ell}_{\delta,n(h)} \qquad \text{if and only if } z_i^*=1.$$

2) the optimal covering of Q.

We consider coverings of Q by families of sets $D^{\delta}(x_i)$, $x_i \varepsilon H$; as H is finite such an n-optimal covering obviously exists: let it be $C^{u}_{\delta,n'(h)}$. As $Q \supset K$ we have $n'(h) \geqslant n^*$.

The covering $C^{u}_{\delta,n'(h)}$ can be constructed as follows:

Let $Q = \bigcup_{j=1}^{r} Q_j$ and $J_j \equiv \{k: D^{\delta}(p_k) \supset Q_j,\ p_k \varepsilon H\}$; to any $p_i \varepsilon H$, $i=1,\dots,r$ a variable z_i is associated which can assume the values 0 and 1; the following ILP problem can be next set up:

$$\min \sum_{i=1}^{r} z_i$$

subject to the constraints

$$\sum_{k \varepsilon J_j} z_k \geqslant 1 \qquad j=1,\dots,s.$$

Let $z^* \equiv (z_1^*, z_2^*, \dots, z_r^*)$ be a solution of the above ILP problem; then we can assume

$$D^{\delta}(p_i) \varepsilon C^{u}_{\delta,n'(h)} \qquad \text{if and only if } z_i^*=1.$$

The question naturally arises if, for $h \to 0$, we have $n(h) \to n^*$ and $n'(h) \to n^*$, i.e., for sufficiently small h, $n(h)=n^*=n'(h)$.

The answer is partly positive, as shown by the following theorem.

Theorem 7. Let $\mathbf{H}_m$ be the regular lattice of side $h_m=\delta/2m$, $H_m=\mathbf{H}_m\cap K$, Q_m the set of squares, carved by $\mathbf{H}_m$ in R^2, with non empty intersection with K, $C^{\ell}_{\delta,n(h_m)}$ and $C^{u}_{\delta,n'(h_m)}$ the n-optimal coverings respectively of H_m and Q_m defined as above; under these assumptions we may prove that

i) $\lim_{m\to\infty} n(h_m)=n^*$;

ii) $\lim_{m\to\infty} n'(h_m)=n^*$, provided that a covering C_{δ',n^*} of K exists, with $\delta'<\delta$.

Proof: i) As $\mathbf{H}_{m+1}\supset\mathbf{H}_m$ we have:

$$n(h_m)\leq n(h_{m+1})$$
$$n'(h_{m+1})\leq n'(h_m).$$

By way of contradiction let $n_\ell=\lim_{m\to\infty} n(h_m)<n^*$.
Then, for sufficiently large m, say $m\geq\bar{m}$, $n(h_m)=n_\ell$. For $m_k\geq\bar{m}$, we can consider a sequence of n_ℓ-tuples $(x_{1,m_k},x_{2,m_k},\dots,x_{n_\ell,m_k})$, $k=1,2,\dots$ with x_{i,m_k}, $i=1,2,\dots,n_\ell$, $k=1,2,\dots$, centres of coverings of H_{m_k} converging, as $k\to\infty$, to the n_ℓ-tuple $(\bar{x}_1,\bar{x}_2,\dots,\bar{x}_{n_\ell})\varepsilon K^n$.
For any such sequence we may write, for $a\varepsilon K$ and $a'\varepsilon H_{m_k}$

$$\min_{i=1,\dots,n_\ell}\rho(x_{i,m_k},a)\leq\min_{i=1,\dots,n_\ell}\rho(x_{i,m_k},a')+\rho(a,a')\leq$$
$$\leq\delta+\rho(a,a')$$

hence

$$\min_{i=1,\dots,n_\ell}\rho(x_{i,m_k},a)\leq\delta+\inf_{a'\varepsilon H_{m_k}}\rho(a',a)\leq\delta+h_{m_k}$$

and therefore

$$\operatorname{Sup}_{a\varepsilon K}\min_{i=1,\dots,n_\ell}\rho(x_{i,m_k},a)\leq\delta+h_{m_k}.$$

As $k\to\infty$, by the continuity of the left term (proved in theorem 1) in the above inequality, it turns out

$$\operatorname{Sup}_{a\varepsilon K}\min_{i=1,\dots,n_\ell}\rho(\bar{x}_i,a)\leq\delta$$

which violates the n-optimality of C_{δ,n^*}. Therefore $n(h_m)\to n^*$ as $m\to\infty$.
To prove ii) assume, by way of contradiction, that $\lim_{m\to\infty} n'(h_m)>n^*$. Therefore, for any n^*-tuple $(x_1,x_2,\dots,x_{n^*})$, $x_i\varepsilon H_m$ we have, for any m

(9) $\quad \mathrm{Sup}_{a'\varepsilon Q_m} \min_{i=1,\ldots,n^*} \rho(x_i,a')>\delta$.

Let $\bar{x}_1,\bar{x}_2,\ldots,\bar{x}_{n^*}$ be the centres of C_{δ',n^*}. By the assumptions, a sequence $\{(x_{1,m},x_{2,m},\ldots,x_{n^*,m})\}$ exists, $x_{i,m}\varepsilon H_m$, such that $x_{i,m}\to\bar{x}_i$, $i=1,\ldots,n^*$, as $m\to\infty$.

For the elements of this sequence we may write, for $a\varepsilon K$ and $a'\varepsilon Q_m$:

$$\min_{i=1,\ldots,n^*}\rho(x_{i,m},a)\geq \min_{i=1,\ldots,n^*}\rho(x_{i,m},a')-\rho(a',a).$$

By the definition of Q_m we have $\inf_{a\varepsilon K}\rho(a,a')\leq h_m$ for any $a'\varepsilon Q_m$. Therefore

$$\mathrm{Sup}_{a\varepsilon K}\min_{i=1,\ldots,n^*}\rho(x_{i,m},a)\geq\min_{i=1,\ldots,n^*}\rho(x_{i,m},a')-h_m,$$

hence

$$\mathrm{Sup}_{a'\varepsilon K}\min_{i=1,\ldots,n^*}\rho(x_{i,m},a')-h_m\leq\mathrm{Sup}_{a\varepsilon K}\min_{i=1,\ldots,n^*}\rho(x_{i,m},a)$$

and finally, by (9)

$$\mathrm{Sup}_{a\varepsilon K}\min_{i=1,\ldots,n^*}\rho(x_{i,m},a)\geq\delta-h_m.$$

Taking the limits of both sides, as $m\to\infty$, it turns out

$$\mathrm{Sup}_{a\varepsilon K}\min_{i=1,\ldots,n^*}\rho(\bar{x}_i,a)\geq\delta$$

which violates the assumptions about C_{δ',n^*}.

Thus $\lim_{m\to\infty}n'(h_m)=n^*$ and the theorem is completely proved.

Remark 1. By the above theorem, once the numbers $n(h)$ and $n'(h)$ coincide, then $n(h)=n^*=n'(h)$ and $C^u_{\delta,n'(h)}$ can be assumed as an n-optimal covering of K.

Remark 2. The assumption about the existence of C_{δ',n^*}, with $\delta'<\delta$, cannot be removed, as shown by the following example.

Let's consider the segment $[\pi,\pi+2]$ and $\delta=1$. Obviously $n^*=1$ and $x_1^*=$ $=\pi+1$. If we now consider a lattice of points $x_i=ih$, $i=1,2,\ldots$ with $h=1/m$ for integer m, it is obvious that, for any h, two points of such lattice are necessary in order to cover $[\pi,\pi+2]$ and hence $n'(h)=2$ for any h.

ACKNOWLEDGEMENTS

The authors are very grateful to Prof. M. Cugiani for fruitful discussion and valuable help in the writing of this paper.

REFERENCES

BRENT R.P.(1973). Algorithms for Minimization without Derivatives. Prentice Hall, Englewood Cliffs, N.J..

DIXON L.C.W., GOMULKA J. and HERSOM S.E.(1975). Reflections on the Global Optimization Problem. The Hatfield Polytechnic, N.O.C. T.R. n°64.

DIXON L.C.W., GOMULKA J. and SZEGO G.P.(1974). Towards Global Optimization Techniques. The Hatfield Polytechnic, N.O.C. T.R. n°61.

DIXON L.C.W. and SZEGO G.P., eds (1975). Towards Global Optimization. North-Holland Publishing Company. Amsterdam.

IVANOV V.V.(1972). Optimal Algorithms of Minimization of Certain Classes of Functions. Cybernetics,8,4.

SHUBERT B.O.(1972). A Sequential Method Seeking the Global Maximum of a Function. SIAM Journal on Numerical Analysis,9,3.

STRONGIN, R.G.(1973). On the Convergence of an Algorithm for Finding a Global Extremum. Engineering Cybernetics,4,pp.549-555.

SUKHAREV A.G.(1971). Optimal Strategies of the Search for an Extremum. U.S.S.R. Computational Mathematics and Mathematical Physics,11,4.

TOWARDS GLOBAL OPTIMISATION 2
L.C.W. Dixon and G.P. Szegö (eds.)

A SEARCH-CLUSTERING APPROACH TO GLOBAL OPTIMIZATION

Aimo A. Törn
Åbo Swedish University School of Economics
Henriksgatan 7
SF-20500 ÅBO 50
FINLAND

A method for finding the local optima of a multimodal function defined in a region $A \subset R^n$ is presented. The method uses a local optimizer which is started from a number of points sampled in A. In order to reduce the number of function evaluations needed to reach the local optima the parallel local search processes are stopped repeatedly, the working points clustered and a reduced number of processes from each cluster resumed. A number of sample problems ($2 \leq n \leq 6$) are solved and the performance of the method used is reported.

1. INTRODUCTION

Given a region $A \subset R^n$ and a function $f: A \rightarrow R$, a global minimum of f in A is to be determined. The objective function f may be multimodal in A so that many local minima may exist. Assuming the existence of a set $M \subset A$, where f obtains its minimum

$$M = \{ x^* \in A : f(x^*) \leq f(x), \ \forall x \in A \}$$

the problem can be stated as "determine a point in M". Because numerical algorithms are used it is usually impossible to expect to determine a point in M but in $M_\varepsilon \supseteq M$ where

$$M_\varepsilon = \{ x \in A : f(x) - f^* < \varepsilon \}, \ f^* = f(x^*) \qquad \text{(A-1)}$$

A point in M_ε can therefore be considered a solution to the problem.

There exist classes of problems for small n (usually $n \leq 2$) which can be solved in a finite number of steps by certain deterministic algorithms. (Cf. Dixon (1975)). For most problems, however no method exists with guaranteed convergence to a solution at feasible cost. For such problems it would be desirable to use a method that would obtain a good estimate at a given computational cost. The decision as to which method is preferred must then be based on the probabilities of obtaining certain levels of the objective function. In a practical situation it is also important to obtain some stopping condition related to the expected quality of the estimate.

One of the most frequently used approaches to global optimization is multiple local searches (MS). The following observation is the motive for our attempts to design a class of better models: When starting from several randomly sampled points (global points), the local optimizer used in MS will normally arrive at the same local solution several times. This means that much of the effort spent on global optimization is unnecessary re-determination of a local solution that has already been found, whereas this effort could be spent on exploring more global points and thus increasing the probability of arriving at a global solution.

In our algorithm a clustering analysis technique is used to prevent multiple determination of a local solution. A few steps of the local optimizer can be expected to gather the points around the local solutions so that clusters of points will emerge. With the aid of the clustering analysis technique these clusters can then be recognized and a reduction of the points leading to the same local solution can be made.

Based on the foregoing reasoning the following model for global optimization is formulated:

<u>Algorithm LC</u> (Multiple local searches with clustering)

<u>LC1</u>	[Global points]	Choose global points.
<u>LC2</u>	[Local search]	Push the points some steps towards the local optima by using a local optimizer.
<u>LC3</u>	[Find clusters]	Find clusters by using a clustering analysis technique. If a tolerance condition is met, stop.
<u>LC4</u>	[Reduce points]	Take a sample of points from each cluster. Return to step LC2.

The performance of a number of global search methods has been experimentally studied by Hartman (1973). He found that MS and the non-local search method of Bocharov and Feldbaum (1962) performed more or less equally well and better than other search methods. The result indicates that it could be worthwhile trying to improve on MS.

Before describing the four steps of Algorithm LC in more detail let us first discuss the probability of finding a local optimum when using a local optimizer.

2. CHARACTERIZATION OF A LOCAL OPTIMUM

The probability of finding a given optimum when a local optimizer is used to find a local optimum from a given starting point is dependent on a number of sets of points around the optimum. First we have the optimum M_i itself, i.e.

$$M_i = \{ x \in A : f(x) = t_i \} , \quad (f^* = t_1 < t_2 < \ldots < t_k)$$

Second, there exists a set $M_{i\varepsilon} \supseteq M_i$ defined analogously to M_ε in Eq. (A-1)

$$M_{i\varepsilon} = \{ x \in A : f(x) \geq t_i \wedge f(x) - t_i < \varepsilon \} \quad \text{(A-2)}$$

Third, there exists a set $T_i \supseteq M_{i\varepsilon}$ such that from every point in T_i there exists a path to some point in $M_{i\varepsilon}$ with decreasing values of the function f and from which no such path exists to a point outside $M_{i\varepsilon}$. Fourth, a still larger set is $W_i \supseteq T_i$ from which some point in T_i can be reached along a non-increasing path but from which also certain other capture sets T_i can be reached. When using a local optimizer to reach $M_{i\varepsilon}$ the points in W_i are of interest. The probability p_i of reaching a point in $M_{i\varepsilon}$ when starting from a point sampled at random in A, assuming that a successful local optimizer is used, is therefore

$$p_i = \mu(T_i)/\mu(A) + p_{wi} \quad \text{(A-3)}$$

where p_{wi} is the probability of reaching a point in T_i when starting from a point in $W_i - T_i$. This probability depends on the local optimizer used. The number p_i could also be interpreted as the measure of a greater hypothetical region T_i', $\mu(T_i') \geq \mu(T_i)$ from which a point in $M_{i\varepsilon}$ will be reached with a probability of one.

For a deterministic local optimizer, such as the gradient trajectory algorithm for example, the region corresponding to T_1', R_1' (the region of attraction of M_ε) can be determined and $R_1' = R_1$.

It is clear that the success in finding a global optimum by means of probabilistic methods is largely dependent on p_1 and a problem where p_1 is large will be relatively easy to solve using these methods.

In a practical situation an estimate $\widehat{f^*}$ of f^* outside M_ε may be obtained as the result of an optimization. This result may also be of some value to a decision-maker. Furthermore, the estimate $\widehat{f^*}$ will, for probabilistic methods, be dependent on some initial conditions such as the choice of starting points and random numbers generated.

This discussion initiates the question of how the performance of different methods should be compared.

3. PERFORMANCE INDEX OF AN OPTIMIZATION METHOD.

Let us assume that $h(t)$ is the function measuring the benefit of obtaining t as a result of estimating f^*. Let us also assume that for a given method $G(t)$ is the distribution function of the estimate $\hat{f}^*$ for a given cost and a random choice of the initial conditions. As a measure of the usefulness of a method the performance index u could then be computed as the Stieltjes integral.

$$u = \int_{t_1}^{\infty} h(t)dG(t) \qquad \text{(A-4)}$$

where

$$0 \le h(t) \le 1 \text{ and } h(t) = 1 \text{ for } t - t_1 < \varepsilon$$

The function $h(t)$ can be assumed to be non-increasing. If $h(t) = 1$ then $u = 1$ for all methods, which means that the methods are considered equally good, an obvious result. Another extreme is the case where

$$h(t) = \begin{cases} 1 & \text{for } t - t_1 < \varepsilon \\ 0 & \text{otherwise} \end{cases} \qquad \text{(A-5)}$$

Here u is the probability P_1 of finding a global optimum.

The function $G(t)$ will be dependent on the problem, cost and method used but serious efforts to reach a solution might be expected to lead to a function $G(t)$, the corresponding frequency function of which will have peaks to the right of the local minima. i.e. $\hat{x}^* \in M_{i\varepsilon}$, $i \in [1, k]$.

If the procedure of a global optimizer, consists of a number N of independent estimations of the global minimum and the distribution function of $\hat{f}^*$ in a single trial is $G_1(t)$, then $G_N(t)$, i.e. the probability that $\hat{f}^*$ in a sample of N estimations is less than t, is given by

$$G_N(t) = 1 - [1 - G_1(t)]^N$$

The MS method consists of N independent local optimizations and

$$G_1(t) = \sum_{i=1}^{k-1} p_i, \quad t_{k-1} \le t \le t_k$$

so that

$$P_1 = 1 - (1 - p_1)^N \qquad \text{(A-6)}$$

(The performance index of some other probabilistic methods is evaluated in Törn (1976b)). If p_1 is known and we require that $P_1 = 1-\varepsilon$, then the number of global points N_1 that, must be sampled in A can be obtained by solving N in Eq. (A-6)

$$N_1 = \lceil \ln \varepsilon / \ln (1 - p_1) \rceil \qquad \text{(A-7)}$$

Let us assume that A can be divided into q identical sized parts. If when using MS the sampling is done by taking N/q points uniformly distributed in one of the parts and by using identically distributed points in the remaining q - 1 parts, then the number of global points N_q that must be sampled in order for P_1 to be $1 - \varepsilon$ is given by

$$N(q, p_1) \le N_q \le qN_1^{\,1}$$

where

$$N(q, p_1) = \left\lceil \frac{q \ln \varepsilon}{\ln(1 - qp_1)} \right\rceil \qquad \text{(A-8)}$$

The upper bound comes from the rather unrealistic case where the region corresponding to p_1 is identically distributed over the q parts. In cases where the region is contained in one of the q parts N_q will be given by the lower bound. From Eq. (A-7) and Eq.(A-8) we see that the smaller p_1 and ε, the greater will be the cost of obtaining a solution. We shall now turn to the question of how to measure this cost.

4. THE COST OF OBTAINING A SOLUTION.

How, then, should the cost of obtaining an estimate $\hat{f}^* = f(\hat{x}^*)$ of f^* be measured? In a study where all algorithms are run on the same computer the computer cost (CPU time) could be used. In this case it would also be possible to take into account such factors as ease of use, memory required and so on. Another important measure of the cost and one which is easier to use when comparing runs performed on different computers would be the number of function evaluations. If, in addition, the time required for one function evaluation and for the whole run were given, this would make comparisons between runs on different computers possible. (See also Shanno and Phua (1975)). For some methods it would also be of interest to give separately the cost of finding a point leading to an estimate of a global optimum and for final local optimization required to achieve a prescribed tolerance.

For constrained problems it may be expensive to obtain feasible points, i.e. points in A.

[1] Note that $N_{q=1}$ is identical with N_1 in Eq. (A-7).

If the sampling of points is performed in a hyperrectangle $H \supset A$, then the choice of H may have a great influence on the performance of a method and therefore H should be explicitly given.

In order to compare the performance of different methods when used with a problem the performance index u in Eq. (A-4) for each method could be computed for different cost levels. If the results could be shown to hold for a class of problems, this could then provide a basis for method selection when estimating a global optimum for a problem belonging to a given class.

We are now in a position to discuss the steps of Algorithm LC in more detail.

5. GLOBAL POINTS.

The goal governing the choice of global points should be to obtain starting points leading to a global optimum. If no a priori information about the location of the global optimum is available all parts of A should be treated as equally important when sampling the global points. The stratified sampling technique described in section 3 will be used. The number of points to be sampled will be determined by specifying ε and using a value p_1' instead of the unknown p_1 in Eq. (A-8). The value $1-\varepsilon$ will then be an estimate of the probability P_1 of obtaining a solution if the additional condition $p_1' \leq p_1$ is satisfied. In order to obtain global points in A the region is circumscribed by a hyperrectangle H. Sampling is then performed in H and the points falling in A are taken as global points. If a priori information as to where the optima are located is available this information can be supplied in form of quesses x_{g_i}. These will be included in the set of global points if $x_{g_i} \in A$.

6. LOCAL SEARCH.

A local optimizer aims to find the local optimum for a given starting point. Our goal is to use the local optimizer to push the N global points towards the local optima so that clusters of points leading to the same local optimum can be found by using the clustering analysis technique.

The local optimizer used for a given problem should be chosen to suit the problem, i.e. an optimizer which is rapid for the given function and region. In order to save function evaluations in the clustering phase the optimizer should be capable of good initial progress.

The choice of local optimizer is critical for MS and LC when the cost of a global optimization is considered. This is because most of the time will be spent in using the local optimizer. A good choice may however require some a priori information about the function f and the region A. If no a priori information about f and A is assumed to be available then a method that works under general conditions could be used. The method implemented here is a method named UNIRANDI. (Cf Järvi (1973)).

UNIRANDI, as used here, consists of the two elements, random search and linear search. Random search means that a line (direction determined at random) is laid through the base point. If one of the two points on that line at distance d from the base point is better than the base point, a linear search is performed in the first promising direction. If two lines are tried without success the distance d is halved and a new random search is started. Linear search means that points at distances 2d, 4d, 8d, ... on the line are tried as long as better and better points are obtained. The best point at distance $2^k d$ from the base point for example, is then taken as the new base point and a new random search is started with $d = 2^{k-1}d$.

The local search processes will be stopped when the step-length is reduced to some given threshold. This threshold will then be the starting step-length for the search processes next time LC2 is performed.

7. FINDING CLUSTERS.

Cluster analysis techniques are used in various scientific areas for dividing a finite set of objects or points into subsets so that each object is more similar to objects within its subset than to the objects outside. We shall assume that the objects can be represented as points in the n-dimensional Euclidean space and that the deviance (square of the Euclidean distance) is used as the dissimilarity measure. Kendall (1972) gives several reasons for using the deviance as the dissimilarity measure when possible.

A class of clustering techniques which does not necessarily require that the whole dissimilarity measure matrix be computed is the technique of growing a cluster from a centre or a seed point. Classification procedures using this technique can be expected to give satisfactory results for hyperspherical and well-separated clusters. One problem that has to be solved when using such a technique is how to locate the seed points. Another problem which is common for most classification procedures is the choice of some condition (threshold distance) for stopping the growth of a cluster. The choice of the threshold distance r is critical because if r is very large, all the points will be assigned to a single cluster and if r is very small, each point will form an isolated cluster.

The preordering of the K points in accordance with their suitablility as seed points is performed by sorting them in ascending order with respect to the corresponding function values. The assumption made is that the local minima would be good seed points to use and that the point with the smallest function value may be taken as an estimate of a local minimum. Therefore when a new seed point is needed the point, among those not yet classified, with the smallest function value is chosen.

The cluster is grown by enlarging a hypersphere around the seed point as long as the point density in the hypersphere remains greater than K/V where the reference region V (the region spanned by the points) is determined by

$$V = \prod_{i=1}^{n'} 4\lambda_i^{\frac{1}{2}} \qquad \text{(A-9)}$$

In Eq. (A-9) λ_i, $i = 1, \ldots, n$ ($\lambda_1 \leq \ldots \leq \lambda_n$) are the roots of the equation $\det(S - \lambda I) = 0$ where S is the covariance matrix corresponding to the points $x_1, \ldots, x_K$. If the points are contained in a subspace ($n' < n$), this will be discovered and taken advantage of in the clustering algorithm.

The number of distances D_s that have to be computed is

$$D_s = c(K - 1) - \sum_{i=1}^{c-1} (c - i) K_i \qquad \text{(A-10)}$$

where c is the number of clusters and K_i is the number of points in the i:th cluster. For $c = K$ Eq. (A-10) gives $D_s = D_o = K(K - 1)/2$, i.e. the same number of distance computations as when the whole distance matrix is used. Usually c is much smaller than K so that $D_s \ll D_o$.

For further details in the cluster analysis algorithm, see Törn (1976a).

8. REDUCTION OF POINTS.

In order that a large number of global points may be used to increase the probability of obtaining a solution, the number of points which are pushed towards the local solutions should be reduced during the iterations of algorithm LC. The number of points should be reduced in such a way that not all the points leading to a solution, if any present, are rejected. Of course, this cannot generally be guaranteed and the probability of success will also depend on the specific problem and on steps one and two in algorithm LC. If the foregoing steps were successful, it would be sufficient to choose one point from each recognized cluster in order to guarantee a solution. In general, at least one point from each cluster should therefore be chosen. If more than one point is chosen from a cluster, the motive for this is to assure that a found cluster really consists of points leading to a unique local solution. The points should therefore be chosen to test this as well as possible. One natural way would be to use cluster analysis for the points in each cluster to obtain subclusters and then choose one point from each subcluster. Another way would be to rank the points in each cluster in ascending order of function values and then choose every second, third or so in the hope that function values corresponding to points leading to a lower local optimum would also be smaller. This technique has been implemented.

9. COMPARISON WITH MULTIPLE LOCAL SEARCHES.

The work, measured in terms of the number of function evaluations, used by multiple local searches w_{MS} is given by

$$w_{MS} = \bar{n}N \qquad \text{(A-11)}$$

where N is the number of global points sampled and $\bar{n}$ is the average number of function evaluations needed to find a minimum.

Consider the application of LC for which the N global points are classified after an average of m steps for each point. Assume that c clusters are obtained and that the best point from each cluster is permitted to continue until the local solutions are reached. The total amount of work used by LC is then

$$w_{LC} < w'_{LC} = mN + c(\bar{n} - m) + v\beta \qquad \text{(A-12)}$$

In Eq. (A-12) v is the work, measured in terms of the number of deviance calculations, required for the classification and β is the ratio between the cost of a deviance calculation and the cost of a function evaluation.

For the best point in a cluster less than $(\bar{n} - m)$ function evaluations would be needed to reach the local minimum and therefore it might be expected that $w_{LC} < w'_{LC}$.

Using the approximation $Nc + Nn + n^2$ (work for clustering, computation of S and λ) for v we obtain

$$\frac{w_{LC}}{w_{MS}} \approx \frac{m}{\bar{n}} + \frac{c}{N} + \beta\left(\frac{n + c}{\bar{n}}\right) \qquad \text{(A-13)}$$

The last term will be small for small β, i.e. problems for which the function evaluations are expensive. If m is large enough, then c will be a good estimate of the number of local minima. Because N will be of the order of 100($\varepsilon = 0.01$, $p_1 = 0.05$ give $N = 92$) the term c/N will be small for problems with a small number of local minima ($c \leq 10$). We obtain

$$\left.\begin{array}{r}\beta\ ,\ c/N<0.1\\ n<10\end{array}\right\} \Rightarrow \frac{w_{LC}}{w_{MS}} < \frac{m+2}{\bar{n}} + 0.1 \qquad \text{(A-14)}$$

which means that $w_{LC}/w_{MS} < 1$ for $m < 0.9\,\bar{n} - 2$. A successful application of LC will normally permit much smaller m to be used. In numerical experiments ratios w_{LC}/w_{MS} smaller than 0.5 have been obtained. For a full application of LC, i.e. when the steps LC2 - LC4 are repeated many times, small m could be used and then still more would be gained by using LC instead of MS.

10. NUMERICAL RESULTS.

In order to determine how many global points N to sample we solve Eq. (A-8) for $q = N_q$ and $p_1' = \varepsilon$. One solution is $N = 64$, $p_1' = 1/65$. This solution means that A is divided into 64 identical parts and that one point is sampled in each part. If at least one of the 64 parts contains a region $T \subset T_1'$ so that $\mu(T)/\mu(A) \geq 1/65 \approx 0.015$, then the probability of obtaining at least one global point leading to the global optimum is no smaller than $64/65 \approx 0.98$. For the given test functions $A = H = [a_1, b_1] \times \ldots \times [a_n, b_n]$ and the 64 parts are obtained by dividing the intervals $[a_i, b_i]$ into parts according to Table 1.

q_i \ i / n	1	2	3	4	5	6
2	8	8				
3	4	4	4			
4	2	4	2	4		
6	2	2	2	2	2	2

Table 1. $A = [a_1, b_1] \times \ldots \times [a_n, b_n]$ is divided into 64 parts by dividing each interval $[a_i, b_i]$, $i = 1 \ldots n$ into q_i parts $\left(q = \prod_{i=1}^{n} q_i\right)$

The algorithm was stopped when two successive clusterings resulted in the same number of clusters. In step LC4 the best point in each cluster was permitted to continue.

The results from the first run for each test function on a UNIVAC 1108 computer are shown in Table 2 and Table 3.

From the column p_1 in Table 2 it can be seen that all the functions are easy to solve when using probabilistic methods. For the worst case, i.e. SHEKEL 10, $p_1 \approx 0.2$, which means that when sampling 14 points at random in A the probability of finding at least one leading to the global minimum is greater than 0.95. For all the problems $p_1 > p_i$, $i = 2, \ldots, k$ i.e. the global minimum is also the easiest local minimum to find. For the test functions SHEKEL 5, SHEKEL 7, RCOS and GOLDPR the clusters determined corresponded to the existing local minima. For SHEKEL 10 the local minimum (1,1,1,1) was not located. For HARTM 3 the local minimum with the approximate location (.47, .44, .75) was not located. For HARTM 6 the six clusters actually represented only two of the local optima, namely the one given in Table 3 and (.40, .91, .82, .57, .077, .036). The CPU-time in Table 2 consists of a part caused by the function evaluations and a part relatively constant for a given N and n. For the test function HARTM 6, 6.4 = 2.6 + 3.8, i.e. the main part was caused by calculations other than function evaluations. If however, the function were 15 or more times costlier to evaluate, then this part would contribute less than 10 per cent to the total CPU-time. A comparison with MS showed that the number of function evaluations required would have been 1.5 to 8 times larger for MS than for LC.

Function	$\hat{p}_1$	n	lo	c	CPU-time s	$f(\hat{x}^*)$	fev
SHEKEL 5	.3	4	5	5	4.1	10.1422	3649
SHEKEL 7	.3	4	7	7	5.1	10.3909	3606
SHEKEL 10	.2	4	10?	9	5.9	10.5300	3874
HARTM 3	.7	3	4?	3	3.3	3.86260	2584
HARTM 6	.65	6	4?	6	6.4	3.31370	3447
RCOS	1.00	2	3	3	1.5	.397901	1558
GOLDPR	.7	2	4	4	1.7	3.00001	2499

Table 2. $\hat{p}_1$ = estimated probability that a point sampled at random will lead to the global optimum

n = dimension of space

lo = number of local optima

c = number of resulting clusters

$f(\hat{x}^*)$ = best value obtained

fev = number of function evaluations performed

Standard time for 1000 evaluations of SHEKEL 5 is 0.41 ± 0.02 s.

Function	$\hat{x}^*$					
SHEKEL 5	4.003	3.998	3.991	3.996		
SHEKEL 7	4.009	4.001	4.000	4.007		
SHEKEL 10	3.995	4.005	4.000	3.996		
HARTM 3	.1310	.5556	.8526			
HARTM 6	.219	.142	.473	.265	.309	.661
RCOS	-3.143	12.28				
GOLDPR	.202E-3	-1.00007				

Table 3. Estimates of x^* corresponding to $f(\hat{x}^*)$ in Table 2.

11. DISCUSSION.

The number of function evaluations required to obtain solutions to the test problems may seem to be large. This number could be reduced by using some a priori information about the problem.

Such information would, for instance, permit the use of a local optimizer suitable for the problem in question. Despite the low accuracy in the results, UNIRANDI used 100-330 function evaluations on average for a point in the final clusters. This number could have been substantially reduced by using a more sophistical local optimizer.

Information about the size of p_1 could bring down the number of global points required for a solution. For the given test functions with $p_1 \geq 0.2$ 5-6 global points might be expected to produce the same $\hat{f}^*$ as obtained here with 64 global points and then the required number of function evaluations would have been only a tenth of those given in Table 2.

As pointed out earlier the value of p_1 is very critical for the success of probabilistic methods like LC, and an assumption regarding the value of p_1 will greatly influence the cost of obtaining a solution. For other techniques related a priori assumptions are made which are critical in the same sense. It is enough to mention the Lipschitz constant in methods like those of Evtushenko and Shubert and the allowable integrating step size in Branin's method.

Information about the number of local optima in A and an interactive use of LC could also reduce the number of function evaluations required. For instance, for RCOS the first iteration in LC requiring 917 function evaluations gave 13 clusters with the number of points in the clusters (ordered according to increasing fuction values of the best point in each cluster given by 14, 16, 14, 2, 2, 8, 1, 1, 1, 1, 2, 1, 1. The three local optima here are easily determined. Also, if the function evaluations are very expensive, then an interactive use and frequent cluster analysis could save much computer time.

12. CONCLUSIONS.

The approach to global optimization presented here has been successfully applied to a number of test problems with minima in the interior of the region of interest A. The algorithm has also been applied to constrained problems Törn (1974) by using the absolute boundary technique for working points that attempts to cross the boundary of A.

An a priori assumption regarding the size of the region of attraction of the global minimum will greatly influence the cost of obtaining a solution. Related critical assumptions apply to other optimization techniques. The local optimizer used here is rather robust but inefficient for most problems. An ideal local optimizer should be rapid for the given problem and should be capable of good initial progress. It might prove advantageous to use separate local optimizers for the first and final iterations in algorithm LC.

Techniques other than local optimization for gathering the points around the local minima could also be considered.

The iterations in LC make interactive use possible. Human intervention between iterations could save much computer time for expensive functions. The results obtained here indicate that LC could be feasible for functions 100-1000 times more expensive to evaluate than the given test functions.

REFERENCES

Bocharov, N. and A.A. Feldbaum (1962). An automatic optimizer for the search of the smallest of several minima, Automation and Remote Control 23, No. 3.

Dixon, L.C.W., J. Gomulka, S.E. Hersom (1975). Reflections on the Global Optimization Problem, Numerical Optimization Centre, Technical Report No. 64, The Hatfield Polytechnic.

Hartman, J.K. (1973). Some Experiments in Global Optimization, Naval Research Logistics Quarterly 20, No. 3, 569.

Järvi, T. (1973). A Random Search Optimizer with an Application to a Max-min Problem. A Special Trajectory Estimation Problem, Inst. of Appl. Math., Un. of Turku, No.3.

Kendall, M.G. (1972). Cluster analysis, Frontiers of Pattern Recognition, (S. Watanabe, ed.) (Academic Press)

Shanno, D.F. and K.H. Phua (1975). Effective Comparison of Unconstrained Optimization Techniques, Management Science 22, No. 3, 321-330.

Törn, A. (1974). Global Optimization as a combination of Global and Local Search, Åbo Swedish University School of Economics, A:13.

Törn, A. (1976a). Cluster Analysis using Seed Points and Density-Determined Hyperspheres with an Application to Global Optimization, Proceedings of the Third International Joint Conference on Pattern Recognition, November 8-11, 1976, Coronado, California, 394-398. IEEE Trans. on Systems, Man and Cybernetics, August 1977.

Törn, A. (1976b). Probabilistic Global Optimization, a Cluster Analysis Approach, Proceedings of the Second European Congress on Operations Research, North-Holland, Amsterdam, 521-527.

TOWARDS GLOBAL OPTIMISATION 2
L.C.W. Dixon and G.P. Szegö (eds.)

A USERS EXPERIENCE WITH TORN'S CLUSTERING ALGORITHM

Dr. J. Gomulka
The Numerical Optimisation Centre
The Hatfield Polytechnic
Hertfordshire, England.

The aim of this paper is to give a user's impressions of two versions of Törn's clustering algorithm. Aimo Törn helped in this study by sending a FORTRAN listing of the algorithm and several explanatory papers, including a user's guide to the N.O.C. His help is gratefully acknowledged.

1. Introduction

In this paper I describe my initial experiences as a user of Törn's clustering algorithm. The algorithm was tested both on a series of standard test functions and on a series of econometric problems. Our experience was that it was consistently reliable and provided considerable information on the nature of the function being tested.

Törn's algorithm consists of the following steps.

Step 0 A sample of points from the domain of an objective function is taken. The points are uniformly distributed over the feasible region.

Step 1 Each of the sampled points is moved towards a local optimum. That is, each point, in turn is taken as an initial approximation for the local optimisation routine; a few iterations of the routine are then performed and the best point obtained replaces the initial point.

Generally speaking, Step 1 transforms a uniform sample into a number of separate "constellations", i.e. groups of points, which have come closer together in the process of converging to the same local optimum.

Step 2 A procedure known in pattern recognition as "clustering" is used to determine groups or "clusters" of points.

If one could be sure that each cluster corresponds to one local optimum, then all but one point from each cluster could be discarded without any essential loss of information. But usually a more cautious course is taken.

Step 3 A certain fraction of points from each cluster is discarded (e.g. only every second or third point is left).

Steps 1 - 3 form an iteration of the algorithm, also referred to as "a clustering", and are repeated in a loop. The process stops when a prescribed number of clusterings has been finished or when a maximum number of function evaluations has been exhausted. Both the maximum number of clusterings and of function evaluations are input parameters to the routine.

A. Törn had kindly made available to N.O.C. a FORTRAN version of his algorithm. We introduced some changes and experimented with both the original and the modified versions. Some comments on individual steps may ease a more general discussion of the results.

Step 0 The algorithm can in principle be used for both the essentially unconstrained problems, i.e. those for which the only bounds are on the size of the region and the solution is expected in an interval point, and for certain constrained problems. For bounds of the form $a_i \leq x_i \leq b_i$, i = 1, ..., n, it is enough to enter the values of a_i, b_i and the points will be sampled in an appropriate box. Other constraints have to be expressed by a feasibility function IRA(x), programmed by the user, such, that IRA(x) = 1 if x is feasible and IRA(x) = 0 otherwise. Then each time a new point is sampled (or, in Step 1, arrived at by the optimisation routine) the value of IRA is checked and infeasible points are rejected. Obviously only inequality constraints defining a domain with interior points can be dealt with in this way and even then it may prove expensive. Our tests were limited to the simplest case of a box-like region.

The output from Step 0 is a list of points, with their coordinates, respective function values and names. The name of a point is simply an integer denoting its place on this initial list. The name does not change when the coordinates of a point are transformed (in Step 1).

Apart from points sampled at random the intial list can contain "guesses", i.e. any points, which the user for any reason would like to include. In particular it can consist of guesses alone.

Step 1 Any local optimisation routine can be used. As Törn noticed in [3], the choice depends on what is likely to work well for a particular problem. In Törn's own code a direct random search routine is used, which doubles or halves the step depending on success or failure in the preceding iteration. When a local minimum is approached the step size tends to 0. Hence the current step size is to a certain degree a measure of progress of the routine. For each clustering a threshold value has to be specified and the local optimisation is stopped as soon as the current step size falls below the threshold.

We substituted for the random search routine one of the N.O.C. optimisation algorithms (OPVM) of the variable metric type, using numerical estimates of the derivatives. Its progress was monitored and controlled by the current gradient value in exactly the same way as the original implementation had been controlled by the stepsize.

It is perhaps worth mentioning that we have not met with any major programming problems while making this and other changes. Törn's algorithm is not only well organised, but has a clear structure and an easy to follow flow of information between subroutines, and a very good and consistent notation.

Step 2 The clustering subroutine used by Törn operates without any outside constants, solely on the basis of the list of points, which are to be divided into clusters. First the dimension of a vector space spanned by the points and the volume of the least box in that space, which contains them, are calculated. The dimension of this space can and often is less than the dimension of the original space, especially when the local minima, towards which convergence in Step 1 occurs, all lie in a subspace. The average distance between two points can then be calculated. Clusters are determined as groups of points with less than average distance from one another (strictly speaking, the distance from the so called "seed point" in each cluster is crucial). (Törn, [4]).

Before clustering, the points are ordered according to the function values, with better points coming first. Therefore the first cluster contains the best point and generally better points come earlier for further processing in this and the next iteration.

In general the routine worked well in our tests. However, the variables have to be scaled so that the input from each coordinate to the euclidean distance, on which the clustering process depends, carries approximately the same weight.

Step 3 This subroutine, which reduces the number of points in each cluster, first prints out the results of Step 2. For each cluster a list of its points is printed, with the name and coordinates of each point, the respective function value and some other information.

In an ideal situation one cluster would correspond to each local minimum and the best point in the cluster would give the approximate position and value of the minimum. The first cluster should correspond to the global minimum. But even when all local minima are eventually located this ideal setup may not be established until some late iteration. At the beginning there are usually changes from one clustering to the next; some clusters split, others merge, individual points pass from one cluster to another.

The Törn algorithm can be regarded as an important and considerable refinement of a popular strategy of repeated local optimisations from different starting points. In fact it includes this strategy as one of its options; initial points can be entered in Step 0 as "guesses" (or, alternatively, can be sampled), the number of clusterings can be put to 1 and the threshold value controlling the local optimisation in Step 1 can be made equal to the desired accuracy of the final results. With this choice of parameters the algorithm is equivalent to repeated local optimisation and the clustering does not play any substantial role.

But in most cases this would not be the best choice. In general the user is faced with the task of striking the right balance between the number of initial points and the number of iterations (i.e. clusterings). The maximum number of function evaluations and the accuracy are usually dictated by cost constraints and by technical requirements and can be regarded as fixed. Let the rate of reduction of points in clusters also be fixed. Then it might be expected, that the more clusterings that are planned the greater the number of initial points that can be allowed, since only a few steps of the local optimisation routine will be performed before reducing the number of points. Hence for the sake of greater confidence, which is associated with denser covering of the region by starting points, one might be inclined to shorten the runs of the local optimisation routine between two consecutive clusterings. However, going too far in this direction proves counterproductive: if the points are not moved far enough, then their distribution remains virtually uniform. Consequently, most "clusters" consist of one point and only very few, if any, points can be discarded.

The right choice of parameters depends on the geometry of the objective function and to some extent on the characteristics of the local optimisation routine. When there is little or no a priori knowledge of either or both these factors, it would be probably best to run the algorithm interactively. It is well suited to such treatment and very little extra programming is needed to make this option available. The computing would have to be stopped in the middle of Step 3, after printing the lists of clusters, and readjustment of parameters made possible.

The interactive mode of operation could be very useful also for other reasons. The method is geared to giving information about many, ideally all, local optima. But in many, perhaps even in most problems only the global optimum and perhaps those local optima which have only slightly worse function values, are of interest. From this point of view it is often evident at an early stage that some clusters offer very little hope and could be discarded altogether or at least pruned very severely. On the other hand some clusters may show signs of impending split, which has not yet been detected by the clustering subroutine, and any reduction should

Function	Local optimiser	No. of global points	Accuracy ϵ	No. of function evaluations	C.U.P. time secs	No. of clusterings	$\rho = \frac{\text{f.e. Törn}}{\text{f.e. MS}}$	$\frac{\text{f.e. Törn}}{\text{f.e. MS}}$ after 3rd clustering (ϵ = .0013)
SHEKEL5	UNIRANDI	40	.0000015	6654	47	7	41%	78%
(f* = 10.1532)	OPVM			7085	53		32%	72%
SHEKEL7	UNIRANDI			6081	43		63%	92%
(-10.4029)	OPVM			6684	65		37%	69%
SHEKEL10	UNIRANDI			6144	56		54%	75%
(-10.5364)	OPVM			7352	65		40%	72%
HARTM3	OPVM		.01	6766	46	3	59%	
HARTM6 (-3.32237)	OPVM			11125	132	3	43%	
GOLDPR	UNIRANDI	20	.001	2006	6	2	90%	
	OPVM		.005	1495	6	2	82%	
RCOS	UNIRANDI		.001	659	4	2	66%	
	OPVM		.005	1318	7	3	66%	

Table 1

preserve points which are likely to go different ways. In such and similar situations a direct look can often easily extract relevant information, which would be difficult to obtain in a purely automatic way.

Results of tests on the standard test function are collected in Table 1. Comparison with the results obtained by Törn [4], with the same local optimiser (UNIRANDI), indicate large discrepancies. There can be several reasons for this. First, the accuracy requirements could have been different in the two runs. Notice that on the SHEKEL functions, for which we used up to 60 - 80% more function evaluations than Törn, our accuracy requirements were rather stringent, and the best function values obtained were slightly better than Törn's. When, on the other hand, these requirements were slackened, as for GOLDPR and RCOS, our numbers for function evaluations are much lower than Törn's. The use of a different number of global points will also have had an effect. On two functions we ran tests, in which only the number of global points was changed; they are given here for comparison with Table 1.

SHEKEL5	UNIRANDI	10 gl. points	1471 f.e.	10 sec C.P.U.
		40 gl. points	6654 f.e.	47 sec C.P.U.
	OPVM	10 gl. points	1829 f.e.	16.5 sec C.P.U.
		40 gl. points	7085 f.e.	53 sec C.P.U.
GOLDPR	UNIRANDI	20 gl. points	2006 f.e.	6 sec C.P.U.
		40 gl. points	4086 f.e.	13 sec C.P.U.

In this range of numbers of global points (10 - 40) the workload seems proportional to the number of points used.

Another explanation may be that Törn was using a slightly different version of his algorithm. In the version which we received, the number of clusterings (iterations) was fixed at the beginning, while in his paper Törn [4] states that he continued until two consecutive clusterings gave the same number of clusters, then took the best point from each cluster and let the local optimiser work on these points only.

This discussion shows that one cannot be too careful when choosing the parameters which control the algorithm. This is hardly a new truth as it applies equally to many other algorithms but it is vitally important nonetheless.

Table 1 shows clearly that on the standard test functions UNIRANDI performed better than OPVM. However, we met practical problems, on which UNIRANDI was unable to achieve the accuracy demanded, while the clustering algorithm with OPVM gave very satisfactory and interesting results. These problems were concerned with fitting the parameters of a mathematical model to data by the least squares method. In a way it was perhaps a more instructive test of the method, since the optimised functions were not built artificially and neither the number nor character of their local minima could be investigated analytically a priori. We shall return to these tests later, but first we shall explain the last two columns of Table 1.

The numbers in the last full column were obtained from the following expression:

$$\rho = \frac{fev_d}{fev_{loc} \times n_{gl}}$$

where fev_d is the number of function evaluations used by the algorithm (= column 5) fev_{loc} is an estimate of the average number of function evaluations needed by the local optimiser (in fact the arithmetic mean of the corresponding numbers for points in the final clustering), and n_{gl} is the number of global points. Thus ρ is the ratio of the work actually done to the estimated work of completing local optimisations from all the global points. The last column, for the SHEKEL functions only, gives the same numbers after the 3rd clustering, when the pattern of clusters was well established but the accuracy much lower (ϵ is the last step-size - or gradient - threshold) than after the final clustering. It must be stressed, that, the numbers are not an accurate estimate, because fev_{loc} is itself a very rough estimate. This is because the number of points in the final iteration is usually small and, moreover, the corresponding numbers of function evaluations, needed for each point, can vary considerably (in extreme cases we had numbers for some points five times greater than for others). However, the figures do give some general idea of the saving that can be expected. In particular, they could aid the user in the choice of the number of initial global points, if cost is a constraint and the number of repeated local minimisations, that could be afforded, can be estimated. A rule of thumb would be: twice the number of local optimisations, if a high accuracy is desired, less for lower levels of accuracy. To finish this discussion, we shall quote values of ρ obtained for the parameter estimating problems (24 - 36 global points, $\epsilon = 10^{-6}$): 66%, 47%, 48%, 56%, 69%, 61%, 75%.

This parameter estimating exercise was connected with an attempt to explain certain peculiarities in the industrial growth of the U.S.S.R. in the postwar period [S. Gomulka, (1977)]. A number of models was considered, with the number of unknown parameters varying between 4 and 7. The models were fitted to statistical data covering 30 years. Perhaps the most interesting experience arose due to a mistake made in specifying some of the models, so that there was a functional dependence between some of their parameters. The algorithm reacted by generating a rather large number of clusters in the final iteration, all with virtually the same function values. For example, in one case 11 clusters were obtained (from

24 points sampled initially). For the first four f = 0.0031033, and x_3 = 0.021, x_4 = 0.95 for all clusters, while the coordinates, x_1 and x_2 were very different for each cluster: x_1 = 4.12, 0.67, 1.85, 3.03; x_2 = 5.41, 0.80, 2.31, 3.98. The remaining seven clusters had the common function value f = 0.01609 and again common values of x_3 and x_4 and widely scattered values of x_1 and x_2. It was a very clear indication of dependence between x_1 and x_2.

This was one example in which benefits from the clustering method could not be fully acknowledged by ρ or some similar "efficiency index". The way in which the whole process was organised and its results displayed mattered even more. One can hope that with more experience in using and perfecting the clustering method it will become possible to spell out its "qualitative" merits better and that they will become more important. In the meantime let me at least add one user's impression: output from the clustering algorithm makes fascinating reading, from which a lot of extra information can be extracted. It is a reliable method which consistently locates the global minimum, but further research is still required to provide a routine less sensitive to the values of the controlling parameters.

Acknowledgement

The author acknowledges the financial support of the S.R.C. while working on this project.

References

1. Gomulka, S. (1977). "Slowdown in Soviet Industrial Growth 1947-75, Reconsidered". European Economic Review (to appear).
2. Törn, A. (1974). "Global Optimisation as a Combination of Global and Local Search". Abo Akademy, Finland.
3. Törn, A. (1976). "Probabilistic Global Optimisation: A Cluster Analysis Approach". Proceedings of EURO II.
4. Törn, A. (1977). "A Search Clustering Approach to Global Optimisation". In "Towards Global Optimisation 2" Eds. L.C.W. Dixon & G.P. Szegö, North-Holland Publishing Company.

TOWARDS GLOBAL OPTIMISATION 2
L.C.W. Dixon and G.P. Szegö (eds.)

A CONTROLLED RANDOM SEARCH PROCEDURE FOR GLOBAL OPTIMISATION

W.L. Price
Department of Engineering
University of Leicester

A new random search procedure is described which, while conceptually simple and easily programmed on a minicomputer, is effective in searching for global minima of a multi-modal function, with or without constraints. The results of trials, using a variety of test problems, are given.

INTRODUCTION

Many methods are available for the optimisation of a given function that is uni-modal within the domain of interest. Gradient methods, such as that of Fletcher and Powell (1963), involve the evaluation of the function and its derivatives at each iteration. Direct methods, involving only function evaluations, are generally less efficient than gradient methods but, because they do not require the calculation of derivatives, are simpler and are applicable to the optimisation of non-differentiable functions. Typical of the direct methods are those of Rosenbrock (1960), Hooke and Jeeves (1969), and Nelder and Mead (1965).

The problem of global optimisation of a multi-modal function has received much less attention. If sufficient is known of the properties of the function to enable the approximate positions of the minima to be estimated, then exploration of the individual modes by established uni-modal methods should lead to the discovery of the global minimum. Without such prior knowledge of the behaviour of the function there would appear to be no alternative to a thorough search over the domain of interest. Having found, with sufficient confidence, the approximate position of a global minimum, a uni-modal method may be used for a final refinement. A systematic search so thorough as to guarantee the discovery of a global minimum would, in general, require so many function evaluations as to be quite impracticable. One must be content with a limited search, systematic or random, and accept the consequent uncertainty as to whether or not a global minimum has been found.

Brooks (1958) has reviewed several random search techniques. The simple random method makes a specified number of trials at points randomly selected in the chosen search domain, V, and accepts, as the optimum, the trial point with lowest function value. The stratified random sampling method divides V into a specified number of sub-domains of equal size and selects, at random, a trial point within each. Such methods represent the best one can do with a limited number of trials if the function itself is quasi-random - i.e. if there is little or no correlation between the values of the function at neighbouring points. In practice however it is usual for correlation to exist - even when the function is discontinuous or the variables discrete some correlation will normally be expected (as in a histogram). It is reasonable, therefore, to suppose that the efficiency of the optimisation procedure can be improved by progressively focussing the search upon those sub-domains which currently tend to contain the points with lowest function values.

Such an improvement on the simple random method is provided by the optimisation routine devised by Becker and Lago (1970). Their procedure begins with a simple random search over the chosen domain, V. Instead of retaining only the point with the lowest function value, as does the simple random method, Becker and Lago retain a pre-determined number of points (those with the lowest function values). If the total number of trials is sufficient then the retained points tend to cluster around minima. A mode-seeking algorithm, such as that developed by Bryan et al (1969), is then used to group the points into discrete clusters and to define the boundaries of sub-domains each embracing a cluster. The clusters are graded, by searching in each for the retained point with the lowest function value, and then rated according to the relative values of the cluster minima. The entire procedure is then repeated using as the initial search region that sub-domain, defined by the mode-seeking algorithm, around the 'best' cluster. The user may choose to examine also the second-best cluster (or indeed all clusters) according to the extent of his doubt as to whether or not the global minimum will be found in the sub-domain defined by the best cluster.

A CONTROLLED RANDOM SEARCH OPTIMISATION PROCEDURE

The new global optimisation procedure about to be described is similar to that of Becker and Lago in that it is a direct, random, method which does not require the function to be differentiable or the variables to be continuous, and which is applicable in the presence of constraints. The controlled random search (CRS) procedure differs from the Becker and Lago method in that it combines the random-search and mode-seeking routines into a single, continuous process. By eliminating both the separate mode-seeking algorithm and the subjective decision making implicit in the choice of clusters for further investigation, CRS provides a simple procedure which, although best used interactively, will function automatically when interactive monitoring is not convenient.

The CRS Algorithm

The essential features of the algorithm are indicated in the flow diagram of fig. 1. An initial search domain, V, is defined by specifying limits to the domain of each of the n variables and a predetermined number, N, of trial points are chosen at random over V, consistent with the additional constraints (if any). The function is evaluated at each trial and the position and function value corresponding to each point are stored in an array, A. At each iteration a new trial point, P, is selected randomly from a set of possible trial points whose positions are related to the configuration of the N points currently held in store (the way in which the set of potential trial points is defined will be explained later). Provided that the position of P satisfies the constraints the function is evaluated at P and the function value, f_p, is compared with f_m, M being the point which has the greatest function value of the N points presently stored. If $f_p < f_m$ then M is replaced, in A, by P. If either P fails to satisfy the constraints or $f_p > f_m$ then the trial is discarded and a fresh point chosen from the potential trial set. As the algorithm proceeds the current set of N stored points tend to cluster around minima which are lower than the current value of f_m. The probability that the points ultimately converge onto the global minimum (minima) depends on the value of N, the complexity of the function, the nature of the constraints and the way in which the set of potential trial points is chosen.

Because the procedure is intended to find global minima, thoroughness of search throughout the current sub-domains is of greater importance than speed of convergence to a potential minimum. Nevertheless, if the procedure is to be more efficient than pure random search the probability of success ($f_p < f_m$) at each trial must be sufficiently high. The CRS procedure achieves a reasonable compromise between the conflicting requirements of search and convergence by defining the set of potential (next) trial points in terms of the configuration of the N points currently stored. At each iteration n + 1

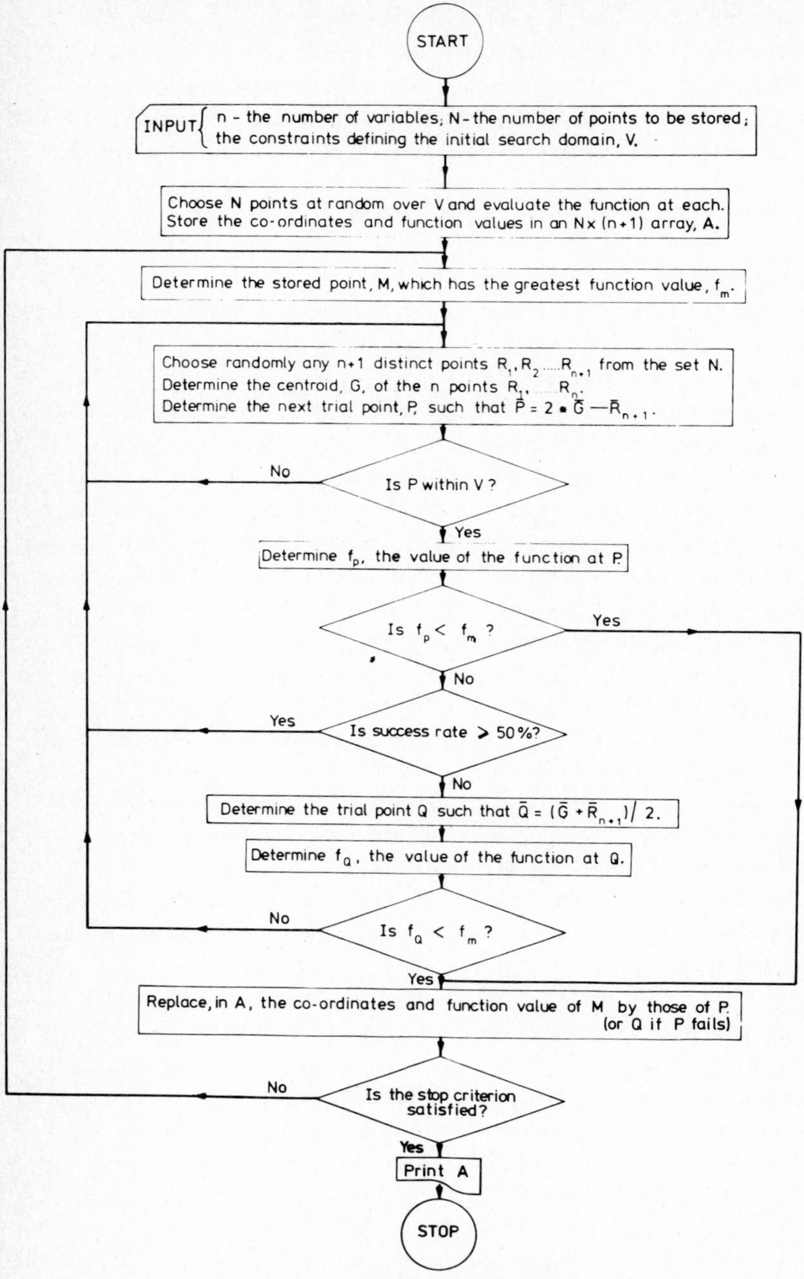

Fig. 1 FLOW DIAGRAM

distinct points, R_1, R_2, ... R_{n+1}, are chosen at random from the N (N>>n) in store and these constitute a simplex of points in n-space. The point R_{n+1} is taken (arbitrarily) as the pole (designated vertex) of the simplex and the next trial point, P, is defined as the image point of the pole with respect to the centroid, G, of the remaining n points.
Thus

$$\bar{P} = 2 \times \bar{G} - \bar{R}_{n+1}$$

where $\bar{P}$, $\bar{G}$ and $\bar{R}_{n+1}$ represent the position vectors, in n-space, of the corresponding points. The number of different ways in which n+1 points can be chosen from N is ${}^{N}C_{n+1}$ and because the choice of R_{n+1} as the pole is arbitrary (a given set of n+1 points may be generated in any random order) the total number of (equi-probable) next trial points associated with the configuration of N stored points is $(n+1) \times {}^{N}C_{n+1}$. The points generated by this procedure will be called <u>primary</u> trial points.

As a simple illustration of this principle fig. 2 shows an assumed configuration of six points at some stage in the search for global minima of a function of two variables. With N = 6 and n = 2 the number of primary trial points generated by this configuration is $3 \times {}^{6}C_{3} = 60$. It will be observed that the disposition of this set of 60 points reflects a general trend in the current configuration and hence the random choice of any one from this set as the next trial point, P, is likely to result in a more efficient search than a procedure based on uncontrolled random search within V. On the other hand the domain of the set is not restricted to the immediate neighbourhood of the configuration. This is conducive to 'out-going' exploration, thus avoiding the dangers, for global search, of too rapid convergence.

The example also shows that while the macroscopic structure of the primary trial set reflects a general trend, the local variations of point density within the set correspond to detailed patterning of the current configuration. Thus the algorithm takes account of the possibility that the current configuration, regarded as two clusters each of three points, might indicate bi-modality. Further, the clusters of primary trial points remote from the centre of the configuration can be interpreted as 'images' of the two clusters within the configuration, each seen through the other, thus promoting the search for other possible modes which might result from periodicities of the function. The algorithm has the merit of simplicity and objectivity (it is not required to judge the modality of a given configuration) but yet demonstrates an elementary pattern recognition capability.

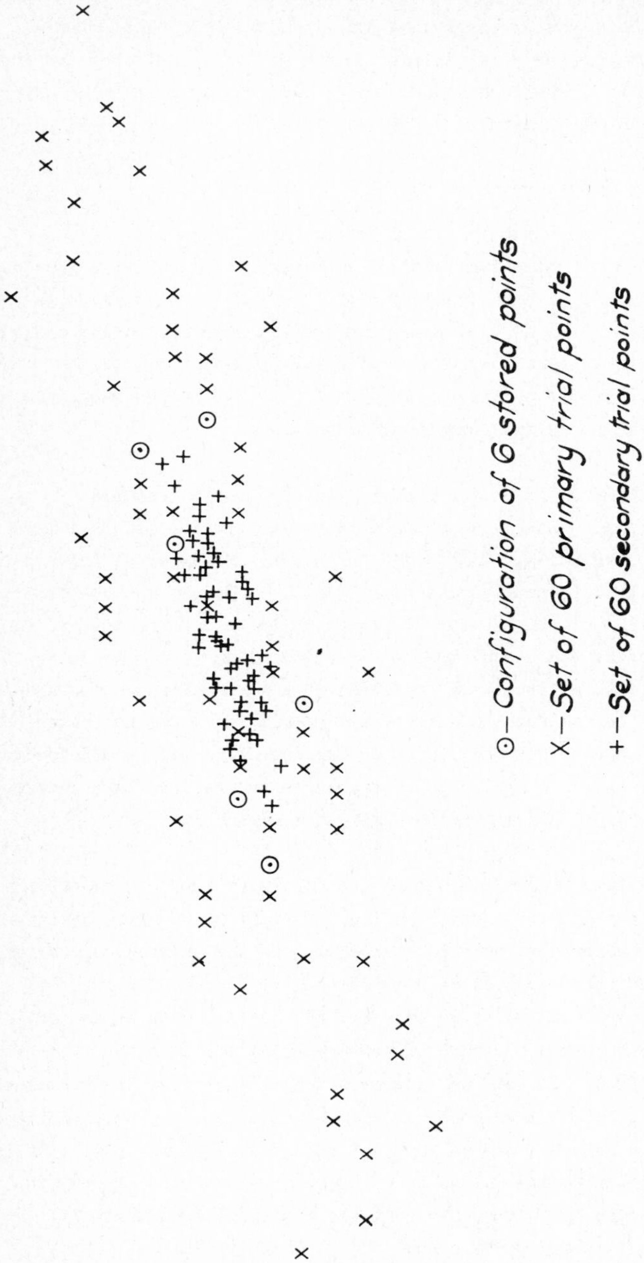

Fig. 2 An illustration of the algorithm for generating trial points in two dimensions

It might appear from the particular example of fig. 2 that, while the algorithm is more efficient than uncontrolled random search, the probability of success ($f_p < f_m$) in a given trial is nevertheless likely to be quite small. In part this arises because, for simplicity of illustration, a configuration of only six points has been assumed. In practice one takes N>>n and, typically, the author has chosen N = 50 for a two-dimensional search. It will be appreciated, however, that a significant proportion of failures is necessary to the effectiveness of the global search because only as a result of such failures can one be assured that the algorithm has reached out to the limits of the currently relevant contours ($f = f_m$). At any stage of the optimisation procedure let the cumulative success rate be defined as the percentage of successes obtained in the total number of trials (involving function evaluations) up to that stage. This success rate is high in the early stages of the optimisation but then falls off in a manner dependent upon the complexity and modality of the function. In the author's experience the success rate of the search based on primary trial points rarely falls below 20% (irrespective of the number of dimensions). It seems desirable, however, to aim for a success rate which does not fall appreciably below 50%. Without significantly reducing the effectiveness of the primary search the efficiency of the procedure can be increased, and the convergence consequently improved, by means of the following extension to the algorithm as so far described. For a given simplex and pole the corresponding primary trial point, P, has already been defined. Using the same notation a <u>secondary</u> trial point, Q, is defined by

$$\bar{Q} = (\bar{G} + \bar{R}_{n+1})/2 .$$

In addition to the primary trial points fig. 2 also shows the secondary trial points generated by the given configuration of six points. Whereas the primary trial points are search oriented it is apparent that the secondary trial points are conducive to convergence, especially if, as is likely in the latter stages of the optimisation, the configuration tends to become unimodal. At any stage of the optimisation procedure if the cumulative success rate is below 50%, then, whenever a primary trial fails, the corresponding secondary point is selected for the next trial. This feature is incorporated in the complete flow diagram shown in fig. 1.

Constraints

The procedure regards all global optimisation problems as being constrained in the sense that the search is confined within the initially prescribed domain, V, defined in terms of upper and lower bounds on each of the variables.

If P falls outside V then the trial is discarded. Additional constraints, restricting further the field of search within V, are handled similarly. When such constraints are imposed then, dependent on their number and complexity, a sufficiently large value of N must be chosen to ensure that a reasonable proportion of potential trial points are valid (all N points of the current configuration necessarily satisfy the constraints).

The Choice of N

Clearly the greater the value of N the more thorough the search and the higher the probability of discovering a global minimum. On the other hand the larger the value of N the greater is the demand on computer storage and the slower the convergence of the algorithm. A feature of the algorithm is that for a given N the number of potential trial points increases rapidly with n, provided that $N \gg n$. Thus high dimensionality does not, per se, demand inordinatedly large storage; it is sufficient that N should increase linearly (rather than exponentially) with n. In a given situation there is likely to be an optimum value of N - dependent on the complexity of the function to be optimised, the nature of the constraints and the volume of the initial search domain - such that the reduction in speed which results from a further increase in N is not justified by the gain in performance. The appropriate choice of N is therefore a matter of experience. An empirical rule adopted by the author is to take $N = 25n$ (i.e. 25 points per dimension) unless there is good reason to choose otherwise.

The Stop Criterion

The flow diagram for the algorithm does not specify a particular stop criterion. The user is free to programme his own criterion - e.g. after a specified number of function evaluations or when the N points fall within a sufficiently small region of n-space.

TRIALS

The CRS procedure has been programmed (in BASIC) so as to run either interactively on a PDP 11/20 minicomputer or in batch mode on a Cyber 72 computer (Standard time 3.6 sec.). The following trials were conducted on the Cyber 72, in each case (except for trial 6) N being chosen as 25 points per dimension. Unless otherwise stated a run was terminated when the function values of all the points in store were identical to within an accuracy of six significant figures. The co-ordinate values quoted as the solution points are those obtained on the last trial. Although these are also stated to six significant figures it is not claimed that the solution point has been obtained to this degree of precision : the spread of co-ordinate values of the N points finally in store indicate that the quoted solutions are, in some instances, accurate only to four significant figures. For each trial the total number of function evaluations (E) and the computer CPU time in seconds (T) are also given.

Trial 1 - Shekel's family

The results for this problem were as follows :

	m = 5	m = 7	m = 10
x_1	4.00003	4.00052	4.00066
x_2	4.00016	4.00078	4.00054
x_3	4.00002	3.99943	3.99955
x_4	4.00011	3.99957	3.99948
f(x)	-10.1532	-10.4029	-10.5364
E	3,800	4,900	4,400
T	50	72	71

Trial 2 - Hartman's family

The results for the two cases (1) m = 4, n = 3, and (2) m = 4, n = 6 were as follows :

	(1)	(2)
x_1	0.114840	0.201643
x_2	0.555515	0.149867
x_3	0.852551	0.477123
x_4	-	0.275331
x_5	-	0.311717
x_6	-	0.657225
f(x)	-3.86278	-3.32237
E	2,400	7,600
T	27	171

Trial 3 - Branin

In the early stages of the optimisation of this function the configuration separates into three clusters, one around each global minimum. Because there is no bias, the function being equally sensitive at each of the minima, the ultimate convergence is dependent upon the particular random sequence used in the search. The result of two trials, using different random sequences, are given. In the first the configuration converges onto one minimum while in the second the configuration remains divided between two global minima.

	(1)	(2)	
x_1	9.42478	-3.14159	3.14159
x_2	2.47500	12.2750	2.27500
f(x)	0.397887	0.397887	
E	1,600	1,800	
T	13	16	

Trial 4 - Goldstein and Price

The results for this function were as follows :

x_1	0.00000
x_2	-1.00000
f(x)	3.00000
E	2,500
T	10

Trial 5

The two-dimensional function defined by

$$f = f_1^2 + f_2^2 ,$$

where $$f_1 = 2x_1^3 x_2 - x_2^3$$

and $$f_2 = 6x_1 - x_2^2 + x_2 ,$$

has three global minima, all having $f = 0$. Defining the initial search volume, V, by $-500 < x_1, x_2 < 500$

the configuration quickly forms clusters around each minimum. After about 1,000 function evaluations however the search concentrates around the global minimum at (0,0) irrespective of the random sequence used. This happens because (unlike trial 3) the minima are not all equally sensitive : the basin centred on (0,0) is much shallower than that around each of the other global minima. In many practical applications of optimisation (e.g. in engineering design where the variables represent physical parameters subject to tolerances) it is clearly desirable that, where more than one global minimum exists, the algorithm should select that minimum which is least sensitive to small changes in the variables.

Trial 6

The presence of exponential terms makes this 9-variable function a particularly difficult one to optimise.

$$f(x_1,\ldots,x_9) = \gamma^2 + \sum_{k=1}^{4} (\alpha_k^2 + \beta_k^2)$$

where

$$\alpha_k = (1-x_1x_2)x_3\left\{\exp[x_5(g_{1k}-g_{3k}x_7 x\, 10^{-3}-g_{5k}x_8 x\, 10^{-3})] -1\right\}-g_{5k}+g_{4k}x_2 ,$$

$$\beta_k = (1-x_1x_2)x_4\left\{\exp[x_6(g_{1k}-g_{2k}-g_{3k}x_7 x\, 10^{-3}+g_{4k}x_9 x\, 10^{-3})] -1\right\}$$

$$-g_{5k}x_1 +g_{4k} ,$$

$$\gamma = x_1x_3 - x_2x_4 ,$$

and the numerical constants g_{ik} are given by the matrix

0.485	0.752	0.869	0.982
0.369	1.254	0.703	1.455
5.2095	10.0677	22.9274	20.2153
23.3037	101.779	111.461	191.267
28.5132	111.8467	134.3884	211.4823

This function provides a 'least-sum-of-squares' approach to the solutions of a set of nine simultaneous non-linear equations which arise in the context of transistor modelling and the primary concern, therefore, is to search for global minima for which $f = 0$. The function has been investigated by Cutteridge (1974) who reduces the dimensionality from 9 to 8 by using the equation $\gamma = 0$ to eliminate one variable.

The CRS procedure was run in batch mode on the Cyber 72 computer using only primary search. A run was terminated if either the total number of function evaluations exceeded 30,000 or $(f_m - f_\ell)/(f_m + f_\ell) < 0.01$, where f_m and f_ℓ are respectively the greatest and least function values of the N points currently in store (this provides a simple, though arbitrary, measure of convergence). For the first run the initial search domain was chosen to be $0 < x_k < 10 \quad k = 1, \ldots 9$, and a storage value of N = 200 was taken. After 20,000 function evaluations (about 12 minutes of computing time) the 200 points were clustered around the point.

$$x_1 = 0.9, \ x_2 = 1.0, \ x_3 = 3.7, \ x_4 = 3.9, \ x_5 = 7.1,$$
$$x_6 = 8.1, \ x_7 = 0.4, \ x_8 = 1.9, \ x_9 = 2.8.$$

The minimum value of f being 22.6.

A second run, again using N = 200, took an initial search domain having the same volume as the first but centred on the above point. This resulted, after 30,000 function evaluations, in the 'best' point at

$$x_1 = .9, \ x_2 = .92, \ x_3 = 5.2, \ x_4 = 5.1, \ x_5 = 4.8$$
$$x_6 = 9.8, \ x_7 = -1.75, \ x_8 = 1.1, \ x_9 = 0.4.$$

having $f = 3.8 \times 10^{-2}$.

The encouraging improvement shown by the second run suggested that this re-start strategy (i.e. re-running the programme with no changes except to shift the centre of the initial search domain to the position of the best point of the previous run) be continued until no further significant improvement is obtained. Through a sequence of six such runs, each involving between 20,000 and 30,000 function evaluations, the function values corresponding to the best points were :

$$22.6, \ 3.8 \times 10^{-2}, \ 1.8 \times 10^{-2}, \ 6.1 \times 10^{-3}, \ 4.7 \times 10^{-4}, \ 3.9 \times 10^{-4}$$

The optimum point for the sixth run was

$$x_1 = 0.8987, \ x_2 = 0.9708, \ x_3 = 10.95, \ x_4 = 10.14,$$
$$x_5 = 3.362, \ x_6 = 6.952, \ x_7 = -8.049, \ x_8 = 1.236,$$
$$x_9 = -0.4278.$$

Refinement, using a gradient method, has shown that this point is close to a global minimum having $f = 0$.

It is not known how many other global minima the function possesses but one occurs close to the point

$x_1 = 0.9$, $x_2 = 0.45$, $x_3 = 1$, $x_4 = 2$, $x_5 = 8$, $x_6 = 8$,
$x_7 = 5$, $x_8 = 1$, $x_9 = 2$.

Although this point lies within the initial search domain for the first run the procedure does not find it. By taking an initial search domain centred on this point and having a range of ± 2 in each variable, clear evidence of convergence towards the point was obtained ($f = 5.7 \times 10^{-2}$ in 44,000 function evaluations) but it would appear that the sensitivity of the function is very much greater in the region of this global minimum than in the region of the minimum found by the CRS procedure.

CONCLUSION

The CRS optimisation procedure while simple and easily programmed on a small computer, is yet capable of global search with an efficiency which compares favourably with that of more sophisticated direct procedures. If the function to be optimised has more than one global minimum then the algorithm tends to converge upon the least sensitive minimum. If the global minima are equally sensitive then the final convergence depends on the particular random sequence used in the search. While high dimensionality does not, of itself, require an inordinate amount of storage it is desirable, when using CRS to optimise complicated functions of many variables, to use as large a value of N as is practicable. Where storage is at a premium one may perform several runs over various search domains either by dividing the given domain into discrete sub-domains or by using a re-start strategy such as that described. Because the CRS algorithm is designed for thoroughness of search rather than for speed of convergence it is desirable, where possible, to refine the discovered optima by recourse to faster, uni-modal methods.

REFERENCES

Becker, R.W. and Lago, G.V. (1970). "A Global Optimisation Algorithm". Proceedings of the Eighth Allerton Conference on Circuits and Systems Theory.

Brooks, S.H. (1958). "A Discussion of Random Methods for Seeking Maxima". Operations Research, 6, 244-251.

Bryan, J.K., Dwyer, S.J., and Lago, G.V. (1969). "Non-Parametric Decision Schemes for EEG Classification". Sixth Annual Rocky Mountain Bio-Engineering Symposium Conference Record, 56-61, University of Wyoming, Laromie, Wyoming.

Cutteridge, O.P.D. (1974). "Powerful 2-Part Program for Solution of Nonlinear Simultaneous Equations". Electronics Letters, Vol. 10, No. 10, 182-184.

Fletcher, R. and Powell, M.J.D. (1963). "A Rapidly Convergent Descent Method for Minimisation". The Computer Journal, 6, 163-168.

Hooke, R. and Jeeves, T.A. (1969). "Direct Search Solution of Numerical and Statistical Problems". Journal of the Association for Computing Machinery, 8, 212-229.

Nelder, J.A. and Mead, R. (1965). "A Simplex Method for Function Minimisation". The Computer Journal, 7, 308-313.

Rosenbrock, H.H. (1960). "An Automatic Method for Finding the Greatest or Least Value of a Function". The Computer Journal, 3, 175-184.

TOWARDS GLOBAL OPTIMISATION 2
L.C.W. Dixon and G.P. Szegö (eds.)

A STOCHASTIC METHOD FOR GLOBAL OPTIMIZATION: ITS STRUCTURE AND NUMERICAL PERFORMANCE(*)

L. DE BIASE, ISTITUTO DI MATEMATICA
UNIVERSITA' DEGLI STUDI DI MILANO
MILANO, ITALY
F. FRONTINI, C.I.S.E. SEGRATE, ITALY

In this paper a method is developed for the minimization of a function $f: S \subset R^N \to R$. The root of a suitable approximation $A(\xi)$ of the level set measure yields an approximation to f^*. $A(\xi)$ is given by a recursive spline technique, smoothing the data obtained by a sequential uniform sampling both on S and on the expected range of function values.

INTRODUCTION

Let $f(x)$ be a continuous real valued function defined on a compact set $S \subset R^N$; our problem is to single out a point x^* such that

$$f^* = f(x^*) \leq f(x), \qquad x \in S.$$

For particular classes of functions many satisfactory algorithms have been exhibited, but in the general case, mainly when the function to be minimized is multiextremal, usual optimization techniques cannot be applied. An interesting stochastic strategy relies upon a one variable function $\psi(\xi)$ defined as the normalized Lebesgue measure of the set

$$E(\xi) = \{x \in S: f(x) \leq \xi\}.$$

If f^* is the essential infimum of $f(x)$ and x^* is an isolated minimum, we have obviously: $\psi(\xi) = 0$ for $\xi \leq f^*$ and $\psi(\xi) > 0$ for $\xi > f^*$. In order to approximate $\psi(\xi)$, $f(x)$ is evaluated at random points out of a uniform distribution in S, yielding a "noisy" pointwise approximation of $\psi(\xi)$. The sampling and the approximation processes are tuned together sequentially in order to get a convergent estimate of f^* at the root

(*) This work was supported in part by G.N.I.M. (Gruppo Nazionale Informatica Matematica) of C.N.R. (Italian National Research Council)

of the approximating function and to control the sample size.

ANALYSIS OF THE METHOD

Let $f(x)$ be a real valued continuous function, defined on a compact set $S\in R^N$. Let $E(\xi)$ be the set

$$E(\xi)=\{x\in S:\ f(x)\leq \xi\} \qquad \xi\in R^1.$$

Let $m(.)$ be a measure proportional to the Lebesgue measure on S, such that $m(S)=1$. We define a function

$$\psi(\xi)=m(E(\xi)) \tag{1.1}$$

mapping R into R^+.

As S is a compact set, $E(\xi)$ belongs to the σ-algebra of Lebesgue sets in S; thus $\psi(\xi)$ is properly defined and $0\leq\psi(\xi)\leq 1$.

Let f^* be the essential infimum of $f(x)$ in S. If $\xi<f^*$, $\psi(\xi)=0$; $m(f^{-1}(f^*))=0$ implies $\psi(\xi)=0$ iff $\xi\leq f^*$.

The function $\psi(\xi)$ has some regularity properties:

Theorem: a) $\psi(\xi)$ is non decreasing in R^1;

b) $\psi(\xi)$ is almost everywhere differentiable;

c) $\psi(\xi)$ is continuous in R^1 provided that no set $H\subset S$ exists such that $f(x)=$const. in H and $m(H)>0$.

The proof of this theorem is in Archetti-Betrò (1976); we give here some hint of it:

a) follows immediately from properties of a Lebesgue measure; to prove b) $m(f^{-1}(.))$ is shown to be a finite Borel measure. After known theorems (Rudin (1970)) this implies that $m(f^{-1}(-\infty,\xi))=\psi(\xi)$ is almost everywhere differentiable; using properties of real monotone sequences right and left continuity are stated and c) is proved.

The behaviour of $\psi(\xi)$ is shown in figure 2 for an $f(x):R\rightarrow R$ (figure 1).

By the properties of $\psi(\xi)$ an approximation to a value β^* such that $\psi(\xi)=0$ for $\xi\leq\beta^*$ and $\psi(\xi)>0$ for $\xi>\beta^*$ may be instrumental in controlling the optimization process: thus we look for an approximation of $\psi(\xi)$ over the interval $[f^*, \|f\|_\infty]$, where $\|f\|_\infty$ is the supremum in S of $f(x)$.

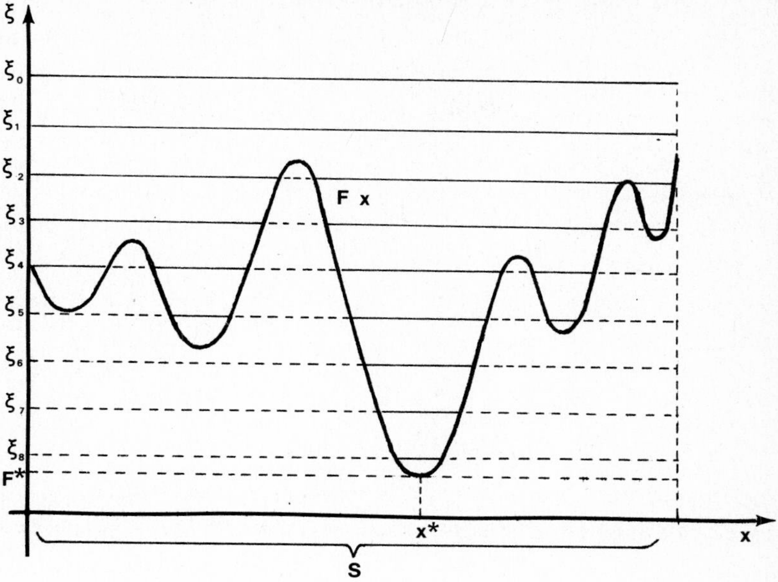

fig. 1

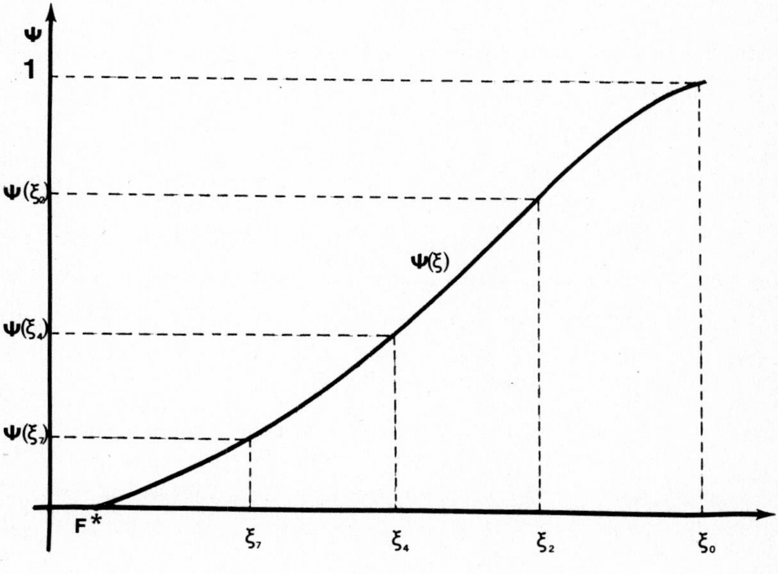

fig. 2

Except for trivial cases, an analytic expression of $\psi(\xi)$ is not available; it is then necessary to set up a stochastic sampling technique to provide an approximation of it in analytical form.
Let $w \in R^N$ be a vector uniformly distributed in S. Let's define

$$P(\xi) = \text{prob} \{ f(w) \leq \xi \}$$

as the probability of hitting the region $E(\xi)$.
Since the distribution is uniform and $m(S)=1$, it follows that

$$\psi(\xi) = m(E(\xi)) = P(\xi).$$

We perform a sample θ_q and evaluate $f(x)$ at every random point w_j of θ_q $(j=1,2,...,q)$.
Let:

$$\theta_\ell = \min_{j=1,2,...,q} f(w_j) \quad \text{and} \qquad (1.2)$$
$$\theta_u = \max_{j=1,2,...,q} f(w_j),$$

and let's choose ξ out of a uniform distribution in $[\theta_\ell, \theta_u]$.
Let p be the number of trial points hitting $E(\xi)$; $\psi(\xi)$ is approximated by $\tilde{\psi}(\xi) = p/q$.
The random variable $\tilde{\psi}(\xi)$ is an unbiased estimator of $\psi(\xi)$.
For the law of large numbers, we have:

$$\text{Prob} \{ \lim_{q\to\infty} p/q = \psi(\xi) \} = 1.$$

The random variable $\tilde{\psi}(\xi)$ follows a binomial distribution with parameter $\psi(\xi)$; the error $|p/q - \psi(\xi)|$ is of the order of $1/\sqrt{q}$; its expected value is 0 and its standard deviation is $\{\psi(\xi)(1-\psi(\xi))/q\}^{1/2}$; thus $\psi(\xi)$ turns out to be an unknown regression function and we look for its root ξ^*, such that $\psi(\xi^*)=0$, $\psi(\xi)>0$ for $\xi > \xi^*$.
Stochastic optimization techniques are likely to fail in this problem because the root we are looking for is located just outside the actually sampled range in the ξ variable: we are thus faced with stochastic extrapolation.

To surmount this difficulty we provide a least squares approximation to $\psi(\xi)$, so that its root may be analytically evaluated.

Let $\phi_\ell(\xi)$ $(\ell=1,2,...,k)$ be our basic functions and $\sum_1^k \lambda_\ell \phi_\ell(\xi)$ the approximation. To calculate the optimal values of the coefficients λ_ℓ, $(\ell=1,2,...,k)$, we consider the following error functional

$$D(\lambda)=\int_{f^*}^{\|f\|_\infty} [\psi(\xi)-\sum_{\ell=1}^{k} \lambda_\ell \phi_\ell(\xi)]^2 d\xi \tag{1.3}$$

and minimize it.

Since $\psi(\xi)$, $\|f\|_\infty$, f^* are not available, $D(\lambda)$ is replaced by the empirical risk functional

$$J^r(\lambda)=\{\sum_{i=1}^{r} \tilde{\psi}_i(\xi_i)-\sum_{\ell=1}^{k} \lambda_\ell \phi_\ell(\xi_i)\}^2 \tag{1.4}$$

where r, $\tilde{\psi}_i$ and ξ_i are defined below:

let $\theta^i_{q_i} \equiv \{f(w^i_{j_i}),\ j_i=1,2,\ldots,q_i;\ i=1,2,\ldots,r\}$ be the i-th sample performed, where $w^i_{j_i}$, $i=1,2,\ldots,r$ and $j_i=1,\ldots,q_i$ have already been described. We define

$$w^r_\ell=\min_i \min_{j_i} \{f(w^i_{j_i})\} \text{ and} \tag{1.5}$$

$$w^r_u=\max_i \max_{j_i} \{f(w^i_{j_i})\}.$$

Consider now a random variable ξ_i uniformly distributed in $[w^r_\ell-\varepsilon, w^r_u+\delta]$, with $\varepsilon,\delta>0$; the following theorem holds (Archetti-Betrò (1976)):

<u>Theorem</u>. Let $m_i=\min\{f(x_j), j=1,2,\ldots,i\}$ and $M_i=\max\{f(x_j), j=1,2,\ldots,i\}$ where x_j are independent random points uniformly distributed in S. Then $m_i\to f^*$ and $M_i\to\|f\|_\infty$ with probability 1 as $i\to\infty$.

This theorem ensures that for r sufficiently large

$$[w^r_\ell-\varepsilon, w^r_u+\delta]\supset[f^*, \|f\|_\infty]$$

with probability 1.

Now we can define an unbiased estimator of $\psi(\xi)$ at every level ξ_i

$$\tilde{\psi}_i=\tilde{\psi}(\xi_i)=p_i/q_i, \tag{1.6}$$

where p_i is the number of trial points hitting the region $E(\xi_i)$ in $\theta^i_{q_i}$. Moreover, as the samples $\theta^i_{q_i}$ are independent, we have in hand an independent sample $\{(\xi_i,\tilde{\psi}_i),\ i=1,2,\ldots,r\}$ for the construction of $\psi(\xi)$.

THE APPROXIMATION OF $\psi(\xi)$

Let $s_k^r(\xi) = \sum_{\ell=1}^{k} \lambda_\ell^{*r} \phi_\ell(\xi)$ (2.1)

where λ^{*r} is such that $J^r(\lambda^{*r}) \leq J^r(\lambda)$ for every λ.

As an approximation of f* we shall consider the value β_k^r, such that

$$s_k^r(\beta_k^r)=0, \quad s_k^r(\xi)>0 \text{ for } \xi>\beta_k^r.$$

We remark that such a β_k^r may fail to exist. In order to ensure its existence (as can easily be shown) we generalize its definition as follows:

$$s_k^r(\beta_k^r)= \varepsilon_k, \qquad s_k^r(\xi)>\varepsilon_k \quad \text{for } \xi>\beta_k^r \tag{2.2}$$

with $\varepsilon_k \to 0^+$ for $k \to +\infty$.

We must of course be sure that our approximation ensures $\beta_k^r \to f^*$ as $r, k \to \infty$, i.e. that every piece of information gained during sampling improves the approximation to $\psi(\xi)$ as well as to f*. This behaviour is ensured if s_k^r converges uniformly to the regression function $\psi(\xi)$ as $r,k\to\infty$, and this happens under mild conditions if $s_k^r(\xi)$ is a spline function of odd degree, as stated in Betrò-De Biase (1976) (*). We now set k=N+2m (the reason will be clear later). Let $s_k^r(\xi)$ be expressed in terms of spline functions of order 2m-1, with N equidistributed knots Z_j, j=1,2,...,N in $[f^*, \|f\|_\infty]$. To find out the N+2m parameters $(\lambda_1, \lambda_2, \ldots, \lambda_{N+2m})$ of our spline function $s_k^r(\xi)$, we minimize the functional

$$J^r(s_k^r)= \sum_{i=1}^{r} [\tilde{\psi}_i - s_k^r(\xi_i)]^2. \tag{2.3}$$

Uniform convergence of the spline functions sequence to $\psi(\xi)$ is ensured by a result of Mikhal'skii's (1974):

(*) We recall that a spline function $s_N(\xi)$ of order 2m-1 with N equidistributed knots Z_i, i=1,2,...,N over a closed interval [a,b] is a piecewise polynomial function of $C^{2m-2}[a,b]$, i.e. it is a function such that $s_N(\xi)$ is a polynomial of degree at most 2m-1 on every interval $(-\infty, Z_1)$, $(Z_N, +\infty)$, (Z_i, Z_{i+1}), i=1,2,...,N-1, with derivatives joining at every knot, up to the (2m-2)-th order.

Theorem: Let $s^r_{N(r)}(\xi)$ be a spline function for which the number $N(r)$ of knots, equidistributed in a closed interval $[a,b]$ increases as $r\to\infty$ Let $f(x)$ be a continuous function defined on $[a,b]$ and $\{(x_i,f_i)$, $i=1,2,...,r\}$ an independent sample such that f_i is an unbiased estimator of $f(x_i)$;
if the sequence $s^r_{N(r)}$ minimizes the empirical risk,
then it converges uniformly to $\psi(\xi)$ provided that the following condition holds:

$$\lim_{r\to\infty}\frac{N^2(r)\ln r}{r}=0.$$

Relying upon all the considerations quoted above about the properties of $\psi(\xi)$ and about the sequential sampling procedure, we can apply this result to the approximation of $\psi(\xi)$.
From now on we shall write s^r_N for the approximation, to stress the **number of knots.**

SPLINE APPROXIMATION

It is shown (Betrò-De Biase (1976)) that a necessary and sufficient condition for the existence of a spline function $s^*_N(\xi)$ of degree $2m-1$ minimizing the empirical risk is the existence of a subset P of the set $\{\xi_i,\ i=1,2,...,r\}$ with $r>N+2m$, such that

$P=\{\xi_{j_i},\ i=1,2,...,N+2m$ with $\xi_{j_1}<\xi_{j_2}<...<\xi_{j_{N+2m}}$ and

$$\xi_{j_i}< Z_i<\xi_{j_{i+2m}}\} . \qquad (3.1)$$

The proof of this theorem relies upon two theorems which provide conditions to ensure that the determinant $|(d_i-b_j)^n_+|$ (*) is positive given any $n>0$ and any numbers $\{d_i,\ i=1,2,...,r\}$, $\{b_j,\ j=1,2,...,r\}$ and that a system $s_N(x_i)=y_i$, $i=1,2,...,N+2m$, and $x_1<x_2<...<x_{N+2m}$ has one and only one solution.
After these theorems it can be shown that the system matrix is positive definite iff a subset like P exists, because the proof of positive definiteness reduces to looking for the kernel of a system with r equations and N+2m variables.
In our algorithm a more restrictive condition than (3.1) is needed,

(*) For the definition of $|(d_i-b_j)^n_+|$ see (3.3).

i. e. at least one point occurs between every two knots.
If this is verified the algorithm proceeds, otherwise other points ξ_i are generated until our requirement is fulfilled.

The number of parameters to be evaluated to determine the spline function is N+2m. Indeed, a polynomial of degree 2m-1 has 2m coefficients and we have N+1 polynomials(we do not consider the bounds of the interval as knots); so we have (N+1)2m parameters. We have also (2m-1)N conditions, since we require our polynomials to join at every knot together with equal derivatives up to order 2m-2. So the number of free parameters is N+2m.

Setting $s_N^{2m-2}(\xi)$= const. on every interval between two knots and integrating with restect to ξ, it is easy to show that every spline function has the following representation:

$$s_N(\xi)=\sum_{i=0}^{2m-1} a_i\xi^i+\sum_{j=1}^{N} c_j(\xi-Z_j)_+^{2m-1} \tag{3.2}$$

where, for every function Γ we define

$$\Gamma_+=\begin{cases}\Gamma & \text{if } \Gamma\geqslant 0\\ 0 & \text{if } \Gamma<0.\end{cases} \tag{3.3}$$

We remark that this representation has

$$1,\ \xi,\ \xi^2,\ldots,\ \xi^{2m-1},\ \{(\xi-Z_j)_+^{2m-1}\ ,\ j=1,2,\ldots,N\} \tag{3.4}$$

as basic functions, i.e. exactly N+2m parameters are needed.

A severe ill-conditioning of the matrix C may arise in the system $C\lambda=b$, where C and b are respectively the matrix and the vector obtained when minimizing $J^r(s_N)$.
In order to avoid ill-conditioning, a proper basis must be chosen. A very good one is composed by the (2m)-th divided differences of the functions

$$(\xi-Z_j)_+^{2m-1}\ ,\qquad j=1,2,\ldots,N+2m. \tag{3.5}$$

This basis was introduced by Schumaker (1969) for spline

interpolation problems and in this case it yielded tridiagonal matrices.When dealing with least squares approximations as in our problem, a matrix is obtained which is no longer tridiagonal, but retains diagonal dominance. Therefore this basis is particularly effective in our problem.

ANALYSIS OF THE ALGORITHM

Let's now outline the algorithm step by step.

Step 1 Sampling range.

We must single out an interval for the sampling of ξ_i, $i=1,2,\dots,r$. As discussed previously, θ_ℓ and θ_u are used as an approximation respectively of f^* and $\|f\|_\infty$.
We perform a sampling of q points in S and evaluate f(x) at each of them. An arithmetic mean of function values obtained in this way will be assumed as an approximation of θ_u.
It is obvious that, when looking for a global minimum, it is not important to extend the range of f(x) upwards, and, on the contrary, in our case, we prefer a higher density of points neighbouring f^*.
A more delicate problem is the choice of θ_ℓ.
We proceed as follows: the amplitude of the interval between θ_u and the minimum value obtained for f(x) is found out; it is multiplied by a proper parameter greater than 1 (depending on the particular function f(x) one is minimizing) and this is assumed as the amplitude of $[\theta_\ell,\theta_u]$.

Step 2 Sampling strategy.

At every sample a random value $\xi_i \varepsilon [\theta_\ell,\theta_u]$ and q random points (q n-tuples of random values) in S are generated; for every ξ_i we evaluate the frequency $\psi(\xi_i)$ (i.e. $m(E(\xi_i))$) by means of $\tilde{\psi}_i = p_i/q$.

Step 3 Updating of frequencies.

To prevent an anomalous sample from influencing the approximation too strongly, the frequencies obtained up to now are updated by the rule:

$$\tilde{\psi}_j^i = \frac{\tilde{\psi}_j^{i-1}(i-1)+p_j/q}{i}, \quad i \geq 2, \tag{4.1}$$

where $\tilde{\psi}_j^i$ is the approximation obtained at sample i for $\psi(\xi_j)$.

Step 4 Spline approximation at the i-th sample.

At the i-th sample we have:

$$s_N^i(\xi) = \sum_{j=1}^{N+2m} \lambda_j^i \phi_j(\xi) \tag{4.2}$$

where $\phi_j(\xi)$, j=1,2,...,N+2m, are the basic functions discussed in previous section.

To calculate coefficients λ_j^i we solve the linear system $C^i\lambda=b^i$ obtained when minimizing

$$J^r(\lambda) = \sum_{\ell=1}^{i} \{\tilde{\psi}_\ell - \sum_{j=1}^{N+2m} \lambda_j \phi_j(\xi_\ell)\}^2. \tag{4.3}$$

Indeed, we must have:

$$\frac{\partial J(\lambda)}{\partial \lambda_k} = 0, \quad k=1,2,...,N+2m.$$

And this means

$$2 \sum_{\ell=1}^{i} \{\tilde{\psi}_\ell - \sum_{j=1}^{N+2m} \lambda_j \phi_j(\xi_\ell)\} \phi_k(\xi_\ell) = 0, \quad k=1,2,...,N+2m. \tag{4.4}$$

This yields:

$$\sum_{\ell=1}^{i} \tilde{\psi}_\ell \phi_k(\xi_\ell) = \sum_{j=1}^{N+2m} \lambda_j \sum_{\ell=1}^{i} \phi_j(\xi_\ell)\phi_k(\xi_\ell), \quad k=1,2,...,N+2m \tag{4.5}$$

Then we have $C^i=\{a_{pq}^i\}$ and $b^i=\{b_p^i\}$, where

$$a_{pq}^i = \sum_{\ell=1}^{i} \phi_p(\xi_\ell)\phi_q(\xi_\ell); \quad b_p^i = \sum_{\ell=1}^{i} \phi_p(\xi_\ell)\tilde{\psi}_\ell. \tag{4.6}$$

We remark that updating of frequencies $\tilde{\psi}_i$ modifies every component of b^i at each sample, but it does not anyway affect C^i.

This is a very important fact because it allows us to reduce computation. Indeed, we do not need inversion of C^i at every sample to evaluate coefficieents λ_j, j=1,2,...,N+2m, until the number N(i) of knots is increased.
As a matter of fact, we have

$$C_N^{i+1} = C_N^i+\Phi(\xi_{i+1})\Phi^T(\xi_{i+1}), \tag{4.7}$$

where

$$\Phi(\xi)=\{\phi_1(\xi),\phi_2(\xi),...,\phi_{N+2m}(\xi)\},$$

and this allows a recursive technique to update $(C_N^{i+1})^{-1}$ on the basis of $(C_N^i)^{-1}$ by means of the formula:

$$(C_N^{i+1})^{-1}=(C_N^i)^{-1}-\frac{(C_N^i)^{-1}\Phi(\xi_{i+1})\Phi^T(\xi_{i+1})(C_N^i)^{-1}}{1+\Phi^T(\xi_{i+1})(C_N^i)^{-1}\Phi(\xi_{i+1})} \tag{4.8}$$

With this device we solve our system with a number of operations of the order of $2(N+2m)^2$ against $(N+2m)^3/3$ of usual solution.

<u>Step 5</u> Termination criteria.

A crucial point in the algorithm is the choice of proper termination criteria. A first test is made on the convergence of the coefficients λ_j, j=1,2,...,N+2m, of the basic spline functions, in order to verify whether, with a fixed number of knots, information deriving from each new sample is completely exploited.
This means that, given a precision η, before changing the number of knots, the condition

$$\frac{|\lambda_j^i-\lambda_j^{i-1}|}{|\lambda_j^{i-1}|} \leqslant \eta \qquad j=1,2,...,N+2m \tag{4.9}$$

must be verified, where λ_j^i denotes the j-th component of the vector λ at the i-th sample.
The roots (in the sense of (2.2)) of the approximation to $\psi(\xi)$ corresponding to every reached convergence of the vectors λ^i are composed in their mean value β_{mean} by the law:

$$\beta^{i+1}_{mean} = \frac{1}{i+1}\{\beta^{i}_{mean}\cdot i + \beta^{i+1}\} \tag{4.10}$$

If we meet for a certain fixed number of times a β^{i+1} which is "far" from β^{i}_{mean} i.e. if

$$\frac{|\beta^{i+1}-\beta^{i}_{mean}|}{|\beta^{i+1}|} \geq 1 \tag{4.11}$$

we give less relevance to β^{i}_{mean} by a new updating formula:

$$\beta^{i+1}_{mean} = \frac{\beta^{i}_{mean}\cdot i' + \beta^{i+1}}{i'+1}$$

where the first value for i' is 1.
The reason for doing this only after several discrepancies between the new nihilating point β^{i+1} and β^{i}_{mean} is that, if it happens only once, possibly the anomalous result is β^{i+1}.

The number β_{mean} is used for a test of convergence of β^{i} to f*.
We check whether

$$\frac{|\beta_{mean}-\theta_{\ell}|}{|\theta_{u}-\theta_{\ell}|} < \varepsilon \tag{4.12}$$

where $\varepsilon>0$ is a fixed precision.
The reason to perform this kind of test (and to wait until it is verified more than once) is that it, somehow, ensures β_{mean}, which is an extrapolation result, to stay within a "small" (with respect to $\theta_{u}-\theta_{\ell}$) neighborhood of the sampled minimum.

Another obvious criterion which leads to termination is verified when the function $J^{r}(\lambda)$ is less than a prefixed precision.
When at least one of the last two exposed criteria is fulfilled, the algorithm performs some local searches, starting from a prefixed number of points (trial number), selected among the best points obtained along sampling, to the extent of calculating the actual value F_{c} and coordinates x* of the global optimum.
If we have

$$\frac{|F_c - \beta^*|}{|\theta_u - F_c|} \leq \alpha \qquad (4.13)$$

where β^* is the last β_{mean} and α a fixed precision, the algorithm terminates and (x_c, F_c) (where x_c is the actually calculated x^*) is accepted as final output.
Otherwise the strategy is started again from step 2 within a fixed number of function evaluations.

COMPUTATIONAL RESULTS

This algorithm was tried on some test functions from the literature and on some technological problems (Archetti-Frontini (1977)).

1) "six hump camel back function" (Branin (1972)):

$$f(x)=4x_1^2-2.1x_1^4+\frac{1}{3}x_1^6+x_1x_2-4x_2^2+4x_2^4.$$

This function has six minima, two maxima and seven saddle points. There are two global minima in which $f(x)=-1.0316$, and they are attained at (0.0898, -0.7126) and (0.7126, -0.0898).
Our minimization is performed over the compact set $-2.5 \leq x_1 \leq 2.5$, $-1.5 \leq x_2 \leq 1.5$.
The results we obtained are shown in Table A.

2) Branin (RCOS) (1972):

$$f(x)=(x_2-\frac{5.1}{4\pi^2}x_1^2+\frac{5}{\pi}x_1-6)^2+10(1-\frac{1}{8\pi})\cos x_1+10.$$

This function has three global minima in the region $-5 \leq x_1 \leq 10$, $0 \leq x_2 \leq 15$.

3) Goldstein&Price (GOLDPR) (1972):

$$f(x)=[1+(x_1+x_2+1)^2\cdot(19-14x_1+3x_1^2-14x_2+6x_1x_2+3x_2^2)].[30+$$
$$+(2x_1-3x_2)^2\cdot(18-32x_1+12x_1^2+48x_2-36x_1x_2+27x_2^2)].$$

This function has 4 local minima in the region

$-2 \leq x_1 \leq 2$, $-2 \leq x_2 \leq 2$.
The global minimum is $f(x)=3$, at $x_1=0$, $x_2=-1$.

A direct comparison to evaluate effectiveness of this procedure can be made by Hartman's method (1973), using some of his test problems. Hartman's method has no global convergence test; this algorithm, on the contrary, spends a number of function evaluations to satisfy internal and external tests even if the problem has already been solved. The objective function is:

$$4) \qquad f(x) = -\sum_{i=1}^{m} c_i \exp\{(x-p_i)^T A(x-p_i)\} ,$$

where c_i , $i=1,2,\ldots,m$ are the levels of the m minima and p_i, $i=1,2,\ldots,m$, their position vectors.
Shape and amplitude of attraction regions are determined by matrix A (n rows and n columns, n being the dimension of the problem), which has to be negative definite.
Also for this test family some trials were performed, with results shown in table A.

Another classical family of test functions was proposed by Shekel (1973), with the form:

$$5) \qquad f(x) = -\sum_{i=1}^{m} \frac{1}{\|x-a^i\|^2 + c_i} .$$

This function has m minima in positions a^i $(i=1,2,\ldots,m)$ with levels c_i.
Note: interaction among minima in Hartman's family acts heavily on the levels of minima but does not affect their positions, while in Shekel's family positions differ a lot from our a^i, $i=1,2,\ldots,m$, but levels are sufficiently preserved.

Regarding this method an important feature becomes evident from the trials performed: the number of function evaluations needed is almost insensitive to the dimension of the problem and to the number of minima.

Function	N	r	q	F.E.	Total F.E.	β^*	F_c	$\left\|\frac{F_c-\beta^*}{\theta_u-F_c}\right\|$	time (prime minutes)
S.H.C.B.F.	2	67	10	670	717	-1.13418	-1.031629	0.024	2.09
Branin RCOS	2	37	15	555	597	0.9433	1.250	0.0061	1.37
Goldstein&Price	2	21	15	315	378	4.6044	2.9997	$0.5*10^{-5}$	1.51
Hartman (4 minima)	3	45	15	675	732	-3.8602	-3.8627	0.0008	1.58
Hartman (4 minima)	6	47	15	705	807	-3.2013	-3.3223	0.04	2.08
Shekel (5 minima)	4	36	15	540	620	-1.534	-10.14	0.9 (*)	2.35
Shekel (7 minima)	4	47	15	705	788	-1.53	-10.399	0.8 (*)	2.07
Shekel (10 minima)	4	54	20	1080	1160	-1.67	-10.43	0.87 (*)	3.03

Table A

(*) Termination reached by the test $J^r(\lambda) \leqslant 10^{-6}$.

The time required for 1000 Shekel 5 evaluations at (4., 4., 4., 4.) was 0.09 prime minutes.

We remark that in table A we quoted two different numbers of function evaluations: the first is the number of function evaluations required by our method without local searches, the second is the total number of them.

In table B we show the behaviour of β_{mean} as r increases, for all the functions described in table A.

r	S.H.C.B.F.	Branin RCOS	Goldst.&Price	Hartman 4;3	Hartman 4;6	Shekel 5;4	Shekel 7;4	Shekel 10;4
15	-1.1622	1.3593	4.6050	-3.6717	-2.6210	-1.393	-1.422	-1.098
20	-1.1622	1.3593	4.6044	-3.6717	-2.6463	-1.393	-1.422	-1.098
25	-1.1622	1.3597		-3.6720	-2.6453	-1.393	-1.490	-1.773
30	-1.1622	0.5197		-3.6945	-2.6453	-1.393	-1.448	-1.792
35	-1.1622	0.9433		-3.6945	-2.6508	-1.534	-1.465	-1.776
40	-1.1622			-3.8602	-2.9810		-1.456	-1.776
45	-1.1622			-3.8602	-3.1512		-1.528	-1.721
50	-1.1622							-1.690
55	-1.1627							-1.668
60	-1.1627							
65	-1.1341							

Table B

It seems on the contrary dependent on the shape of the function. Indeed, if the function has very sharp minima the number of function evaluations increases noticeably; if it is very smooth very few function evaluations are needed, even for high dimension problems. Comparing our algorithm with Hartman's method A3, we can notice that for cases A, B, F, H our number of function evaluations is of the order of 50% of the average number needed by Hartman's method to reach the global minimum.

We remark that even if it is not explicitly stressed in table A, all trials reached the position of the global minimum.
We point out, however, that our main interest in testing this method was in obtaining a good prediction of the level of global optima; this will be clear from the close checks introduced into the algorithm described at the end of the previous section.
If our goal were a dependable answer after a small number of function evaluations, this total number could significantly be reduced.
It is worth noticing that this control on the goodness of the prevision was violated an insignificant number of times during all the trials we performed.

ACKNOWLEDGEMENTS

We thank Professor M. Cugiani for reading the draft of this paper. We are also grateful to Dr. J. Gomulka and Dr. L.C.W. Dixon for sending test functions and for fruitful discussions.

REFERENCES

Archetti, F. (1975). A sampling technique for global optimization; in: Towards Global Optimization, Szego & Dixon eds., North-Holland Publishing Company.

Archetti, F. and Betrò, B. (1976). Recursive stochastic evaluation of the level set measure in global optimization problems;quaderni del Dipartimento di Ricerca Operativa e Scienze Statistiche, A21, Università di Pisa.

Archetti, F. and Frontini, F. (1977). A global optimization method and its application to technological problems. This book.

Betrò, B. and De Biase, L. (1976). A recursive spline technique for uniform approximation of sampled data; quaderni del Dipartimento di Ricerca Operativa e Scienze Statistiche, A31, Università di Pisa.

Branin, F.H. (1972). A widely convergent method for finding multiple solutions of simultaneous equations; Tech. Rept. 21466, I.B.M. S.D. D.L., I.B.M. Journal on Research & Development.

Christensen, T. and Hartman, K. (1974). A strategy of optimization based on random search; Proceedings of I.F.A.C. Symposium on Stochastic Control, Budapest.

Frontini, F. (1975). Un metodo automatico per l' ottimizzazione globale in problemi di progettazione; C.I.S.E. Rept. N-172.

Hartman, K. (1973). Some experiments in global optimization; Naval Postgraduate School, Tech. Rept. NPS55HH72051/A.

Hartman, K. (1973). A new method for global optimization; Naval Postgraduate School, Tech. Rept. NPS55HH73041/A.

Mikhal'skii, A.J. (1974). The method of averaged splines in the problem of approximating dependencies on the basis of empirical data ; Automation and Remote Control, Vol. 35, N° 3.

Opacic, J. (1973). A heuristic method for finding most extrema of a nonlinear function; IEEE Transactions on Systems, Man and Cybernetics; 102, 107.

Rudin, W. (1970). Real and complex analysis, Mc. Graw Hill.

Schumaker, L.L. (1969). Theory and applications of spline functions; Greville ed. (Academic Press, New York).

TOWARDS GLOBAL OPTIMISATION 2
L.C.W. Dixon and G.P. Szegö (eds.)

A MIXED STOCHASTIC-DETERMINISTIC TECHNIQUE FOR GLOBAL OPTIMISATION

Enrico Fagiuoli (°) - Paolo Pianca (°) - Marco Zecchin (°°)

ABSTRACT. A space-covering method for global optimisation is presented, based on the construction of a stochastic automaton. Three possible implementations of the method are suggested. Some numerical results are given for two of these algorithms.

ACKNOWLEDGEMENTS. This study has been partially done during the visit of E. Fagiuoli at the Numerical Optimisation Center of Hatfield Polytechnic, in the framework of the Italo-British research project on global optimisation techniques sponsored by S.R.C. and C.N.R.

(°) University of Venice.
(°°) University of Trieste.

1. Introduction.

A space-covering method for the solution of a global optimisation problem was proposed by McMurtry and Fu [5]: this method uses a stochastic automaton as a controller for the search, adjusting its search probabilities according to past experience.
A different procedure was proposed by Hartman [2] and, later, by Dal Castello, Mereau and Pacquet [3]. A random search was performed until a favourable point (on which the function value is greater than a stored one) was found; then a deterministic local search was done, and the value relative to the local maximum was stored. The basic idea was to search for favourable points, avoiding the effort of a search from points unlikely to give improvements.
In what follows we describe a class of global optimisation methods that aim to construct a more efficient computational technique by including both McMurtry and Hartman features.
We define a space-covering method in which the stochastic automaton supervising the choice between cells is updated more efficiently than in the McMurtry and Fu method: at the same time, through the space-covering description, the method allows a more precise representation of the objective function.
These results are achieved by using an estimate obtained by the statistics of extrema as the action probabilities for the stochastic automaton.
Some numerical tests have been made by using the algorithms presented in this paper: they show that the method converges to the global optimum in a reasonable number of function evaluations.

2. Definition of the problem.

Let us define the following global optimisation problem:

$$\max_{\underline{x}} f(\underline{x})$$

over $R : a_r \leq x_r \leq b_r$, $r = 1, \ldots, n$, $f : R \rightarrow R^1$, $R \subset R^n$, i.e. the essentially unconstrained global optimisation problem and let the function f have a finite number of maximum points over R.

3. A class of space-covering methods.

Define a space-covering of the set R by dividing each side of R in h_r parts, of length $l_r = (b_r - a_r) / h_r$.
The center of the j-th subdivision is given by

$$x_{rj} = l_r(j - 1/2) + a_r \ , \ j = 1, \ldots, h_r, \ r = 1, \ldots, n,$$

and the index of the cells generated by the above subdivision is chosen as

$$i = j(x_1) + \sum_{s=2}^{n} \left\{ \left[j(x_s) - 1 \right] \prod_{t=1}^{s-1} h_t \right\},$$

where $x_s = x_{sj}$, $j = 1, \ldots, h_s$ and $j(x_s) = j(x_{sj}) = j$, $j = 1, \ldots, h_s$.
Therefore the set R is divided in cells C_i, $i = 1, \ldots, N$, each cell having volume $V = \prod_{r=1}^{n} l_r$.
Let us define the following stochastic algorithm:

1) choose a cell, according to a search probability distribution

$$\{d_i\} \quad i = 1, \ldots, N.$$

2) Within the cell, choose a point $\underline{x}^* = \{ x_1^*, \ldots, x_n^* \}$, according to the uniform distribution

$$g(x_r) = 1/l_r, \quad x_{rj} - l_r/2 \le x_r \le x_{rj} + l_r/2,$$

$r = 1, \ldots, n$.

3) If $f(\underline{x}^*)$ is less or equal to a previously stored value, go to 1). If $f(\underline{x}^*)$ is greater than the previously stored value, start a deterministic maximisation algorithm within the cell. Let $\underline{z}$ be the result of the maximisation process: change the stored value to $f(\underline{z})$.
Go to 1).

A step is performed successfully when a value $f(\underline{x})$ is found which is better than the previously computed value. Therefore the probability of a successful step is given by

$$P(f(\underline{x}) > W) \ , \ \underline{x} \in C_i \ , \ W = \text{stored value}.$$

Provided the functions $F_i(W) = P(f(\underline{x}) > W)$, $\underline{x} \in C_i$, are known, the performance of the algorithm can be easily evaluated.
The functions $F_i(y)$ are related to Chichinadze Ψ-transform [1]:

$$\Psi_i(W) = \text{meas}\{\underline{x} : f(\underline{x}) > W, \underline{x} \in C_i\} .$$

Then $F_i(W) = \Psi_i(W)/V$.
In section 6. a method based on approximations to F_i will be proposed.

4. Description of the algorithm as a stochastic automaton in time-varying environment.

For the following we shall suppose that the functions F_i are known. The optimisation process can be viewed as an interaction among the following components:

a) An environment, which is characterized after s-1 successful steps, by a vector of success probabilities:

$$\begin{bmatrix} p_1^s \\ p_2^s \\ \vdots \\ p_N^s \end{bmatrix}$$

where p_i^s is the probability of a favourable answer from the environment (i.e. $f(\underline{x}) > W$) when $\underline{x} \in C_i$. Therefore $p_i^s = F_i(W)$.
Due to the algorithm's structure, the environment is varying during the process. Suppose that the current probability vector is $(p_1^s, p_2^s, \ldots, p_N^s)^T$, when an improved point is found. Suppose that the deterministic maximisation algorithm starting from this point leads to the global maximum in the λ -th cell.
Then the new probability vector is given by

$$p_\lambda^{s+1} = F_\lambda(\max_{\underline{x} \in C_\lambda} f(\underline{x}))$$

and therefore the new vector is

$$(p_1^{s+1}, p_2^{s+1}, \ldots, p_{\lambda-1}^{s+1}, 0, p_{\lambda+1}^{s+1}, \ldots, p_N^{s+1}) .$$

The environment can be described by a set of matrices, each corresponding to the location of the maxima within a cell.
The k-th matrix, corresponding to the maxima in the k-th cell, is:

$$P^k = \begin{bmatrix} p_1^{k,1} & p_1^{k,2} & \dots & p_1^{k,m_k+1} \\ p_2^{k,1} & p_2^{k,2} & \dots & p_2^{k,m_k+1} \\ \dots & \dots & \dots & \dots \\ \dots & \dots & \dots & \dots \\ p_j^{k,1} & p_j^{k,2} & \dots & p_j^{k,m_k+1} \\ \dots & \dots & \dots & \dots \\ p_N^{k,1} & p_N^{k,2} & \dots & p_N^{k,m_k+1} \end{bmatrix}$$

where m_k is the number of maxima in the k-th cell.

Thus the environment is described by a matrix P of dimensions N x M, where $M = \sum_{i=1}^{N} (m_i + 1)$:

$$\tilde{P} = (P^1, P^2, \dots, P^N) .$$

Each of the column vectors $\underline{p}^{k,r}$, $k = 1, 2, \dots, N$, $r = 1, \dots, m_k+1$ of the matrix $\tilde{P}$ will define an environment state. The vectors $\underline{p}^{k,r}$ have some components equal to zero, corresponding to cells where maxima have already been located.

A vector which has all its components equal to zero corresponds to the location of the global maximum on R.

Note: The vector $\underline{p}^s$ used above to characterize the environment is therefore a column vector of the matrix $\tilde{P}$: it depends on the particular sequence of previously identified maxima.

b) A stochastic automaton, whose input is the environment state s, and whose output is a point $\underline{x} \in C_j$, where C_j is chosen according to the distribution $\{d_i^s\}$, and $\underline{x} \in C_j$ is chosen according to a uniform distribution.

c) A deterministic automaton, whose input is the output from b) and whose output, which results from the maximisation process, is fed-back into the environment, and induces a state change in the environment.

5. Limit properties.

Let us now reorder the cells in such a way that the cell with the lowest global maximum is numbered 1, the cell with the next lowest global maximum is numbered 2, and so on.

Let us reorder the environment states as follows:

$$\begin{array}{ccccccccccccc}
\text{state} & 1 & 2 & \dots & m_1 & m_1+1 & m_1+2 & \dots & m_1+m_2+2 & \dots & M \\
 & \begin{bmatrix} p_1^1 \\ p_2^1 \\ p_3^1 \\ \cdot \\ \cdot \\ p_N^1 \end{bmatrix} & \begin{bmatrix} p_1^2 \\ p_2^2 \\ p_3^2 \\ \cdot \\ \cdot \\ p_N^2 \end{bmatrix} & \dots & \begin{bmatrix} p_1^{m_1} \\ p_2^{m_1} \\ p_3^{m_1} \\ \cdot \\ \cdot \\ p_N^{m_1} \end{bmatrix} & \begin{bmatrix} 0 \\ p_2^{m_1+1} \\ p_3^{m_1+1} \\ \cdot \\ \cdot \\ p_N^{m_1+1} \end{bmatrix} & \begin{bmatrix} 0 \\ p_2^{m_1+2} \\ p_3^{m_1+2} \\ \cdot \\ \cdot \\ p_N^{m_1+2} \end{bmatrix} & \dots & \begin{bmatrix} 0 \\ 0 \\ p_3^{m_1+m_2+2} \\ \cdot \\ \cdot \\ p_N^{m_1+m_2+2} \end{bmatrix} & \dots & \begin{bmatrix} 0 \\ 0 \\ 0 \\ \cdot \\ \cdot \\ 0 \end{bmatrix}
\end{array}$$

Then the transition between the environment states is a Markov chain, whose transition matrix is:

$$[q_{ij}] = \begin{bmatrix} q_{11} & 0 & 0 & 0 & \cdot & \cdot & \cdot & 0 \\ q_{21} & q_{22} & 0 & 0 & \cdot & \cdot & \cdot & 0 \\ q_{31} & q_{32} & q_{33} & 0 & \cdot & \cdot & \cdot & 0 \\ \cdot & \cdot & \cdot & \cdot & \cdot & \cdot & \cdot & \cdot \\ \cdot & \cdot & \cdot & \cdot & \cdot & \cdot & \cdot & \cdot \\ q_{M1} & q_{M2} & q_{M3} & \cdot & \cdot & \cdot & q_{M,M-1} & 1 \end{bmatrix}$$

where

$$\begin{cases} q_{ii} = \sum_{k=1}^{N} d_k^i (1 - p_k^i) & i = 1, \dots, M-1 \\ q_{ij} = 0 \quad \text{per } i < j \\ \sum_{i=j+1}^{N} q_{ij} = \sum_{k=1}^{N} d_k^j \, p_k^j . \end{cases}$$

Provided that q_{ii}, $i = 1, \dots, M-1$, are less than 1, the chain has only one absorbing state, the one corresponding to the location of the global maximum in R: therefore the system reaches the final state with probability 1 in a finite number of steps.

6. Specific algorithms.

Consider the following three specifications for d_i^j, given the knowledge of $F_i(y)$, and therefore of the elements p_i^j of the environment matrix:

A.1 : $$d_i^j = \frac{1}{N} \qquad i = 1, \ldots, N$$

A.2 : $$d_i^j = \frac{p_i^j}{\sum_{i=1}^{N} p_i^j}$$

A.3 : $$d_i^j = 1, \text{ i such that } p_i^j = \max_k p_k^j; \quad d_r^j = 0, \ r \neq i.$$

The ranking between the three algorithms, in terms of the expected value of the probability of success in one step is:

$$A.1 \leqslant A.2 \leqslant A.3 \ .$$

Proof.

A.1 $\leqslant$ A.3 :

$$E_1(p^j_{success}) = \sum_{i=1}^{N} d_i^j \, p_i^j = \frac{1}{N} \sum_{i=1}^{N} p_i^j \leq \frac{\sum_{i=1}^{N} (\max_i p_i^j)}{N} = \max_i p_i^j$$

$$E_3(p^j_{success}) = \sum_{i=1}^{N} d_i^j \, p_i^j = \max_i p_i^j \Rightarrow E_1(p^j_{success}) \leq E_3(p^j_{success}) \, .$$

A.2 $\leqslant$ A.3 :

$$E_2(p^j_{success}) = \frac{\sum_{i=1}^{N} p_i^j \, p_i^j}{\sum_{i=1}^{N} p_i^j} \leqslant \frac{\sum_{i=1}^{N} p_i^j \, (\max_i p_i^j)}{\sum_{i=1}^{N} p_i^j} = E_3(p^j_{success}) \ .$$

A.1 $\leqslant$ A.2 :

The proof of this statement is based on the fact that the quadratic form

$$\underline{p}^T A \underline{p} = N \sum_{i=1}^{N} p_i^j \left\{ \frac{\sum_{i=1}^{N} (p_i^j)^2}{\sum_{i=1}^{N} p_i^j} - \frac{\sum_{i=1}^{N} p_i^j}{N} \right\}$$

is related to the variance of a random variable $\{p_i^j\}$, $i = 1, \ldots, N$.

Remark 1

It can be shown that all three choices for d_i^j satisfy the condition needed in section 5., namely $q_{jj} < 1$, $j = 1, \ldots, M-1$.
To prove this statement consider:

$$q_{jj} = \sum_{i=1}^{N} d_i^j \,(1 - p_i^j) .$$

q_{jj} is less or equal to 1.
Now we prove that q_{jj} is less than 1.
For A.1 we have

$$q_{jj} = \frac{1}{N}\left[\sum_{i=1}^{N} (1 - p_i^j)\right] = 1 - \frac{\sum_{i=1}^{N} p_i^j}{N} ,$$ which is equal to 1

only if $p_i^j = 0 \;\forall\, i$, i.e. only if $j = M$.

For A.2 we have

$$q_{jj} = \frac{\sum_{i=1}^{N} (1 - p_i^j) p_i^j}{\sum_{i=1}^{N} p_i^j} = 1 - \frac{\sum_{i=1}^{N} (p_i^j)^2}{\sum_{i=1}^{N} p_i^j} ,$$

which is never equal to 1.

For A.3 : $q_{jj} = 1 - p_i^j$, i such that $p_i^j = \max_k p_k^j$.

Therefore $q_{jj} < 1$, unless $\max_k p_k^j = 0$, which can happen only if $j=M$.

Remark 2

The algorithms A.2 and A.3 are based on the knowledge of the values p_i^j, which is not available in practical problems; an approximate algorithm can be implemented by making use of results due to Clough [4], in the framework of the statistics of extrema. s_i samples, each of size n_i are taken for the values of $f(\underline{x})$ in each cell. From the values of $f(\underline{x})$, as $\underline{x}$ is chosen at random, we have an estimate of the probability of maximum z_i in the i-th cell being greater than an arbitrary value L, where

$$L = z^i_{max}\left(\frac{s_i + c}{s_i - 1}\right) - z^i_{med}\left(\frac{1 + c}{s_i - 1}\right).$$

In this expression c is an arbitrary parameter: if $z^i_{k_i}$, $k_i=1,\dots,s_i$, is the maximum value of $f(\underline{x})$ for $\underline{x}$ belonging to the sample k_i, z^i_{max} is defined as

$$\max_{k_i = 1, \dots, s_i} z^i_{k_i}$$

and z^i_{med} as

$$\frac{1}{s_i}\sum_{k_i=1}^{s_i} z^i_{k_i}.$$

The probability that the maximum z_i be greater than the arbitrary value L is:

$$P(z_i > L) = \left(\frac{s_i - 1}{s_i + c}\right)^{s_i-1}$$

i.e.

$$P(z_i > L) = \left(\frac{z^i_{max} - z^i_{med}}{L - z^i_{med}}\right)^{s_i-1}$$

This expression is an approximation for $F_i(L)$; it can therefore be used as an approximation to success probabilities p^j_i.

Remark 3

We can relax the condition 3 of section 3 by dropping the requirement that the deterministic maximization algorithm is constrained to act within the cell in which the starting point has been chosen. Since

$$\bar{m}^*_i = \max_{\underline{x}\in R} f(\underline{x}) \geq \max_{\underline{x}\in C_i} f(\underline{x}) = m^*_i,$$

when the maximization is performed from the same starting point the environment is changed to a state in which both m^*_i and $\bar{m}^*_i$ have been located, and the convergence process is faster.

The algorithms by Hartman [2] and Dal Castello, Mereau and Pacquet [3] belong to the algorithms proposed in section 3.; these are obtained by choosing $h_r = 1$, $r = 1, \ldots, n$.

7. Implementation of the algorithms.

The algorithms A.2 and A.3 have been implemented, with the probability estimation suggested in remark 2 of section 6.
After the construction of initial samples in each cell, the probabilities of success are computed; in algorithm A.2, a random search is made according to the success probabilities, until a favourable point is found.
In algorithm A.3, the choice is made within the cell with largest success probability, until a favourable point is found.
Then, a deterministic maximization procedure is started, and the new function value is used in the probability estimation formula; the process continues by starting a new search procedure.
The deterministic maximization algorithm used in this implementation is the REQP package, produced by the Numerical Optimisation Center of Hatfield Polytechnic.
The iterative process is stopped when the probability of further improvement becomes lower than a given value.

8. Numerical results.

For the algorithm A.2 tests have been performed on a set of test functions with the results shown in table 1.
Remark 1: SQRIN ($n = 4$, $m = 5$) was used for testing the algorithm A.2, with n° cells = 1. This algorithm is equivalent to Hartman's method. The algorithm did not find the minimum after 20,000 function evaluations.
Remark 2: In all the tests convergence to the minimum was achieved within the initial sampling. These results are valid for the algorithm A.3 too, since A.2 and A.3 behave differently only out of the initial sample.

Further tests were performed for A.2 and A.3 on different test functions. For this set, no stopping rule was applied. The results are shown in table 2.

The functions used for test in table 2 are defined as follows:

Chichinadze:

$$f(\underline{x}) = x_1^2 - 12x_1 + 11 + 10\cos\left(\frac{\pi}{2}x_1\right) + 8\sin(5\pi x_1) -$$

$$- \frac{1}{\sqrt{5}} \exp\left(-\tfrac{1}{2}(x_2 - \tfrac{1}{2})^2\right), \quad -30 \le x_1 \le 30, \quad -10 \le x_2 \le 10;$$

Six hump camel back:

$$f(\underline{x}) = -4x_1^2 + 2.1x_1^4 - \frac{1}{3}x_1^6 - x_1 x_2 + 4x_2^2 - 4x_2^4 ,$$

$$-3 \le x_1 \le 3 , \quad -1.5 \le x_2 \le 1.5 .$$

Note

The time for the standard computation (1000 evaluations of SQRIN 5) on this machine was 0.041 sec.

Function	minimum point	minimum value	nº cells	nº samples and sample size	nº of function evaluations to achieve the minimum	nº of function evaluations to achieve convergence in probability (.90)	computational time (CDC 7600)
SQRIN n=4 m=5	(4. ,4. 4. ,4.)	10.153	5;5;5;5	2; 2	2514	2514	0.27 sec.
SQRIN n=4 m=7	(4. ,4. 3.9995,3.9996)	10.403	5;5;5;5	2; 2	2519	2519	0.359 sec.
SQRIN n=4 m=10	(4.0007,4.0006 3.9997,3.9995)	10.536	5;5;5;5	2; 2	2518	2518	0.519
Hartman n=3 m=4	(.11458,.55565, .85255)	3.8628	5;5;5	2; 2	513	2060	0.189
Hartman n=6 m=4	(.40465,.88244, .84610,.57399, .13893,.03496)	3.2032	3; 3; 3; 3; 3; 3	2; 2	2916	10970	4.157
GOLDPR n=2 m=2	$(-.149 \cdot 10^{-8}; -1.)$	-3.	3; 3	4; 4	158	1076	0.029
RCOS	(9.4248; 2.25)	.39789	10; 10	4; 4	1600	6012	0.215

Tab. 1

Function	extremum point	extremum value	nº cells	nº samples and sample size	nº of function evaluations to achieve the minimum	convergence within initial sample
A.2						
Chichinadze	(5.6235, 0.5) (max)	40.964 (max)	4; 4	2; 2	839	no
Six hump camel back	(0.089842;.71266) (min)	1.0316 (min)	4; 4	2; 2	71	yes
A.3						
Chichinadze	(5.6235, 0.5) (max)	40.964 (max)	5; 5	2; 2	132	no
RCOS	(−3.1416,12.25) (min)	0.39789 (min)	4; 4	2; 2	71	yes
Six hump camel back	(0.089842,.71266) (min)	1.0316 (min)	4; 4	2; 2	72	yes

Tab. 2

9. Conclusions.

In the above paragraphs we have described a class of global methods, within the framework of stochastic automata in time-varying environment.

Some particular implementations have been suggested with the aid of results from statistics of extrema.

The methods share the troubles common to all the space-covering methods, i.e. inapplicability in high dimensions, due to the high number of cells needed for a precise description.

Tests have been performed on a number of test functions, showing the efficiency of the methods in finding global optima.

References

[1] Chichinadze, V.K.; "The Ψ-transform for solving linear and nonlinear programming problems". Automatica (1969) vol. 5, pp.347-355.

[2] Hartman, J.K.; Some experiments in global optimisation. Naval Postgraduate School - Monterey - Cal. - Tech. Rep., May 1972.

[3] Dal Castello, G.; Mereau, P.; Pacquet, J.G.; "Methode numerique pour la recherche d'un optimum global". INFOR, June 1975, pp. 185-196.

[4] Clough, D.J.; "An asymptotic extreme value sampling theory for estimation of a global maximum". CORS J., July 1969, pp.102-115.

[5] McMurtry, G.J.; Fu, K.S.; "Variable structure automaton used as a multimodal searching technique". IEEE Autom. Contr., AC-11, n. 3, July 1966, pp. 379-387.

TOWARDS GLOBAL OPTIMISATION 2
L.C.W. Dixon and G.P. Szegö (eds.)

THE APPLICATION OF BAYESIAN METHODS FOR SEEKING THE EXTREMUM

J. Mockus, V. Tiešis, A. Žilinskas
Institute of Mathematics and Cybernetics
Academy of Sciences of the Lithuanian SSR
Vilnius

The purpose of this paper is to describe how the Bayesian approach can be applied to the global optimization of multiextremal functions. The function to be minimized is considered as a realization of some stochastic function. The optimization technique based upon the minimization of the expected deviation from the extremum is called Bayesian. The implementation of Bayesian methods is considered.

The results of the application to the minimization of some standard test functions are given.

INTRODUCTION

Many well known methods for seeking the extremum have been developed on the basis of quadratic approximation. In some problems of global optimization the function to be minimized can be considered as a realization of some stochastic function. The optimization technique based upon the minimization of the expected deviation from the extremum is called Bayesian.

The description of such methods is given in [1, 2, 3]. However, to make this paper reasonably complete a brief definition of the Bayesian methods will be given.

DEFINITION OF BAYESIAN METHODS

Assume the function to be minimized is a realization $f(x, \omega)$ of some stochastic function $f(x)$, where $x \in A \subset R^n$ and $\omega \in \Omega$ is some fixed but unknown index.

The probability distribution P on Ω is defined by the equalities:

$$F_{x_1, \dots, x_m}(y_1, \dots, y_m) = P\left\{ \omega : f(x_1, \omega) < y_1, \dots, f(x_m, \omega) < y_m \right\} \quad (1)$$

where P is a priori probability of an event:

$$\left\{ \omega : f(x_1, \omega) < y_1, \dots, f(x_m, \omega) < y_m \right\} \quad (2)$$

The observation is the evaluation of function f at some fixed point x_i. The vector

$$z_m = (f(x_1), \dots, f(x_m), x_1, \dots, x_m)$$

contains the information gained in all the observations from 1 to m.

A decision function is the measurable vector function $d = (d_0, \dots, d_N)$, which expresses the dependence between the point of the next observation and the results of the previous observations

$$x_{m+1} = d_m(z_m), \quad m = 0,1,\dots,N. \tag{3}$$

The decision function d^0 is called the Bayesian method for seeking the minimum, if it minimizes the expected deviation from the extremum

$$\min_d E\left\{ f(x_{N+1}) - f_0 \right\} \tag{4}$$

where E defines the expectation and $f(x_{N+1})$ is the value of function f at the point of final decision x_{N+1}.

The criterion (1) is satisfied under some con ditions [2 - 3] by the solution of the following recurrent equations

$$u_N(z_N) = \min_{x \in A} E\left\{ f(x) / z_N \right\}$$

$$u_{m-1}(z_{m-1}) = \min_{x \in A} E\left\{ u_m(z_{m-1}, f(x), x) / z_{m-1} \right\} \quad m = N, \dots, 2 \tag{5}$$

$$u_0 = \min_{x \in A} E\left\{ u_1(f(x), x) \right\}$$

where $E\left\{ f(x)/ z_N \right\}$ is a conditional expectation of the random variable $f(x)$ with respect to the random vector z_m.

In accordance with the definition (4) the Bayesian method depends on a priori probability distribution P. The conditions under which the Bayesian methods converge to the minimum of any continuous function are given in [4]. For example, those conditions are satisfied for a Markov process.

THE ONE-STAGE METHOD

One of the simplifications for the solution of the equations (2) is "one-stage" method [1] when at each stage it is assumed that the next observation is the least one. In such a case the sequence of observations is defined by the equations

$$E\left\{ u(z_m, f(x_{m+1}), x_{m+1}) / z_m \right\} = \min_{x \in A} E\left\{ u(z_m, f(x), x) / z_m \right\} \tag{6}$$

where

$$u(z_{m+1}) = \min_{x \in A} E\left\{ f(x) / z_{m+1} \right\}, \quad m = 0,1,\dots,N. \tag{7}$$

The one-stage Bayesian method converges to the minimum of any continuous function under the same conditions as eq. (5).

SELECTION OF AN A PRIORI DISTRIBUTION FUNCTION

The selection of a priori distribution function (1) is the only way to adjust the Bayesian method to a given class of optimization problems.

For simplicity it will be supposed that the set A is an n-dimensional cube

$[-1,1]^n$. One of the weakest conditions to satisfy the continuity of realizations of a stochastic function is the independence of n-th differences [6], which are a discrete approximation of an n-th derivatives. It is known [6] that the continuity of realizations and independence of the n-th differences implies the Gaussian probability distribution. Furthermore it is very reasonable to suppose that the stochastic model is homogeneous on A. All those conditions (continuity, independence of n-th differences and homogeneouity on A) are satisfied in a case of the Gaussian stochastic function of n variables with the expectation μ and the covariance [7]

$$K_{x_j x_k} = \sigma^2 \prod_{i=1}^{n} \left(1 - \frac{|x_j^i - x_k^i|}{2}\right) \tag{8}$$

where

$$x_j^i,\ x_k^i \in [-1,1].$$

The covariance (8) can be expressed in the following form

$$K_{x_j x_k} = \frac{1}{2^n} \sum_{l=1}^{2^n} K_{x_j x_k}(x_l)$$

where x_l, $l=1,\dots,2^n$ are the vertices of an n-dimensional cube $-1,1^n$

$$K_{x_j x_k}(x_l) = \sigma^2 \prod_{i=1}^{n} |v_i|, \text{ and } v_i = \begin{cases} \min(x_j^i - x_l^i,\ x_k^i - x_l^i), & \text{if } x_j^i \geqslant x_l^i,\ x_k^i \geqslant x_l^i \\ \min(x_l^i - x_j^i,\ x_l^i - x_k^i), & \text{if } x_j^i < x_l^i,\ x_k^i < x_l^i . \end{cases}$$

Consequently, when $\mu = 0$ the stochastic function with covariance (8) is the sum of 2^n of Wiener fields [6] with the origins on the vertices of the cube $[-1,1]^n$.

When $n=1$ it is the sum of two indepen dent Wiener processes running in the opposite directions: the origin of the first is at -1, the origin of the second is at $+1$.

In such a way one can define a priori distribution functions correct up to some unknown parameters: μ and σ^2. The estimation of those parameters can be realized on the basis of some number M of additional observations. It is convenient to distribute those observations uniformly with the equal probability on A.

The unbiased maximum likelihood estimations $\bar{\mu}$ and $\bar{\sigma}^2$ of the parameters μ and σ^2 are correspondingly:

$$\bar{\mu} = \frac{\sum_{i,k=1}^{M} \rho^{-1}_{x_i x_k} f(x_k)}{\sum_{i,k=1}^{M} \rho^{-1}_{x_i x_k}}$$

and

$$\bar{\sigma}^2 = \frac{1}{M-1} \sum_{i,k=1}^{M} \rho^{-1}_{x_i x_k} \bar{z}_i \bar{z}_k$$

where $\bar{z}_i = f(x_i) - \bar{\mu}$ and $\rho^{-1}_{x_i x_k}$ is an element of the inverse correlation matrix with the elements $\rho_{x_i x_k} = K_{x_i x_k} / \sigma^2$.

The maximum likelihood estimation $\bar{\sigma}^2$ of the only parameter σ^2 in a special case of a Wiener process on the unit interval where the distance between the additional observations are equal is very simple:

$$\bar{\sigma}^2 = \sum_{i=1}^{M} (f(i/M) - f((i-1)/M)^2$$

The computer simulation had demonstrated [8] that the effectiveness of the one-step Bayesian method (6) can be improved if the estimation of the variance σ^2 is increased by the factor $\alpha > 1$. In the case of the Wiener process the recommended [8] value of $\alpha = 7$. Apparently, it is because the increasing of the estimation of $\bar{\sigma}^2$ makes a search of the extremum more "global" and so in some way approximates the influence of the consequent steps which are neglected in the case of the one-step method.

The Gaussian field with the expectation μ and covariance

$$K_{x_i x_k} = \sigma^2 e^{-\sqrt{\sum_{i=1}^{M} c_i^2 (x_j^i - x_k^i)^2}} \qquad (9)$$

also were investigated [5, 9]. This function is interesting because in a special case when n=1 it corresponds to the stationary Markov process.

THE RELATION OF EFFECTIVENESS OF BAYESIAN SEARCH ON A PRIORI DISTRIBUTION

The arbitrariness of an a priori distribution makes it necessary to investigate the behaviour when a Bayesian method designed to minimize the realizations of one stochastic function actually is minimizing the realization of another quite different stochastic function. In [10] the one-step Bayesian method intended for the minimization of realizations of the Wiener process is applied to the minimization of realizations of the stationary Markov process with zero expectation, unit variance and exponential correlation.

The results (when the scale factor c=1) are given in the first row of Table 1. In the second row of this table the results are given for the one-stage Bayesian method designed for the minimization of the Markov process. Similarly in the third row the results of the Monte-Carlo method with random uniformly distributed observations are given.

Table 1
The dependence of the results of the Bayesian method on the a priori distribution

m	15	30	45	60	75
1	-0.929	-0.984	-1.009	-1.020	-1.026
2	-0.927	-0.989	-1.0011	-1.0022	-1.1027
3	-0.848	-0.907	-0.933	-0.951	-0.956

As one can see that in spite of considerable difference between the a priori distributions corresponding to row 1 and row 2 (the Wiener process is not even a stationary one), the observed results of the Bayesian methods designed on the basis of true and false stochastic models are very close. So it is possible to suppose, that the efficiency of Bayesian methods does not depend very much on the a priori distribution.

ONE-DIMENSIONAL SEARCH

For the solution of some multidimensional problems of optimization it is convenient to use the coordinate optimization technique, when along each coordinate a one-dimensional global search is carried out [11]. The one-dimensional search also can be applied to multidimensional optimization using the condition

$$\min_{x_1} \min_{x_2} \dots \min_{x_n} f(x_1, \dots, x_n) \tag{10}$$

The condition (10) is convenient in the two-dimensional case. Therefore, methods of one-dimensional search are of special interest.

The one-dimensional one-step Bayesian method is relatively simple [8]. In the case when the a priori distribution corresponds to the Wiener process on the interval [0,1] the coordinates m+1-th observation are

$$x_{m+1} = \arg \max_{0 \leqslant x \leqslant 1} w_{m+1}(x)$$

where

$$w_{m+1}(x) = \sigma_m(x) \int_{-\infty}^{\frac{f(x_{0m}) - \mathcal{M}_m(x)}{\sigma_m(x)}} \Pi(z)\,dz, \quad m=0,1,\dots,N$$

$$\mathcal{M}_0(x)=0, \quad \sigma_0^2(x) = \sigma^2 x, \quad f(x_{0m}) = \min_{1 \leqslant i \leqslant m} f(x_i), \quad f(x_{00}) = 0$$

and $\Pi(z)$ is a probability integral. Let $x_{i_1}, \dots, x_{i_m}$ be the coordinates of observations ordered by the increasing values. Then

$$\mu_m(x) = \frac{f(x_{i_{k+1}})(x - x_{i_k}) + f(x_{i_k})(x_{i_{k+1}} - x)}{x_{i_{k+1}} - x_{i_k}}$$

and

$$\sigma_m^2(x) = \frac{(x - x_{i_k})(x_{i_{k+1}} - x)}{(x_{i_{k+1}} - x_{i_k})}$$

if $x_{i_k} < x < x_{i_{k+1}}$. If $x = x_{i_k}$, $k = 1, \ldots, n-1$, then $\mu_n(x_{i_k}) = f(x_{i_k})$ and $\sigma_m^2(x_{i_k}) = 0$.

The function $w_m(x)$ is unimodal on the interval $x_{i_k} \leqslant x \leqslant x_{i_{k+1}}$ and so it can be minimized using usual methods of one-dimensional optimization.

COMBINATION OF THE BAYESIAN METHODS AND THE METHODS OF LOCAL OPTIMIZATION

Such properties of stochastic functions as the independence of n-th differences usually do not contradict a priori notions about the "global" behaviour of real functions. Unfortunately, some important local characteristics of the realizations are very different. For example, the realizations of the Wiener process are not differentiable at almost every x. This makes the adequacy of the stochastic models (8) or (9) very doubtful when short distances are considered. The local inadequacy of the stochastic models (8) and (9) is one reason which can explain the decline in efficiency which occurs near the termination of the Bayesian procedures of optimization when the distances between the subsequent observations are usually small. Therefore, it would be very useful to detect in advance such regions where the general stochastic model no longer describes the behaviour of the function to be minimized sufficiently well because the distances between the nearest observations have become too small. In such regions it is quite reasonable to use the quadratic approximation approach which is widely used in some very efficient methods of local optimization.

In the case when the Wiener process was used as a general stochastic model [12], the following rule for the detection of intervals of local inadequacy was used. If at the k-th observation for some l_k, $4 \leqslant l_k \leqslant k-3$ the condition

$$f(x_{i_k - 1}) > f(x_{i_k}), \quad i_k = l_k - 2, \ldots, l_k \tag{11}$$

and

$$f(x_{i_k}) < f(x_{i_k + 1}), \quad i_k = l_k, \ldots, l_{k+2}$$

is satisfied, then the interval $(x_{l_k + 2}), (x_{l_k - 2})$ is considered as the interval of local inadequacy.

The probability of inequality (11) is less than 0.016 in the case of the Wiener process when $x_{i_k}, x_{i_k} = l_k - 2, \dots, l_k$ are fixed. This inequality is quite natural, if the function $f(x)$ is unimodal at the interval $(x_{l_{k-2}}, x_{l_{k+2}})$. Therefore when such an interval is detected the corresponding local minimum is calculated correct up to $\varepsilon > 0$ using the algorithm of the parabolic approximation.

In the multidimensional case the direct detection of the regions where the function is unimodal with a given probability is more complicated. So far in such case the simplified procedure is used. The procedure is based upon the assumption, that local minima usually are in the neighbourhood of such points x_l where observed values of $f(x_l)$ are significantly lower than conditional expectation. Such points can be detected using the following inequality

$$\delta_l \leq \delta \qquad (12)$$

where

$$\delta_l = \frac{f(x_l) - E\left\{f(x_l) / z_{l-1}\right\}}{\sigma\left\{f(x_l) / z_{l-1}\right\}},$$

$\sigma\left\{f(x_l) / z_{l-1}\right\}$ is the conditional standard deviation and δ is some fixed number which depends on a significance level. In some situations it is more convenient to fix not some significance level but the number L, of local minima. In such cases the local minimization is carried out from the L starting points with lowest values of δ_l.

NUMERICAL EXAMPLES

For the minimization of seven standard test functions, three algorithms were considered. The first algorithm was Bayesian of the type (6), (7), (8), (12) with the fixed numbers of global observations N and local minimizations L. For the local minimization with simple constraints the modification of a variable metric method [13], [14] was used. The local minimization was terminated if the norm of the gradient was less than 5.10^{-4} or if the value of the function was decreasing less than $5 \cdot 10^{-4}$ in k iterations, where $k = 2$ if $n = 2$ and $k = 4$ if $n > 2$. The number of observations performed for the local minimization will be denoted N_1 and the point, that was found $x_p = (x_p^1, \dots, x_p^n)$.

The second algorithm was a combination of the Monte Carlo method with local minimization. The first N_2 observations were random and uniformly distributed. From the best of those observations the local minimization was carried out by the same method as in the case of the first algorithm. The procedure was replated K times.

The third algorithm was of the type (10) using the Bayesian algorithm (12) for one-dimensional minimization. The same accuracy level of 0.001 was fixed for both the coordinates and the function.

The first test function considered was Branin's function, which has three equal local minima:

$$f(x^1, x^2) = a(x^2 - b(x^1)^2 + (x^1 - d) + l(1 - f)\cos x^1 + l \qquad (13)$$

$$a = 1, \quad b = 5.1/4\pi^2, \quad c = 5/\pi, \quad d = 6, \quad l = 10, \quad f = 1/8\pi,$$

$$-5 \leqslant x^1 \leqslant 10, \qquad 0 \leqslant x^2 \leqslant 15.$$

The results of the minimization of function (13) by the Bayesian algorithm when $N = 70$, $L = 4$ are given in Table 2.

Table 2
The results of the optimization of Branin's function by the Bayesian algorithm

l	δ_l	x_l^1	x_l^2	x_p^1	x_p^2	N_1
19	-2.30	9.457	2.379	9.425	2.475	22
70	-2.09	-3.034	15.000	-3.142	12.275	39
39	-1.78	3.330	1.358	3.142	2.275	32
55	-1.48	3.087	2.520	3.142	2.275	26

In the case of function (13) and the Monte-Carlo algorithm when $N_2 = 1$, $K = 300$ the average number of observations was 45. The point of the first local minimum (-3.142, 12.275) was found in 35,5 % of the cases, the point of the second local minimum (3.142, 2.275) - in 39,5 % of the cases, the point of the third local minimum (9.425, 2.475) - in 25 % cases.

The third algorithm was terminated after 1021 observations of function (13). The results are given in Table 3.

Table 3
The local minima of Branin's function which were found by the third algorithm

x^1	-3.14159	3.14159	9.42478
x^2	12.27500	2.27500	2.47500
f	0.39879	0.39879	0.39879

The second test function was that of Goldstein and Price with four local minima:

$$f(x^1, x^2) = \left[1 + (x^1 + x^2 + 1)^2 (19 - 14x^1 + 3(x^1)^2 - 14x^2 + 6x^1x^2 + 3(x^2)^2)\right] \times$$

$$\times \left[30 + (2x^1 - 3x^2)(18 - 32x^1 + 12(x^1)^2 + 48x^2 - 36x^1x^2 + 27(x^2)^2)\right] \qquad (14)$$

$$-2 \leqslant x^1, \; x^2 \leqslant 2.$$

The results of the minimization of the function (14) by the Bayesian algorithm when $N = 70$, $L = 4$ are given in Table 4.

Table 4
The results of the minimization of the Goldstein and Price function by the Bayesian algorithm

l	δ_l	x_l^1	x_l^2	x_p^1	x_p^2	N_1
59	-6.02	-0.025	-0.944	0.000	-1.000	55
26	-2.06	-0.455	-0.579	-0.600	-0.400	84
44	-1.81	1.098	-1.367	0.000	-1.000	83
45	-1.42	-0.608	-0.493	-0.600	-0.400	70

The results of the minimization of function (14) by the Monte-Carlo algorithm with $K=200$ are given in Tables 5 and 6, where β_1 is the percentage of the cases when the global minimum was found, N_3 is the average number of observations, β_2 is the percentage of the cases when the result of the optimization was in the region of a singular point (x_s^1, x_s^2).

Table 5
The results of the optimization of the Goldstein and Price function by the Monte-Carlo algorithm

N_2	1	5	10	30	50	70
β_1	50	64.5	68	81.5	89	89
N_3	92	82	79	94	112	131

Table 6
The percentage of the cases when the results of optimization by Monte-Carlo method was in the region of the singular point of the Goldstein and Price function (x_s^1, x_s^2)

x_s^1	0.000	-0.600	1.800	1.200	-0.398
x_s^2	-1.000	-0.400	0.200	0.800	-0.602
β_2^2	50	25.5	15.5	8	1

The third algorithm was terminated after 520 observations of function (14). One local minimum (0.00001, -1.00000) where $f=3$ was calculated with given accuracy.

The third, fourth and fifth test functions from Shekel's family were given by the formula

$$f(x) = -\sum_{i=1}^{k} \frac{1}{(x-a_i)^T (x-a_i) + c_i} \qquad (15)$$

$$x = (x^1, \dots, x^n)^T, \quad a_i = (a_i^1, \dots, a_i^n)^T, \quad c_i > 0$$

$$0 \leqslant x^j \leqslant 10, \quad j = 1, \dots, n.$$

The values of the parameters used are those specified in the introduction to this volume.

The results of the minimization of function (15) by the Bayesian algorithm $N = 70$, $L = 5$ and Monte-Carlo algorithm ($K = 200$) are given in Tables 7 and 8, correspondingly.

Table 7
The results of the minimization of Shekel's function by the Bayesian algorithm

k = 5			k = 7			k = 10		
l δ_l N_1	x_l	x_p	l δ_l N_1	x_l	x_p	l δ_l N_1	x_l	x_p
51	0.713	1.000	61	4.398	4.994	27	5.009	4.001
-7.51	0.815	1.000	-11.6	4.397	4.995	-4.08	2.798	4.001
117	0.570	1.000	171	3.275	3.006	216	4.831	4.000
	1.487	1.000		2.586	3.007		3.034	4.000
14	5.938	5.999	29	4.053	4.994	32	5.285	4.995
-3,68	6.803	6.000	-6.56	4.566	4.995	-2.41	6.148	4.994
137	5.398	5.999	189	2.907	3.006	243	3.654	3.008
	5.734	6.000		3.146	3.007		3.213	3.007
15	4.378	4.000	12	6.393	5.998	57	6.196	6.006
-2.98	3.370	4.000	-2.15	5.404	6.000	-2.01	1.351	2.010
227	2.892	4.000	164	6.253	5.997	261	8.077	6.004
	4.434	4.000		4.654	5.999		1.481	2.009
70	4.057	3.002	63	2.730	4.001	40	1.402	3.001
-1.99	7.739	6.998	-2.15	5.186	4.001	-1.45	6.696	7.000
155	3.358	3.002	359	6.153	3.999	189	3.909	3.001
	7.499	6.998					6.890	7.000
67	5.700	3.001	42	0.836	3.001	39	2.799	3.001
-0.45	3.037	6.998	-0.97	9.595	7.001	-0.67	8.785	7.000
468	8.856	3.001	326	0.501	3.000	230	0.617	3.001
	2.394	6.998		9.807	7.000		6.477	7.000

Table 8
The results of the minimization of Shekel's function by the Monte-Carlo algorithm where N_3 is the average number of observations, β_1 - percentage of the cases when the global minimum was found

	N_2	1	5	10	30	50	70
k = 5	β_1	38.5	38	32.5	32.5	24.5	28.5
	N_3	353	306	288	265	263	263
k = 7	β_1	44	39.5	32	33	22	32.5
	N_3	369	308	288	272	266	285
k = 10	β_1	38	34	34	30.5	21	27.5
	N_3	343	295	280	267	265	275

The sixth and seventh test functions from Hartman's family were given by the formula

$$f(x) = -\sum_{i=1}^{\bar{K}} c_i \exp\left(-\sum_{j=1}^{n} \alpha_{ij} (x^j - p_{ij})^2\right) \tag{16}$$

$$x = (x^1, \dots, x^n), \quad \alpha_i = (\alpha_{i1}, \dots, \alpha_{in}), \quad p_i = (p_{i1}, \dots, p_{in})$$

The standard parameters values were used.

The results of the minimization of function (16) by the Bayesian algorithm ($N=70$, $L=5$) and Monte-Carlo algorithm ($K=200$) are given in Tables 9, 10, correspondingly, where N_3 is the average number of observations and β_1 is the percentage of cases when the global minimum was found.

Table 9
The results of the minimization of Hartman's function by the Monte-Carlo algorithm

	N_2	1	5	10	30	50	70
n = 3	β_1	60	71	89.5	92.5	95.5	99.5
	N_3	103	97	97	118	126	155
n = 6	β_1	64	71	74	64	51	57
	N_3	321	268	259	270	278	303

Table 10
The results of the minimization of Hartman's functions by the Bayesian algorithm

n = 3				n = 6			
1 δ_1	x_1	x_p	N_1	1 δ_1	x_1	x_p	N_1
37 -4.84	0.0000 0.5668 0.8297	0.1146 0.5556 0.8525	87	65 -3.01	0.0692 0.4275 0.5566 0.3245 0.2232 0.8545	0.2017 0.1500 0.4768 0.2753 0.3117 0.6573	246
28 -4.45	0.8398 0.4609 0.8904	0.1146 0.5556 0.8525	110	5 -2.00	0.5467 0.3877 0.8051 0.4761 0.4542 0.8605	0.2017 0.1500 0.4768 0.2753 0.3117 0.6573	257
48 -2.95	0.9573 0.5447 0.9278	0.1146 0.5556 0.8525	91	15 -1.66	0.6265 0.9441 0.9181 0.7578 0.0802 0.2041	0.4047 0.8824 0.8462 0.5740 0.1389 0.0385	221
24 -1.38	0.4527 0.0000 0.1854	0.3687 0.1176 0.2675	78	58 -1.18	0.2167 0.8614 1.0000 0.9419 0.9640 0.0000	0.4077 0.8824 0.8462 0.5740 0.1387 0.0385	202
34 -1.38	0.5003 0.8212 0.5331	0.1093 0.1093 0.5641	77	46 -0.68	0.4026 1.0000 0.0000 0.9010 0.9800 0.3206	0.4047 0.8824 0.8461 0.5740 0.1388 0.0385	236

CONCLUSIONS

The Bayesian algorithms require significantly less observations (function evaluations) than the Monte-Carlo algorithms under the conditions which were considered in this paper. However, applying the Bayesian algorithms it is necessary to perform a supplementary calculations to find the best point at the next observation. It is not so important when the observations are complicated or expensive. But when the function evaluations are simple it is reasonable to simplify the Bayesian algorithms using some approximation of the equations (6) (7).

REFERENCES

1. Mockus J.B.(1972). On Bayesian Methods for Seeking the Extremum. Automatics and Computers, No. 3, 53-62 (Russian).

2. Mockus J.B. (1974). On Bayesian Methods for Seeking the Extremal Point, Kybernetes, Vol. 3, 103-108.
3. Mockus J.B. (1975). On Bayesian Methods of Optimization, Towards Global Optimization, North-Holland, Amsterdam, 166-181.
4. Mockus J.B. (1975). On Bayesian Methods for Seeking the Extremum, Lecture Notes in Computer Science, 27, Springer Verlag, Berlin, Heidelberg, New York, 400-404.
5. Žilinskas A.G., Mockus J.B. (1970). On a Bayesian Method for Seeking the Minimum, Automatics and Computers, No. 3, 42-44 (Russian).
6. Katkauskaitė A.J. (1972). The Random Fields with Independent Increments, Lithuanian Mathematical Transaction, No. 4, 75-85 (Russian).
7. Mockus J.B. (1977). On Bayesian Methods for Seeking the Extremum and Their Application, Information Processing 77, B. Gilchrist (ed.), North-Holland, Amsterdam, 195-200.
8. Žilinskas A.G. (1975). The One-step Bayesian Method for Seeking the Extremum of the Function at One Variable, Cybernetics, No. 1, 139-144 (Russian).
9. Šaltenis V.R. (1971). On the Methods for the Multiextremal Optimization, Automatics and Computers, No. 3, 34-38 (Russian).
10. Žilinskas A.G., Timofeev L.L. (1976). On the Relation of the Efficiency of Bayesian Methods for Seeking the Extremum and the a Priori Distribution, Numerical Methods of Nonlinear Programming, Naukova Dumka, Kiev, 186-189.
11. Mockus J.B. (1976). Multiextremal Problems in Design, Nauka, Moscow (Russian).
12. Žilinskas A.G. (1974). On a Simplified Bayesian Algorithm for Seeking the Minimum, The Computer-aided Design of Engineering Units and Technological Processes, Gorky State University, Gorky, 54-57(Russian).
13. Goldfarb B.D. (1969). Extension of Davidson's Variable Metric Method to Minimization Under Linear Unequality and Equality Unstrain't, SIAM Journal Applied Mathematics, Vol. 17, No. 4, 733-764.
14. Biggs M.C. (1971). Minimization Algorithms Making use of Nonquadratic Properties of the Objective Function, Journal of the Inst. of Mathematics and Applications, Vol. 8, No. 3, 315-327.

TOWARDS GLOBAL OPTIMISATION 2
L.C.W. Dixon and G.P. Szegö (eds.)

GLOBAL OPTIMIZATION AS AN EXTENSION OF INTEGER PROGRAMMING

by

E.M.L. Beale & J.J.H. Forrest

(Scicon Computer Services Ltd)

SUMMARY

Branch and Bound methods can be used to find global optima of functions that are defined as sums of products of functions of single arguments. These methods can be incorporated as extensions to the Integer Programming facilities in general Mathematical Programming Systems. We discuss their implementation using Special Ordered Sets, and a new extension known as Linked Ordered Sets. Appropriate formulations for the four families of test problems suggested for this study are given, as are computational results using Scicon's new mathematical programming system SCICONIC.

1. INTRODUCTION

Branch and Bound methods for integer programming can be extended to find arbitrarily close approximations to the global optimum value of a wide range of continuous functions. The fundamental requirement is that the function be explicitly defined. But we make the stronger assumption that it is defined as a finite sum of products of a finite number of uniformly continuous functions of single arguments, and that each of these arguments can be limited a priori to a finite range. In fact we can allow a finite number of discontinuities in these otherwise continuous functions, but this is hardly relevant to the present paper.

A convenient way to do this is to use Special Ordered Sets. These were introduced by Beale and Tomlin (1970) and revised by Beale and Forrest (1976). In their published form, they apply only to sums of nonlinear functions of single arguments. They can be adapted to products by using logarithms or other transformations. But an extension of Special Ordered Sets, known as Linked Ordered Sets, allows us to treat products more directly.

The theory of Special Ordered Sets and Linked Ordered Sets is outlined in Section 2. Then in Section 3 we discuss their application to the four families of problem suggested for this study.

2. THEORY

Suppose that the nonlinear function $f(z)$ occurs in either the objective function or a constraint of an otherwise linear programming problem, and suppose that the argument z can take only a finite number of possible values, say Z_k for $k = 0,\ldots K$. Then we can introduce a set of nonnegative variables λ_k for $k = 0,\ldots K$ and write

$$\sum_k \lambda_k \qquad = 1$$

$$\sum_k Z_k \lambda_k \quad -z \quad = 0.$$

The nonlinear function $f(z)$ is then represented by the linear function $\sum_k f(Z_k)\lambda_k$, if we impose the further restriction that not more than one of the λ_k may be nonzero. And if we solve the linear programming problem without this further restriction we obtain an upper bound on how good the solution can be.

Furthermore, in order to sharpen this upper bound we can carry out branch and bound operations on the set of λ-variables allowed to take nonzero values. Specifically, if λ_{k_1} is any member of the set, we note that in any valid solution

<u>Either</u> $\qquad \lambda_0 = \ldots = \lambda_{k_1} \quad = 0$

<u>or</u> $\qquad \lambda_{k_1+1} = \ldots\ldots = {}_K \quad = 0.$

So we can replace our original linear programming problem by two new linear programming subproblems that include all valid solutions to the original problem while excluding the solution to the previous linear programming approximation. We can therefore solve the problem by a branch and bound process.

In the above formulation, the set of λ-variables is known as a Special Ordered Set of Type One, or <u>S1 set</u>, since only one member may ultimately be nonzero. But we may relax the formulation and allow two adjacent members of the set to be nonzero. This is then known as a Special Ordered Set of Type Two, or <u>S2 set</u>. It amounts to permitting linear interpolation to $f(z)$ between adjacent given values of the argument Z_k. The modification to the branching rule is simply that

Either $\lambda_0 = \ldots = \lambda_{k_1-1} = 0$

or $\lambda_{k_1+1} = \ldots = \lambda_K = 0.$

Now suppose that we wish to represent $u_j f_j(z)$ for $j=1\ldots J$, where the u_j are other variables of the problem, or possibly linear functions of such other variables. We suppose again that z must take one of the values Z_k for $k=0,\ldots K$, and that $U_{MINj} \leq u_j \leq U_{MAXj}$.

We now introduce 2J sets of nonnegative variables λ_{j1k} and λ_{j2k} for $k=0\ldots K$, $j=1\ldots J$ and write

$$\sum_k \lambda_{j1k} + \sum_k \lambda_{j2k} = 1$$

$$\sum_k Z_k\lambda_{j1k} + \sum_k Z_k\lambda_{j2k} - z = 0$$

$$\sum_k U_{MINj}\lambda_{j1k} + \sum_k U_{MAXj}\lambda_{j2k} - u_j = 0$$

Then if $z = Z_{k_1}$ we can represent the nonlinear function $u_j f_j(z)$ by the linear function

$$\sum_k U_{MINj} f_j(Z_k)\lambda_{j1k} + \sum_k U_{MAXj} f_j(Z_k)\lambda_{j2k},$$

if we impose the further restriction that $\lambda_{jvk} = 0$ whenever $k \neq k_1$ for all j and v. We therefore treat the sets of variables λ_{jvk} as Linked S1 Sets, and carry out branch and bound operations on the first and last values of k for which the variables may take nonzero values for all j and v simultaneously.

Beale and Forrest (1976) show how Special Ordered Sets can be used when z is a truly continuous variable. Provision must then be made for introducing new variables λ_k while solving linear programming subproblems. Such automatic column-generation techniques are essentially part of the Decomposition Principle introduced by Dantzig and Wolfe (1960). There remains the global optimization problem of choosing a value of z for which the corresponding λ_k is within ε of the most negative reduced cost. Beale and Forrest (1976) define a finite algorithm for this using ideas similar to those used by Brent (1973). They require that each f(z) can be written in the form

$$f(z) = \sum_i f_i(z),$$

where

(a) each $f_i(z)$ is twice differentiable

(b) $f_i(z), f_i'(z)$ and $f_i''(z)$ can be calculated for any z

and

(c) the range of possible values of z can be divided a priori into a finite number of intervals such that $f_i''(z)$ is monotonic within each interval.

Some additional code must be added to the Mathematical Programming System for each family of possible functions $f_i(z)$ requiring automatic interpolation. At the time of writing SCICONIC allows functions of the form z^θ or exp(z), but not cos z or $\exp(-z^2)$. So, while we have solved all the test problems using a finite grid of possible values for all nonlinear variables, we have solved only some of them using automatic interpolation. It is perhaps worth adding that this part of the System is under active development. In particular, we hope that the computing time using automatic interpolation will be greatly reduced in future by making more use of parametric programming.

3. FORMULATIONS FOR THE TEST PROBLEMS

It will be noted that our approach requires a more fundamental analysis of the mathematical structure of the function to be maximized than, say, a hill-climbing method for local optimization. One could argue about the nature of the difference between the work involved in the mathematics and programming of this analysis and the mathematics and programming for, say, function and derivative evaluation. But we hope that the reader will agree after reading this section that it is fairly straightforward when the functions have simple forms.

3.1 Shekel's Family

$$\underset{x}{\text{Maximize}} \quad \sum_{i=1}^{m} \{\sum_{j=1}^{n} (x_j - a_{ij})^2 + c_i\}^{-1} \quad \text{when } 0 \le x_j \le 10.$$

This problem requires only Special Ordered Sets.

Write $f(\theta) = \theta^2, \quad g(\theta) = 1/\theta$.

Then the problem is to maximize$_{x,s,z}$ $\sum_{i=1}^{m} g(z_i)$, subject to the constraints

$$z_i - s + 2\sum_j a_{ij}x_j^2 - c_i = \sum_j a_{ij}^2 \qquad (i=1,\ldots m)$$

$$s - \sum_j f(x_j) = 0.$$

Since $0 \le x_j \le 10$, we note that

$$c_i \le z_i \le c_i + \sum_j \max(a_{ij}^2, (10-a_{ij})^2).$$

The explicit formulation for the Mathematical Programming System is therefore as follows:

Maximize x_0 subject to
x,μ,s,λ

$$x_o - \sum_{i=1}^{m} \sum_k Z_{ik}^{-1} \mu_{ik} = 0$$

$$\sum_k \mu_{ik} = 1 \qquad (\forall i)$$

$$\sum_k Z_{ik}\mu_{ik} - s + 2\sum_j a_{ij}x_j = c_i + \sum_j a_{ij}^2 \qquad (\forall i)$$

$$s - \sum_j \sum_k X_{jk}^2 \lambda_{jk} = 0$$

$$\sum_k \lambda_{jk} = 1 \qquad (\forall j)$$

$$-x_j + \sum_k X_{jk}\lambda_{jk} = 0 \qquad (\forall j).$$

The problem therefore have m+n S2 sets, and 2(m+n)+1 constraints. There is no temptation to overestimate s, so we will never need to branch on any of the λ-sets. But we will have to branch on the m μ-sets.

3.2 Hartman's Family

$$\underset{x}{\text{Maximize}} \sum_{i=1}^{m} c_i \exp(-\sum_{j=1}^{n} \alpha_{ij} (x_j-p_{ij})^2) \quad \text{when } 0 \le x_j \le 1.$$

This problem is interesting, in that it can be solved using only Special Ordered Sets, or using Linked Ordered Sets. The Special Ordered Set formulation is as follows:

Write $f(\theta) = \theta^2$, $g(\theta) = \exp(-\theta)$.

Then the problem is to maximize $\underset{z,x,y}{} \sum_{i=1}^{m} c_i g(z_i)$,

subject to the constraints

$$z_i - \sum_{j=1}^{n} \alpha_{ij} (y_j - 2p_{ij}x_j + p_{ij}^2) = 0 \quad (i=1,\ldots m)$$

$$y_j - f(x_j) = 0 \quad (j=1,\ldots n).$$

Since $0 \leq x_j \leq 1$ and $0 \leq p_{ij} \leq 1$, we note that

$$0 \leq z_i \leq \sum_j \alpha_{ij} \max(p_{ij}^2, (1-p_{ij})^2.$$

The complete formulation is:

Maximize x_0 subject to
x, μ, y, λ

$$x_0 - \sum_{i=1}^{m} \sum_k c_i \exp(-Z_{ik}) \mu_{ik} = 0$$

$$\sum_k \mu_{ik} = 1 \qquad (\forall i)$$

$$\sum_k Z_{ik}\mu_{ik} - \sum_j \alpha_{ij}y_j + 2\sum_j \alpha_{ij}p_{ij}x_j = \sum_j \alpha_{ij}p_{ij} \qquad (\forall i)$$

$$y_j - \sum_k X_{jk}^2 \lambda_{jk} = 0 \qquad (\forall j)$$

$$\sum_k \lambda_{jk} = 1 \qquad (\forall j)$$

$$-x_j + \sum_k X_{jk} \lambda_{jk} = 0 \qquad (\forall j).$$

This problem therefore has m+n S2 sets, and 2m+3n constraints. There is no temptation to overestimate y_j, so we again never need to branch on any of the λ-sets. But we will have to branch on the m μ-sets.

The formulation using linked ordered sets is rather more complex. But it is certainly preferable if m >> n, since branching operations are restricted to 1 S1 (or S2) set μ_k defining the value of x_1 and n-1 sets of linked S1 sets λ_{jivk} defining the values of x_j for j = 2,...n.

If we write

$$h_{ij}(x_j) = \exp(-\alpha_{ij}(x_j - p_{ij})^2),$$

the problem is to maximize $\sum_{i=1}^{m} \prod_{j=1}^{n} h_{ij}(x_j)$;

or if $u_{ij} = \prod_{j_1=1}^{j} h_{ij_1}(x_{j_1})$,

the problem is to maximize $\sum_{i=1}^{n} u_{in}$, where

$$u_{i1} = h_{i1}(x_1) \quad \text{and} \quad u_{ij} = u_{i,j-1}h_{ij}(x_j) \quad \text{for } j=2,\ldots n.$$

Note that $U_{MINij} \leq u_{ij} \leq U_{MAXij}$, where

$$U_{MINij} = \prod_{j_1=1}^{j} \min(h_{ij_1}(0), h_{ij_1}(1)), \text{ and } U_{MAXij} = 1.$$

Further, if $\sum_k \mu_k = 1$, $\sum_k X_k\mu_k - x_1 = 0$, then $u_{i1} = \sum_k h_{i1}(X_{1k})\mu_k$,

and if

$$\sum_k \lambda_{ji1k} + \sum_k \lambda_{ji2k} = 1$$

$$\sum_k X_{jk}\lambda_{ji1k} + \sum_k X_{jk}\lambda_{ji2k} = 0$$

$$\sum_k U_{MINi,j-1}\lambda_{ji1k} + \sum_k U_{MAXi,j-1}\lambda_{ji2k} - u_{i,j-1} = 0,$$

then $u_{ij} = \sum_k U_{MINi,j-1}h_{ij}(X_{jk})\lambda_{ji1k} + \sum_k U_{MAX,j-1}h_{ij}(X_{jk})\lambda_{ji2k}$.

We can now eliminate the variables u_{ij} from the problem, and the problem becomes:

Maximize$_{x,\lambda,\mu}$ x_0 subject to

$$x_0 - \sum_{i=1}^{m}\sum_k c_i U_{MINi,n-1}h_{in}(X_{nk})\lambda_{ni1k} - \sum_{i=1}^{m}\sum_k c_i U_{MAX,n-1}h_{in}(X_{nk})\lambda_{ni2k} = 0$$

$$\sum_k \mu k = 1$$

$$\sum_k X_{1k}\mu_k - x_1 = 0$$

$$\sum_k \lambda_{ji1k} + \sum_k \lambda_{ji2k} = 0 \quad (\forall i, j=2,\ldots n)$$

$$\sum_k X_{jk}\lambda_{ji1k} + \sum_k X_{jk}\lambda_{ji2k} - x_j = 0 \quad (\forall i, j=2,\ldots n)$$

$$\sum_k U_{MINi1}\lambda_{2i1k} + \sum_k U_{MAXi1}\lambda_{2i2k} - \sum_k h_{ij}(X_{ik})\ {}_k = 0 \quad (\forall i)$$

$$\sum_k U_{MINij}\lambda_{j+1,i1k} + \sum_k U_{MAXij}\lambda_{j+1,i2k}$$

$$-\sum_k U_{MINij-1}h_{ij}(X_{jk})\lambda_{ji1k} - \sum_k U_{MAXij-1}h_{ij}(X_{jk})\lambda_{ji2h} = 0 \quad (\forall i, j=2,\ldots n)$$

3.3 Branin's Family

Minimize $a(x_2 - bx_1^2 + cx_1 - d)^2 + e(1-f)\cos x_1 + e,$

where $-5 \leq x_1 \leq 10, \quad 0 \leq x_2 \leq 15.$

This problem requires just two S2 sets.

Write $f(\theta) = \theta^2, \quad g(\theta) = \cos\theta.$

Then the problem is to minimize $af(z) + e(1-f)g(x_1) + e$, subject to the constraint

$$z - x_2 + b\, f(x_1) - cx_1 + d = 0$$

Since $f(z)$ is a convex function to be minimized, it does not matter if we exaggerate its range. So we merely note that

$$-100b - 5c - d \leq z \leq 15 + 10c - d.$$

The complete formulation is:

Maximize$_{x,\mu,\lambda}$ x_0 subject to

$$x_0 + \sum_k aZ_k^2\mu_k + \sum_k e(1-f)\cos X_k\lambda_k = -e$$

$$\sum_k \mu_k = 1$$

$$\sum_k Z_k\mu_k - x_2 + b\sum_k X_k^2\lambda_k - cx_1 = -d$$

$$\sum_k \lambda_k = 1$$

$$- x_1 + \sum_k X_k \lambda_k = 0.$$

3.4 Goldstein and Price's Family

We treat Goldstein and Price's problem as one of minimizing a general polynomial of degree 8 in two variables x_1 and x_2, each required to lie between -2 and +2. So the problem is to minimize $\sum_i \sum_j a_{ij} x_1^{\,i} x_2^{\,j}$.

Write $h_i(x) = \sum_j a_{ij} x^j$.

Then the problem is to minimize $h_o(x_2) + \sum_{i=1}^{7} h_i(x_2) x_1^{\,i} + a_{80} x_1^{\,8}$.

We use μ_k as an S1 set defining the value of x_1, and λ_{ivk} (for $v = 1,2$) as a set of linked S1 sets defining the value of x_2 and the functions $h_i(x_2)x_1^{\,i}$.

If $\sum_k \mu_k = 1$, $\sum_k X_{1k}\mu_k - x_1 = 0$, then $x_1^{\,i} = \sum_k X_{1k}^{\;\,i}\mu_k$.

Further, if

$$\sum_k \lambda_{i1k} + \sum_k \lambda_{i2k} = 1$$

$$\sum_k X_{2k}\lambda_{i1k} + \sum_k X_{2k}\lambda_{i2k} - x_2 = 0$$

$$\sum_k V_{MINi}\lambda_{i1k} + \sum_k V_{MAXi}\lambda_{i2k} - \sum_k X_{1k}^{\;\,i}\mu_k = 0,$$

then $h_i(x_2)x_1^{\,i} = \sum_k V_{MINi} h_i(X_{2k})\lambda_{i1k} + \sum_k V_{MAXi} h_i(X_{2k})\lambda_{i2k}$,

provided that $V_{MINi} \le x_1^{\,i} \le V_{MAXi}$.

$$V_{MINi} = 0 \text{ if } i \text{ is even}$$

$$= -2^i \text{ if } i \text{ is odd.}$$

$$V_{MAXi} = 2^i.$$

The complete formulation is therefore:

Maximize x_o subject to
x, λ, μ

$$x_o + \sum_k h_o(X_{2k})\,\lambda_{11k} + \sum_k h_o(X_{2k})\lambda_{12k} + \sum_{i=1}^{7}\sum_k V_{MINi} h_i(X_{2k})\lambda_{i1k}$$

$$+ \sum_{i=1}^{7}\sum_k V_{MAXi} h_i(X_{2k})\lambda_{i2k} + \sum_k a_{80} X_{1k}^{8}\mu_k = 0$$

$$\sum_k \mu_k = 1$$

$$\sum_k X_{1k}\,\mu_k - x_1 = 0$$

$$\sum_k \lambda_{i1k} + \sum_k \lambda_{i2k} = 1 \quad (i=1,\dots 7)$$

$$\sum_k X_{2k}\lambda_{i1k} + \sum_k X_{2k}\lambda_{i2k} - x_2 = 0 \quad (i=1,\dots 7)$$

$$\sum_k V_{MINi}\lambda_{i1k} + \sum_k V_{MAXi}\lambda_{i2k} - \sum_k X_{1k}^{i}\mu_k = 0 \quad (i=1,\dots 7)$$

An alternative approach is to use Chebychev polynomials in x_1 rather than simple powers. Since we are interested in the range $-2 \le x_1 \le +2$, the relevant polynomials $C_i(x)$ are defined as follows:

$$C_o(x) = 2$$
$$C_1(x) = x$$
$$C_2(x) = x^2 - 2$$
$$C_3(x) = x^3 - 3x$$
$$C_4(x) = x^4 - 4x^2 + 2$$
$$C_5(x) = x^5 - 5x^3 + 5x$$
$$C_6(x) = x^6 - 6x^4 + 9x^2 - 2$$
$$C_7(x) = x^7 - 7x^5 + 14x^3 - 7x$$
$$C_8(x) = x^8 - 8x^6 + 20x^4 - 16x^2 + 2.$$

These functions all lie between -2 and +2 if $-2 \le x \le 2$.

The objective function $\sum_i \sum_j a_{ij} x_1^{\,i} x_2^{\,j}$ can be written as $\sum_i (\sum_{j=o}^{8} b_{ij} x_2^{\,j}) C_i(x_1)$, and the coefficients b_{ij} can be computed from the formulae

$$
\begin{aligned}
b_{8j} &= a_{8j} \\
b_{7j} &= a_{7j} \\
b_{6j} &= a_{6j} + 8b_{8j} \\
b_{5j} &= a_{5j} + 7b_{7j} \\
b_{4j} &= a_{4j} + 6b_{6j} - 20b_{8j} \\
b_{3j} &= a_{3j} + 5b_{5j} - 14b_{7j} \\
b_{2j} &= a_{2j} + 4b_{4j} - 9b_{6j} + 16b_{8j} \\
b_{1j} &= a_{1j} + 3b_{3j} - 5b_{5j} + 76_{7j} \\
b_{0j} &= \tfrac{1}{2}a_{0j} + b_{2j} - b_{4j} + b_{6j} - b_{8j}.
\end{aligned}
$$

Now write

$$g_i(x) = \sum_j b_{ij}x^j.$$

The problem is then to minimize $2g_o(x_2) + \sum_{i=1}^{7} g_i(x_2)C_i(x_1) + b_{80}C_8(x_1)$, and the explicit formulation is:

maximize x_o subject to:
x,λ,μ

$$
\begin{aligned}
x_o &+ \sum_k 2g_o(X_{2k})\lambda_{11k} + \sum_k 2g_o(X_{2k})\lambda_{12k} + \sum_{i=1}^{7}\sum_k C_{MINi}g_i(X_{2k})\lambda_{i1k} \\
&+ \sum_{i=1}^{7}\sum_k C_{MAXi}g_i(X_{2k})\lambda_{i2k} + \sum_k b_{80}C_8(X_{1k})\mu_k = 0
\end{aligned}
$$

$$\sum_k \mu_k = 1$$

$$\sum_k X_k\mu_k - x_1 = 0$$

$$\sum_k \lambda_{i1k} + \sum_k \lambda_{i2k} = 1 \quad (i=1,\ldots 7)$$

$$\sum_k X_{2k}\lambda_{i1k} + \sum_k X_{2k}\lambda_{i2k} - x_2 = 0 \quad (i=1,\ldots 7)$$

$$\sum_k C_{MINi}\lambda_{i1k} + \sum_k C_{MAX}\lambda_{i2k} - \sum_k C_i(X_{1k})\mu_k = 0 \quad (i=1,\ldots 7),$$

where $C_{MINi} \leq C_i(x_1) \leq C_{MAXi}$, i.e. where $C_{MINi} = -2$ and $C_{MAXi} = +2$ for all i.

4. COMPUTATIONAL RESULTS

All the problems have been solved on a Univac 1108 computer using the Mathematical Programming System SCICONIC. Computer times are quoted in seconds of CPU time. They exclude the times taken to run the Matrix Generator and Input to SCICONIC: but these times are all small. The problems were run with 21 values and with 101 values of the argument of each nonlinear function. These values were equally spaced for simplicity, although this spacing is not logical unless the function is quadratic.

Tabular results for each problem and each formulation give the following information:

1. No. of Rows.
2. No. of Columns.
3. No. of Nonzero elements.

   These are a crude measure of the sizes of the linear programming problems. The numbers of columns and nonzero elements are meaningless with automatic interpolation and are not quoted.
4. The continuous LP optimum. This is the first lower bound on the value of the objective function.
5. The value of the First Solution. A comparison of this with the optimum solution is a measure of the efficiency of the Branch and Bound process.
6. The final optimum value of the objective function of the linear programming problem.
7. The number of branches explored.
8. The cumulative number of LP iterations used to solve the continuous optimum and all necessary branches.
9. The CPU time in seconds.
10. The optimum values of the variables under the current approximation to the problem. When using fixed grids of possible values, the optima are quoted to the precision of the grids. When using automatic interpolation the optima are quoted to 6 decimal places in case it is of interest to compare these with the solutions found by other methods, although the code does not attempt to solve the problems to this degree of precision.

4.1 Shekel's Family

The standard data outlined in the introduction to this volume was used, i.e.

The Data: $n = 4$, $m = 5,7,10$

	a_{ij}				c_i
	j=1	j=2	j=3	j=4	
i=1	4.0	4.0	4.0	4.0	0.1
i=2	1.0	1.0	1.0	1.0	0.2
i=3	8.0	8.0	8.0	8.0	0.2
i=4	6.0	6.0	6.0	6.0	0.4
i=5	3.0	7.0	3.0	7.0	0.4
i=6	2.0	9.0	2.0	9.0	0.6
i=7	5.0	5.0	3.0	3.0	0.3
i=8	8.0	1.0	8.0	1.0	0.7
i=9	6.0	2.0	6.0	2.0	0.5
i=10	7.0	3.6	7.0	3.6	0.5

Results

$m = 5$

	21 Point Grids	101 Point Grids	Interpolation
Rows	21	21	21
Cols	212	932	
Elements	624	2784	
Continuous Optimum	-22.854055	-22.869828	-22.869954
First Solution	-10.155971	-10.154045	-10.153243
Optimum Solution of LP	-10.155971	-10.154103	-10.153243
No. of Branches	62	52	155
No. of Iterations	205	350	1348
CPU Time (Secs.)	7	31	31
x_1	4.0	4.0	4.001965
x_2	4.0	4.0	4.000600
x_3	4.0	4.0	3.999867
x_4	4.0	4.0	4.000656
Corresponding F(x)	-10.06903	-10.06903	-10.15320

GLOBAL SOLUTION $F(x^*)$ - 10.15320

m = 7

	21 Point Grids	101 Point Grids	Interpolation
Rows	25	25	25
Cols	258	1138	
Elements	766	3046	
Continuous Optimum	-27.484770	-27.494089	-27.494350
First Solution	-11.774769	-10.410592	-8.488798
Optimum Solution of LP	-11.776349	-10.410613	-10.403004
No. of Branches	179	180	689
No. of Iterations	430	850	8214
CPU Time (Secs.)	20	86	205
x_1	4.0	4.0	4.000216
x_2	4.5	4.0	4.001133
x_3	4.0	4.0	3.999686
x_4	4.0	4.0	3.999535
Corresponding F(x)	-3.24211	-10.40282	

GLOBAL SOLUTION F(x*) -10.4029

m = 10

	21 Point Grids	101 Point Grids	Interpolation
Rows	31	31	31
Cols.	327	1447	
Elements	979	4339	
Continuous Optimum	-32.291721	-32.306712	-32.306678
First Solution	-11.915683	-10.544371	-8.383578
Optimum Solution of LP	-11.925474	-10.544375	-10.536605
No. of Branches	790	824	1094
No. of Iterations	1690	2538	20911
CPU Time (Secs.)	94	353	556
x_1	4.5	4.0	4.000923
x_2	4.0	4.0	4.000677
x_3	4.0	4.0	3.999645
x_4	4.0	4.0	3.999187
Corresponding F(x)	-3.462930	-10.53628	

GLOBAL SOLUTION F(x*) - -10.5364

The odd results quoted with 21 Point Grids arise from the fact that an equally spaced 21 Point Grid is quite inadequate to represent the reciprocal functions arising in this problem. The problem is an intrinsically easy one for branch and bound methods, since one local minimum is so much better than all others.

4.2 Hartman's Family

First Problem: $n = 3$, $m = 4$.

	α_{ij} j=1	j=2	j=3	c_i	p_{ij} j=1	j=2	j=3
i=1	3.0	10.0	30.0	1.0	0.3689	0.1170	0.2673
i=2	0.1	10.0	35.0	1.2	0.4699	0.4387	0.7470
i=3	3.0	10.0	30.0	3.0	0.1091	0.8732	0.5547
i=4	0.1	10.0	35.0	3.2	0.03815	0.5743	0.8828

Second Problem: $n = 6$, $m = 4$.

	α_{ij} j=1	j=2	j=3	j=4	j=5	j=6	c_i
i=1	10.0	3.0	17.0	3.5	1.7	8.0	1.0
i=2	0.05	10.0	17.0	0.1	8.0	14.0	1.2
i=3	3.0	3.5	1.7	10.0	17.0	8.0	3.0
i=4	17.0	8.0	0.05	10.0	0.1	14.0	3.2

	p_{ij} j=1	j=2	j=3	j=4	j=5	j=6
i=1	0.1312	0.1696	0.5569	0.0124	0.8283	0.5886
i=2	0.2329	0.4135	0.8307	0.3736	0.1004	0.9991
i=3	0.2348	0.1451	0.3522	0.2883	0.3047	0.6650
i=4	0.4047	0.8828	0.8732	0.5743	0.1091	0.0381

Results

First Problem Using Special Ordered Sets.

	21 Point Grids	101 Point Grids	Interpolation
Rows	19	19	19
Cols.	167	727	
Elements	492	2172	
Continuous Optimum	-7.739467	-7.741731	-7.741819
First Solution	-4.047685	-3.786536	-3.845759
Optimum Solution of LP	-4.047685	-3.881084	-3.863225
No. of Branches	36	75	103
No. of Iterations	117	553	1032

CPU Time (Secs.)	4	39	21
x_1	0.15	0.12	0.111554
x_2	0.55	0.56	0.555859
x_3	0.85	0.85	0.854464

GLOBAL SOLUTION $F(x^*)$ = -3.86278

Second Problem Using Special Ordered Sets.

	21 Point Grids	101 Point Grids	Interpolation
Rows	28	28	28
Cols.	242	1042	
Elements	720	3120	
Continuous Optimum	-7.546336	-7.549444	-7.549570
First Solution	-3.299720	-3.326780	-3.204061
Optimum Solution of LP	-3.426354	-3.330978	-3.322984
No. of Branches	60	85	123
No. of Iterations	332	1016	1291
CPU Time (Secs.)	12	85	32
x_1	0.20	0.20	0.200137
x_2	0.15	0.15	0.150293
x_3	0.50	0.49	0.480312
x_4	0.25	0.27	0.274474
x_5	0.30	0.31	0.312131
x_6	0.65	0.66	0.656848

GLOBAL SOLUTION $F(x^*)$ = -3.32237

First Problem Using Linked Ordered Sets

	21 Point Grids	101 Point Grids
Rows	28	28
Cols.	394	1754
Elements	1530	7130
Continuous Optimum	-5.983085	-6.029093
First Solution	-0.885272	-1.000064
Optimum Solution of LP	-3.860990	-3.861419
No. of Branches	24	77
No. of Iterations	311	1254
CPU Time (Secs.)	16	152
x_1	0.10	0.12
x_2	0.55	0.56
x_3	0.85	0.85

GLOBAL SOLUTION $F(x^*)$ = -3.86278

Second Problem Using Linked Ordered Sets

	21 Point Grids	101 Point Grids
Rows	64	64
Cols.	949	4229
Elements	3630	16910
Continuous Optimum	-5.287467	-5.314402
First Solution	-3.051300	-3.263650
Optimum Solution of LP	-3.291596	-3.321284
No. of Branches	98	511
No. of Iterations	1383	4446
CPU Time (Secs.)	113	1293
x_1	0.20	0.20
x_2	0.15	0.15
x_3	0.45	0.48
x_4	0.30	0.28
x_5	0.30	0.31
x_6	0.65	0.66

GLOBAL SOLUTION $F(x^*) = -3.32237$

It is interesting to note that the solutions using Linked Ordered Sets are substantially slower than the corresponding solutions using only Special Ordered Sets, in spite of the fact that the Continuous Optimum is substantially more realistic. This is partly because for these problems the number of sets over which branching and bounding must take place is not significantly reduced by using Linked Ordered Sets. It is perhaps also worth noting that the results with 21 Point Grids are much more realistic with this formulation, so the comparisons are not entirely fair.

4.3 Branin's Family

$a = 1.000000$, $b = 0.129185$, $c = 1.591549$ $d = 6.000000$,
$e = 10.000000$, $f = 0.039789$

Results	21 Point Grids	101 Point Grids	999 Point Grids
Rows	6	6	6
Cols.	48	208	2004
Elements	156	716	7002
Continuous Optimum (LP)	1.415346	0.403913	0.397999
No. of Branches	0	0	0
No. of Iterations	13	15	17
CPU Time (Secs.)	0.2	0.5	3.6

x_1	3.250000	9.399998	9.428841
x_2	1.211624	2.509663	2.484055

This problem requires no branching or bounding.

GLOBAL SOLUTION F(x*) = 0.397887

4.4 Goldstein and Price's Family

a_{ij}

	j=0	j=1	j=2	j=3	j=4	j=5	j=6	j=7	j=8
i=0	600	720	3060	12288	14346	1944	-4428	-648	729
i=1	720	-4680	-19296	-23616	-11880	-1188	1944	972	
i=2	1260	7344	7776	5040	8730	3672	-1458		
i=3	-1072	5784	9840	1240	-3480	-1836			
i=4	-2454	-7680	-5370	-4080	1305				
i=5	1344	-168	2592	1224					
i=6	952	1344	-648						
i=7	-768	-288							
i=8	144								

Results

Using Ordinary Polynomials

	21 Point Grids	101 Point Grids	Interpolation
Rows	25	25	25
Cols.	347	1547	
Elements	1439	6719	
Continuous Optimum	-1525699.3	-1525951.2	-1526077.6
First Solution	2690.916	1960.617	9115.250
Optimum Solution of LP	3.000000	2.999905	2.663142
No. of Branches	40	194	2810*
No. of Iterations	750	4902	71964
CPU Time (Secs.)	29	469	3540
	0.0	0.0	0.000000
	-1.0	-1.0	-1.008944

GLOBAL SOLUTION F(x*) = 3.00000

* Search not completed. Note that the optimum objective function is over-optimistic because of the standard tolerances used.

Using Chebychev Polynomials

	21 Point Grids	101 Point Grids	Interpolation
Rows	25	25	
Cols.	347	1547	
Elements	1439	6719	
Continuous Optimum	-293026.26	-296841.70	
First Solution	2690.923	1960.621	
Optimum Solution	3.000000	2.996	
No. of Branches	40	182	
No. of Iterations	797	4692	
CPU Time (Secs.)	31	485	
	0.0	0.0	
	-1.0	-1.0	

GLOBAL SOLUTION $F(x^*) = 3.00000$

At the time of writing, the full results using automatic interpolation are not to hand.

It is interesting that the use of Chebychev Polynomials seems to have little effect on the results, although it produces a somewhat less absurdly optimistic continuous optimum solution.

REFERENCES

E.M.L. Beale and J.J.H. Forrest (1976) "Global Optimization Using Special Ordered Sets" Mathematical Programming 10 pp 52-69.

E.M.L. Beale and J.A. Tomlin (1970) "Special facilities in a general mathematical programming system for non-convex problems using ordered sets of variables" in Proceedings of the Fifth International Conference on Operational Research Ed. J. Lawrence pp 447-454 (Tavistock Publications, London).

R.P. Brent (1973) Algorithms for Minimization without derivatives (Prentice-Hall Inc., Englewood Cliffs, New Jersey).

G.B. Dantizg and P. Wolfe (1960) "Decomposition principle for linear programming" Operational Research 8 pp 101-111.

TOWARDS GLOBAL OPTIMISATION 2
L.C.W. Dixon and G.P. Szegö (eds.)

TWO IMPLEMENTATIONS OF BRANIN'S METHOD: NUMERICAL EXPERIENCE

Dr. J. Gomulka
The Numerical Optimisation Centre
The Hatfield Polytechnic
Hertfordshire, England

In this paper the numerical experience gained in applying two implementations of Branin's method for the global optimisation problem are described.

1. NOTATION AND GENERAL ASSUMPTIONS

f, $f(x)$ objective function, at least twice continuously differentiable on an open set containing the feasible set;

$x = (x_1, \ldots, x_n)$ argument of f, vector of an n-dimensional Euclidean space E_n;

D feasible set; usually D is a box $a_i \leq x_i \leq b_i$, $i = 1, n$, but it can be any closed bounded region in E_n with piecewise regular boundary;

g, $g(x)$ gradient of f, at a point x;

H, $H(x)$ hessian matrix of f at x;

x^* a stationary point of f, i.e. $g(x^*) = 0$. It is assumed, that f has only isolated stationary points in D, none of them on the boundary of D, and that $H(x^*)$ is always nonsingular.

In matrix expressions vectors are identified with n x 1 matrices:

$$x = [x_1, \ldots, x_n]^T = \begin{bmatrix} x_1 \\ \vdots \\ x_N \end{bmatrix}.$$

2. INTRODUCTION

A set C_{go} may be defined as all feasible x, at which gradient is parallel to g_o, i.e.

$$g(x) = \lambda g_o \qquad (1)$$

for a suitable scalar λ. The study of these curves introduced by Branin (1970, 1972), has been based on differential equations, whose trajectories satisfy (1).

The first one is

$$\dot{g} = -g \tag{2}$$

with solutions $g = g_o \exp(-t)$ tending to $g = 0$ for $t \to +\infty$. In regions of x-space, where the hessian H is nonsingular, (2) is equivalent to

$$\dot{x} = -H^{-1}g \tag{3}$$

Using the relation $H^{-1} = A/\det H$, where A is the adjoint matrix of H, one can see that (3) is closely related to

$$\dot{x} = -Ag \tag{4}$$

Indeed, wherever both (3) and (4) are defined their right-hand sides differ by a scalar factor det H, which means that their trajectories are geometrically identical and can differ only in parameterisation. However, (4) is defined everywhere in D, without the restriction det H = 0, and is therefore easier to investigate.

A study of (4) was made in Gomulka (1975) with the following results:
System (4) has singular points of two types:
(1) $g = 0$; so called essential singularities; these are stationary points x^* of f. Under assumptions listed in the previous section they are isolated and are degenerate nodes. That is, all trajectories from a neighbourhood of x^* tend to x^* either for $t \to +\infty$ (if $\det H(x^*) > 0$) or for $t \to -\infty$ (if $\det H(x^*) < 0$) and each trajectory is tangent at x^* to a different direction.

(2) $g \neq 0$, $Ag = 0$; these are called extraneous singularities [Branin]. Generally speaking, they form sets of dimension $n - 2$. For $n = 2$ they are isolated points and can be either centres, with a nest of closed trajectories around them, or saddle points.

For $n = 2$ system (4) can be shown to have no spiralling trajectories. Therefore, each trajectory followed in any direction, if it is not a closed curve, must either approach an essential singularity asymptotically or leave D. Exceptions are half trajectories which enter extraneous saddles; they are isolated and have little significance, apart perhaps from dividing regions with different behaviour.

Similar conclusions can be drawn for the case $n > 2$ from the results given in Gomulka (1975). They were not formulated explicitly there and we shall sketch the argument here. Let x_o be an ordinary point of (4), i.e. $Ag \neq 0$ at x_o, and let $g_n(x_o)$ be a nonvanishing coordinate of $g(x_o)$. Since the trajectory T_{xo} through x_o satisfies (1), it follows that $g_n(x) \neq 0$ for any $x \in T_{xo}$. Hence the functions

$$\psi_\alpha(x) = g_\alpha(x)/g_n(x), \qquad \alpha = 1, \ldots, n-1 \tag{5}$$

are well defined in a neighbourhood of every point on T_{xo} and, because of (1), are constant on T_{xo}. Proposition 3 of Gomulka (1975) implies that grad ψ_α, $\alpha = 1, \ldots, n-1$, are linearly independent. Hence T_{xo} is at each of its points a proper intersection of smooth hypersurfaces

$$\psi_\alpha(x) = c_\alpha \,, \qquad \alpha = 1, \ldots, n-1 \tag{6}$$

This excludes all phenomena which, like spiralling, involve having infinitely many disjoint arcs of one trajectory in a neighbourhood of some ordinary points. Therefore each trajectory either approaches a singular point asymptotically or leaves the region. The same proposition 3 can be used to show that in general (i.e. under some regularity assumptions involving second derivatives of ψ_α) only exceptional trajectories tend to extraneous singularities and, as in the case n = 2, almost all non-closed trajectories lead either to the boundary of D or to one of the essential singularities.

It is now clear that the complete solution of (1) for a fixed g_o and all k, which was denoted by C_{go}, consists of one or more trajectories of (4). It follows, that if $x \in C_{go}$ and x is a nonsingular point of (4), then C_{go} is a smooth arc in a neighbourhood of x and λ is a smooth function of the points of this arc. Smoothness of C_{go} near an essential singularity x* can be derived from the Implicit Function Theorem, which also implies that λ is smooth and changes sign at x*.

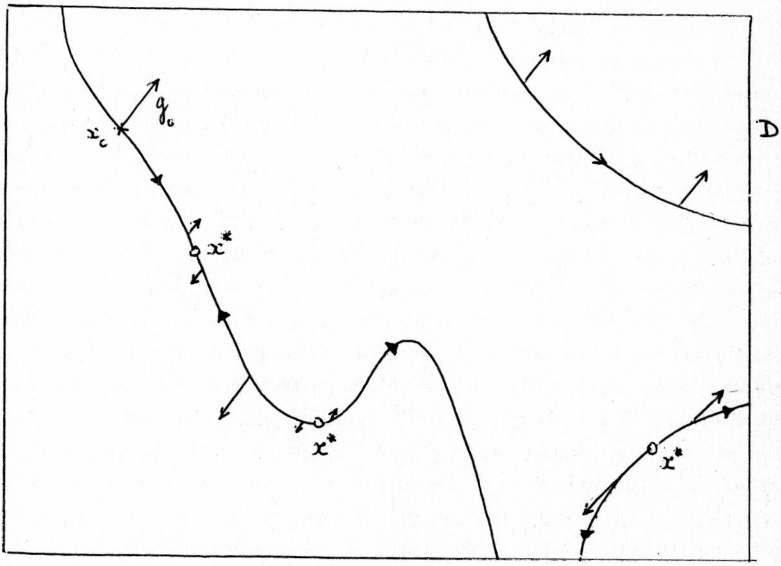

Fig. 1: A Typical C_{go}

3. ALGORITHMS FOR COMPUTING ONE BRANCH OF C_{go}

3.1 AN INTEGRATION ALGORITHM

Earlier attempts to construct an algorithm for computing a complete component of a C_{go} were based on straightforward integration of a differential equation derived from (3) or (4).

For use in the algorithm, equation (3) was normalised and fitted with a sign factor thus reading

$$\dot{x} = -s_H \, H^{-1}g/(\| H^{-1}g \|), \quad s_H = \pm 1 \tag{7}$$

With equation (7), the integration step in terms of the parameter is at the same time the distance between two consecutive points on a trajectory and the choice of s_H determines the direction of the integration process: $s_H = 1$ corresponds to decreasing length of g, generally leading towards a stationary point, while for $s_H = -1$ the length of g is increasing (direction of g of course remains constant).

Equation (7), after an adjustment of the sign factor, is identical with a normalised form of (4), namely

$$\dot{x} = -s \, Ag/(\| Ag \|), \tag{8}$$

where

$$s = s_H \, \mathrm{sign} \, (\det H) \tag{9}$$

For constant value of s, a single trajectory of (8) goes from one stationary point of f to another (unless it leaves D). To obtain the same movement by integrating (7), one has to switch s_H each time that det H changes sign. We shall define the term <u>trajectory direction</u>, as the direction of increasing parameter for trajectories of (4) or, equivalently, for trajectories of (8) with s = 1. This direction is pointed by the arrows along trajectories on the Fig. 1. On the other hand, the direction of integration in the algorithm is chosen to give a continuous run along a whole connected part of C_{go} on one side of an initial point. Consequently, if at the beginning the direction of integration coincides with the trajectory direction, then after crossing the first stationary point these directions will be opposite to each other, they will agree again after crossing the next stationary point etc. In other words, the sign of s in (8) has to be changed each time a stationary point is crossed. At the same moment the direction of g along the curve is reversed, i.e. the sign of λ in (1) is also changed (see the end of the previous section). Therefore during the whole integration process

$$s.\mathrm{sign}\,\lambda = s_H.\mathrm{sign}\,(\det H)\,.\mathrm{sign}\,\lambda = \mathrm{const} \tag{10}$$

We shall now describe in general terms an integration algorithm for tracing a component of C_{go}.

3.1.1 INITIAL STAGE

An initial point x_o has to be chosen, and also a direction of integration, i.e. the sign of s in (8) or of s_H in (7). To trace a whole component of C_{go} two runs from x_o are needed, one in each direction. The unit vector of the gradient direction, which is constant along the trajectory,

$$g_o = g(x_o)/\| g(x_o) \| \tag{11}$$

is calculated and stored. In the algorithm there is also a provision for starting from a stationary point x_o, when a vector g_o has also to be given. Instead of starting the main loop the program will then immediately cross a Stationary Point part, to seek a nearby point with the gradient parallel to g_o and start from there.

The size h_o of the integration step has to be specified. The stepsize will be constantly adjusted in the main loop, but h_o will be used again for restarting a trajectory after crossing each stationary point.

In addition to x_o, h_o and possibly g_o, other values which have to be chosen on entry to the program are a maximal stepsize h_{max}, accuracy parameters ϵ, tol_x, tol_g, the bounds on x defining the feasible region and the maximum number of iterations. Subroutines computing an objective function, its gradient and its hessian at any feasible point have to be included.

3.1.2 THE MAIN LOOP

A current point x and a current integration step h are needed to start an iteration. The sign of h is that of -s in (8), hence it changes only after crossing a stationary point.

1. PREDICTION (INTEGRATION) STEP

To predict the next point x_{new}, one step of a Runge-Kutta integration routine is applied to

$$\dot{x} = \text{sgn}\,(\det H)\, H^{-1}g/(\| H^{-1}g \|) \tag{12}$$

with the initial point x and the step h.

3.1.3 CORRECTION STEP

There can be various ways of estimating the deviation of x_{new} from C_g, and the second algorithm in this section is an example of a different approach. Here $g(x_{new})$ is compared with the "expected" gradient g_{exp},

$$g_{exp} = \| g(x) \|\, g_o \exp\,(h_s) \tag{13}$$

where

$$h_s = h\,\text{sgn}\,(\det H)/\| H^{-1}\, g \| \tag{14}$$

To justify this definition, let us notice that if x lies exactly on C_{go}, then the right hand side of (13) reads $g(x) \exp(h_s)$, which is a solution of $\dot{g} = g$, corresponding to the initial value $g(x)$ and step h_s of the integration parameter.

From the deviation of gradient,

$$dev_g = g(x_{new}) - g_{exp} \tag{15}$$

a deviation of x_{new} itself from C_{go} is calculated. If x_{corr} denotes a point such that $g(x_{corr}) = g_{exp}$, then, approximately,

$$x_{corr} = x_{new} - dev_x \tag{16}$$

where

$$dev_x = H^{-1} dev_g \tag{17}$$

The aim is to find a point for which

$$\| dev_g \| / (\| g_{exp} \|) < tol_g \text{ and } \| dev_x \| < tol_x \tag{18}$$

There are three alternatives:

a) Condition (18) holds true for x_{new}. Then h is doubled - unless it would then exceed h_{max}, in which case $h \rightarrow h_{max}$ sign h , $x \rightarrow x_{new}$ and the next iteration starts.

(b) Otherwise x_{corr} given by (16) is substituted for x_{new} in (15) and the test (18) is applied again. If (18) is satisfied this time then $x \rightarrow x_{corr}$ and h remains unchanged.

(c) If both x_{new} and x_{corr} fail the test, then $h \rightarrow h/2$ and the main loop is restarted from the beginning, i.e. from the integration step, with the same x.

The main loop is thus repeated with the same x and decreasing h until either a new point satisfies (18), or $|h|$ falls below ϵ, which causes an emergency exit from the program.

3.1.4 CROSSING A STATIONARY POINT

Each new point x_{new} is tested for the condition

$$\| g \| < \epsilon \tag{19}$$

If (19) is satisfied, then $x \rightarrow x_{new}$ and the program jumps to the stationary point procedure. This includes printing $f(x)$, g and eigenvalues and eigenvectors of H and then finding a point x_{beg}, which will serve as the beginning of a new arc of C_{go}, composed of a trajectory, which forms a smooth extension of the arc pursued until now across the stationary point x. As mentioned before, along this new arc g will be pointing in the opposite direction. In order to retain $g = \lambda g_o$, $\lambda > 0$, a change $g_o \rightarrow -g_o$ is made.

At first a prediction of x_{beg} aims at a point at the distance $\bar{h} = h_o$ from x,

$$x_{new} = x + \bar{h} \cdot H^{-1} g_o / (\| H^{-1} g_o \|) \tag{20}$$

Deviation of $g(x_{new})$ is measured with respect to its projection on g_o, namely

$$g_{exp} = (g(x_{new}) \cdot g_o)\, g_o \tag{21}$$

followed by (15) and (17). Conditions (18) are then checked. If they are not satisfied, then $\bar{h} \rightarrow \bar{h}/2$ and the sequence (20), (15), (17), (18) is repeated. This process is repeated until either (18) is satisfied, in which case $x \rightarrow x_{new}$, $h \rightarrow \bar{h}$ sgn (det H) and the main loop is reentered, or $\| g\ (x_{new}) \| < 10.\epsilon$, which again causes an emergency exit from the program.

3.1.5 STOPPING CONDITIONS

The algorithm stops, when the first point is found, which satisfies (18) and lies outside D.

Some additional emergency exits have been mentioned before. A bound on the number of iterations provides another safeguard. There is, however, no separate safeguard against running in circles along a closed component of C_{go}. If there is a single closed trajectory, then no defence seems possible at a reasonable price. In other cases it would be perhaps enough to store the first stationary point discovered, compare it with each new stationary point and stop the algorithm if the first point occurs again.

3.2 AN EXTRAPOLATION ALGORITHM

This is a later version of the trajectory algorithm, suggested by S.E. Hersom, in which the next point on C_{go} is predicted by extrapolation from four previous points. Let $x(l)$, where l is the arclength, be a parameterisation of a component of C_{go} and let $x^i = x(l_i)$ be four known points of this curve. Let $l_4 = 0$ and denote $l_{i-1} - l_i = h_i$, $i = 2, 3, 4$, so that

$$\begin{aligned} x^4 &= x(l_4) = x(0) \\ x^3 &= x(l_3) = x(l_4+h_4) = x(h_4) \\ x^2 &= x(l_2) = x(l_3+h_3) = x(h_4+h_3) \\ x^1 &= x(l_1) = x(l_2+h_2) = x(h_4+h_3+h_2) \end{aligned} \tag{22}$$

Now let

$$x^o = x(l_1+h_1), \tag{23}$$

where h_1 is fixed, be the next point to be determined. An extrapolation formula, based on Taylor expansion, has the form

$$x^o = \sum_{i=1}^{4} s_i(h)\, x^i \tag{24}$$

where $s_i(h) = s_i(h_1, h_2, h_3, h_4)$. When x^o is predicted on the basis of only two or three previous points, the extrapolation formula is similar to (24) but with only two or three terms on the right hand side and simpler expressions s_i. Those are used at the beginning of the trajectory and also when the four point extrapolation process breaks down.

3.2.1 INITIAL STAGE

Information required on entry to the program is almost the same as for the previous algorithm, DEVX. There is one real parameter less, namely tol_g, and there is an additional integer parameter max_{cor} (maximum number of corrections in one iteration). All other data and all subroutines are the same as for DEVX.

Two characteristics of the point x_o are calculated and stored, namely

$$g_o = g(x_o)/\| g(x_o) \|, \quad det_o = sign\ (det\ H(x_o)) \tag{25}$$

Then $x \rightarrow x_o$ and the next point is predicted,

$$x_{pred} = x + h \,.\, H^{-1}g/\| H^{-1}g \| \tag{26}$$

where $h = s_o h_o$, s_o denotes the direction parameter equal +1 or 1. The tangent vector to the trajectory at x, directed forward, is

$$v_{rhs} = s_o H^{-1} g \tag{27}$$

and the scalar l with the property

$$g(x) = l g_o \tag{28}$$

is $\lambda = \| g_o \|$. A flag is set for the extrapolation denoting that only two points will be available.

3.2.2 MAIN LOOP

Data on entry: current point x, scalar λ satisfying (28), vector v_{rhs} tangent to the trajectory at x and directed forward, predicted next point x_{pred}, extrapolation matrix W containing up to four previous points, including x, with the corresponding steps h_4, h_3, h_2 (see (22)) and also the step h_1, which was used for computing x_{pred}.

(i) CORRECTION

Only x_{pred} is used in this procedure. The deviation of x_{pred} is measured by its distance from the nearest point x_{corr} of the trajectory. Therefore,

$$x_{corr} = x_{pred} + dev_x \tag{29}$$

$$(dev_x, H^{-1}g_o) = 0 \tag{30}$$

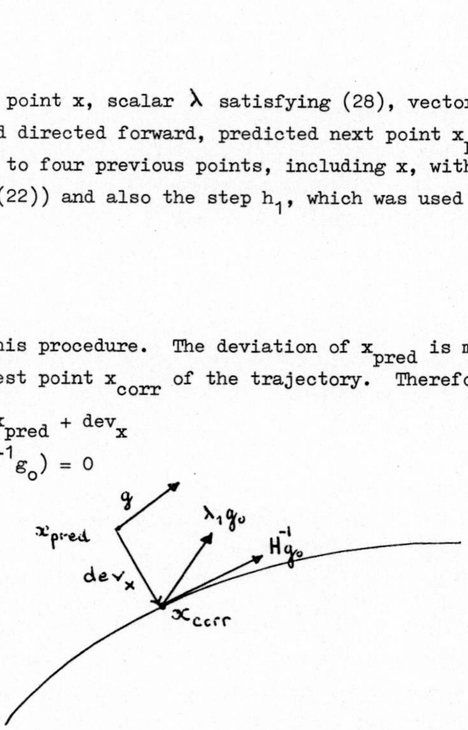

Fig. 2

Denoting the gradient at x_{corr} by $\lambda_1 g_o$

$$\lambda_1 g_o - g = H\ dev_x \tag{31}$$

where $g = g(x_{pred})$. From (30) and (31) we obtain

$$\lambda_1 = (H^{-1}g,\ H^{-1}g_o)\ /\ \|\ (H^{-1}g_o,\ H^{-1}g_o)\ \| \tag{32}$$

$$dev_x = H^{-1}(\lambda_1 g_o - g) \tag{33}$$

The correction loop works as follows. For given x_{pred} values of λ_1 and dev_x are computed from (32), (33). Then the condition

$$\|\ dev_x\ \| < tol_x \tag{34}$$

is checked. If it is satisfied, then $x_{corr} \rightarrow x_{pred}$ which ends the correction process. If (34) fails, then $x_{pred} \rightarrow x_{pred} + dev_x$ and the sequence (32) - (34) is repeated. In the following repetitions, when necessary, the current value of $\|\ dev_x\ \|$ is also compared with the previous one and if no progress has been made, the correcting sequence is modified. Altogether no more than max_{corr} corrections are allowed.

The correction procedure, when successful, returns the following values: a point x_{corr} satisfying (34), corresponding to x_{corr} values of det H, $H^{-1}g$ and λ_1 (from (32)) and the number n_{cor} of corrections actually performed to obtain x_{corr}.

(ii) EVALUATION OF PARAMETER INCREMENT FROM x TO x_{corr}

The absolute value of the increment of trajectory parameter l is approximately equal to the distance between the points. If $\|\ x_{corr} - x\ \|$ significantly exceeds h_1, on which the prediction of a new point was based, then x_{corr} is discarded and a recovery procedure is put in motion. This is because of a suspicion, that x_{corr} may then lie on a distant - in terms of the parameter - part of the trajectory, which by a roundabout path returned near x, or even on an altogether different component of C_{go}. If, on the other hand, the value of

$$\Delta\ x = x_{corr} - x$$

is acceptable, then the parameter increment Δ l is estimated as

$$\Delta\ l = \text{sign}\ (\Delta\ x,\ v_{rhs})\ \|\ \Delta\ x\ \| \tag{35}$$

Here the direction of Δ x is compared with v_{rhs} to determine whether x_{corr} follows or precedes x on the trajectory. Although a move forward is always intended and most of the time this is what really happens, an occasional step backward does occur. Usually it is soon followed by continued steady progress, but sometimes it is a first sign of breakdown.

(iii) TEST FOR A STATIONARY POINT / ACCEPTING THE NEW POINT

This stage is reached if (i) and (ii) were successful, i.e. x_{corr} satisfies (34) and is not too distant from x.

Let us recall that, approximately,

$$g(x) = \lambda g_o, \quad g(x_{corr}) = \lambda_1 g_o \quad .$$

If

$$\lambda . \lambda_1 < 0 \tag{36}$$

then there must be a stationary point between x and x_{corr}. In this case a procedure is started, which makes further tests and prints out particulars of the stationary point. In general this procedure does not influence the main flow of computation, but, when one of the hessians H(x), $H(x_{corr})$ is positive definite and the other is not, then recovery procedure is started.

The point x_{corr} now becomes the new current point x and values of other variables are changed accordingly. In particular,

$$v_{rhs} = s_c H^{-1} g, \tag{37}$$

where

$$s_c = s_o \det_o \ \text{sign}\ (\det(H)) \ \text{sign}\ \lambda_1 \tag{38}$$

Also, $h_2 \rightarrow \Delta l, \quad \lambda \rightarrow \lambda_1$.

<u>(iv) PREDICTION</u>

The prediction step h_1 is increased or decreased, if the number of corrections n_{corr} is respectively below, or above, certain limits. However, h_1 cannot exceed h_{max}. A new extrapolation matrix W is prepared by shifting the rows downwards,

$$W_{i,j} = W_{i-1,j} \ , \ i = 2,3,4; \ j = 1, \ldots, n+1, \tag{39}$$

and putting the new step h_1 and the point $x^1 = x$ into the first row. Expressions $s_i(h)$ occurring in (24) are evaluated and a point x_{pred} is computed according to the formula (24) for x^o.

The main loop is now restarted from (i).

<u>3.2.3 RECOVERY PROCEDURE</u>

This procedure is started, when conditions considered as a breakdown of the method have occurred at some point of the main loop. An attempt is then made to restart the trajectory from the current point x, without relying on the preceding points. It is thought that breakdowns are often caused by rapid changes in function and derivative values. In such a case past information may be misleading and a short step from the current point x, guided by derivatives at x, may result in a more successful prediction. Judging by our experiments, this tactic is very often successful.

The procedure is very similar to the initial stage of the algorithm. First, the prediction step h_1 is halved and the current direction coefficient is computed from (31), where λ is substituted for λ_1. The step h is defined by

$$h = s_c h_1 \tag{40}$$

and x_{pred} is computed according to (26). Only the first row is left in the matrix W, so that at the end of the following interation only two points will be used for the extrapolation.

If h_1 falls below ϵ an emergency exit from the program occurs.

3.2.4 STOPPING CONDITIONS

All remarks made concerning the stopping conditions for DEVX apply here.

4. NUMERICAL EXPERIENCE WITH DEVX AND EXTRAP

The efficiency of EXTRAP is substantially higher than that of DEVX. Perhaps the most important reason is that EXTRAP needs less function evaluations per iteration than DEVX. However, it also seems to use less iterations for a comparable task.

One iteration of DEVX requires at least 5 evaluations of the gradient and hessian, namely 4 for the Runge-Kutta step used for the prediction and at least 1 in the correction loop. EXTRAP also requires at least 1 gradient-hessian evaluation for the correction, but none for the prediction, so that its lower limit is 1 evaluation per iteration. The average number of gradient-hessian evaluations for one iteration of EXTRAP was in our experiments remarkably steady, ranging between 1.8 and 2.1 for most runs, with extreme values of 1.7 and 2.2. In earlier experiments with DEVX only iterations were counted, the gradient-hessian count was only added in the last version. In one run of this version there were on average 9.3 evaluations per iteration and by all other indicators (C.P.U. time, rate of progress along the trajectory etc.) it seemed a typical example.

There is much evidence to suggest that progress of DEVX is to a large degree slowed down in the neighbourhood of stationary points. When a stationary point is being approached the stepsize is cut down, usually by a factor of 0.1 - 0.05, occasionally more. This process, triggered off by small gradients, involves an increase in both the number of corrections per iteration and the number of iterations. Restarting a trajectory after a stationary point brings similar problems. EXTRAP, on the other hand, jumps over stationary points without slackening the pace. In a straightforward comparison, on the same trajectory and under similar accuracy requirements, the following numbers were obtained: 107 iterations and 1000 gradient-hessian evaluations for DEVX and 93 iterations and 173 evaluations for EXTRAP on a two-dimensional problem; 250 iterations of DEVX and 136 iterations of EXTRAP on a four-dimensional function of the Shekel family. One certainly should not make too much of the exact numbers, which can be influenced by peculiarities of the problems, but taken in connection with all our experience of trajectory methods these results left little doubt that EXTRAP is a far better version of the Branin method than DEVX or any of its predecessors.

The success of EXTRAP depends very much on a good choice for two of its parameters, namely HMAX and TOLX. If HMAX is too large, then it may happen that the trajectory is not followed in an orderly, quasi continuous way; large parts of it can be missed altogether, others can be repeated cyclically. This is by no means a purely theoretical danger, but one that is only too easily demonstrable in practice.

The proper value of HMAX is very problem-dependent. For the standard test functions HMAX was taken between 0.5 and 1.

Values of TOLX for the standard test functions were between 0.01 and 0.001. Previous studies, on DEVX, had shown, that the change of TOLX from 0.01 to 0.001 increased the number of function evaluations by 5%; further change to 0.001 brought further increase by 40%. A change in the opposite direction, to TOLX = 0.1, caused breakdown. A similar pattern emerged when TOLG was changed, but this time breakdown occurred for both too large and too small values. Influence of TOLX on performance of EXTRAP was not investigated in detail, but it seems likely that it is not very different. Values which were good for DEVX on a given function seemed to suit EXTRAP as well.

EXTRAP was applied to the standard test functions of the Shekel family. In each case two runs were made from the point (5., 5., 5., 5.), which is the centre of the feasible region, and proceeding in opposite directions, i.e. one complete component of a C_{go} curve was traced.

The results are as follows:

SQRN5	275 gradient and hessian evaluations
	20 sec C.P.U. time
	Five local minima, including the global one, were found
SQRN7	251 evaluations
	23.6 sec C.P.U. time
	Five local minima, including the global one, were found
SQRN10	243 evaluations,
	26.6 sec C.P.U. time
	Five local minima, including the global one, were found.

It is a matter of pure chance that in each of these cases the global minimum was found without additional searches. Examples of trajectories containing only a few local minima or even no stationary points at all appeared frequently in our experiments.

ACKNOWLEDGEMENT

The author acknowledges the financial support of the S.R.C. while working on this project.

REFERENCES

1. Branin, F.H. (1972). "A method for finding multiple extrema of a function of n variables" In "Numerical Methods of Nonlinear Optimisation" Ed. F. Lootsma, Academic Press, London, 1972.
2. Branin, F.H. (1972). "A widely convergent method for finding solutions of simultanious non-linear equations". I.B.M.C. New York, TR 21.4.66 January 1972.
3. Gomulka, J. (1975). "Remarks on Branin's method for solving nonlinear equations" Paper 4 in "Towards Global Optimisation" Eds. L.C.W. Dixon & G.P. Szegö, North Holland, 1975.

TOWARDS GLOBAL OPTIMISATION 2
L.C.W. Dixon and G.P. Szegö (eds.)

A GLOBAL DESCENT OPTIMISATION STRATEGY

GIULIO TRECCANI

UNIVERSITY OF GENOA, ITALY

ABSTRACT

Two approaches are described which attempt to solve the global optimization problem by descending from a local minimum. Such methods would only have to locate a small percentage of the local minima of the function. Unfortunately while both methods perform reasonably satisfactorily on a few test functions, neither can be guaranteed to find the global minimum in every situation.

1. INTRODUCTION

Various approaches have been suggested for solving the following global optimization problem: given a real-valued function $f: D \subseteq R^n \longrightarrow R$, which satisfies some regularity conditions, find a point $x^* \in D$ such that $f(x) \geq f(x^*)$ for any $x \in D$.

Apart from "grid methods" in which the function is evaluated at all points of some discrete subset of D and some estimation of a global Lipschitz constant is made, no deterministic method has been suggested which can always locate the neighbourhood of the global minimum of a function, even when every restrictive regularity conditions are imposed on the function. The "grid methods" are known to be impractical for $n > 3$ as they involve evaluating the function at a very large number of points, even when a good estimate of the Lipschitz constant is given. Other deterministic methods, such as the trajectory method proposed by Branin which tries to locate the global minimum by finding all the stationary points of f(x), have regions of nonconvergence{1}, and moreover can be impractical when there are a large number of stationary points.

Goldstein and Price {2} suggested the alternative approach of descending from a local minimum. They modified the deflation technique which is so successful on polynomial functions, but their adaptation is not rigorous on more general functions or in large dimension.

Two new results are discussed in this work: it is shown that

(i) the global optimisation problem for some function f(x) is equivalent to the local problem for a family of related functions, and

(ii) it is also equivalent to finding a global minimum point of a function,

whose global minimum value is known.
Even though these two problems seem to be particular cases of the global one, since the general global problem can, under reasonable assumptions on the function, be reduced to either of these particular problems, then if we could solve these we would be able to solve the global problem. Two particular strategies for solving the global problem are suggested by these results and will be illustrated in the following. Neither of these strategies can theoretically solve the problem, but both can be implemented in algorithms which will be helpful in finding the global minimum of most standard numerical test problems.
The main practical advantage of solving the global optimization problem by the two approaches introduced here is that in both cases we do not actually find all the local minima of the function to be optimised, as for example in Branin's method, but only a decreasing finite sequence of local minima; this will be illustrated in the following section.

2. ANALYSIS.

In this section we consider a real valued function $f(x)$, $f: R^n \longrightarrow R$, which is continuously differentiable as many times as we need.
We denote by g, G, H, G^{-1} the gradient, the hessian matrix, the adjoint and the inverse of the hessian respectively. We assume that $f(x)$ satisfies the following conditions:
2.1 There exists an open, connected and bounded set $D \subseteq R^n$, which contains all the local minima of $f(x)$ and is such that $x_1 \in D$, $x_2 \notin D$ implies $f(x_1) < f(x_2)$.
2.2 G is nonsingular at the local minima, whose number is finite.
2.3 If y_1, y_2 are local minima of $f(x)$, then $y_1 \neq y_2$ implies $f(y_1) \neq f(y_2)$.
These conditions, together with the continuity assumptions, imply that the local minima are isolated, and that there exists a unique point $x^* \in D$, which is the global minimum of $f(x)$ in R^n.
Assumption 2.3 is restrictive, since it is not true when special symmetry properties hold, but it is necessary to ensure the prevention of cycling with our descent method.
The global descent method consists of the following steps:

2.4 Starting from some initial point x_o, which is not a local minimum of $f(x)$, find a local minimum $y_o \in D$ of $f(x)$, such that $f(y_o) < f(x_o)$.

2.5 Find a point $x_1 \in D$ such that $f(x_1) \leq f(y_o)$ and such that x_1 is not a local minimum of $f(x)$. If this is impossible $y_o = x^*$ and the search stops.

2.6 Set $x_o = x_1$ and return to 2.4.

From 2.1, 2.2 and 2.3, it follows that after a finite number of iterations the global minimum point x^* must be reached.

Step 2.4 never presents difficulties, since the function $f(x)$ is bounded from below and the initial condition x_o is never a minimum, unless $y_o = x^*$.

It will be shown that step 2.5 can be performed by solving a special global minimisation subproblem.

Now assume that at any iteration step 2.4 has been performed, giving us a local minimum $y_o \in D$ such that $y_o \neq x^*$. By the translation:

2.5.1 $$\tilde{f}(x) = f(x+y_o) - f(y_o)$$

we get a function $\tilde{f} : R^n \longrightarrow R$ which has the same continuity properties of f and satisfies 2.1, 2.2 and 2.3 for the translated set:

2.5.2 $$\tilde{D} = \{z \in R^n : \exists\, x \in D \text{ s.t. } z = x - y_o\}$$

which is a bounded neighbourhood of the origin. The origin is a local minimum point of $\tilde{f}$ such that $\tilde{f}(0) = 0$.

Then we consider the two following real-valued functions:

2.5.3 $$\beta(x) = \tilde{f}^2(x)\,(x^T \tilde{G}_o x)^{-3}$$

2.9.4 $$w(x) = \left[\tilde{f}(x) + \varepsilon\right]^2$$

where $\tilde{G}_o$ is the hessian matrix of $\tilde{f}$ at 0 and $\varepsilon > 0$ is such that, according to 2.3, there does not exist a minimum point $y_1 \in D$ of $f(x)$ such that

$$f(y_o) - \varepsilon \leq f(y_1) < f(y_o).$$

$\beta(x)$ is $R^n \setminus \{0\} \longrightarrow R^+$, while $w(x)$ is $R^n \longrightarrow R^+$.

The global minima of $\beta(x)$ are those points other than y_o such that $f(x) = f(y_o)$, while if y_1 is a global minimum of $w(x)$, then

$$f(y_o) - \varepsilon \leq f(y_1) \leq f(y_o).$$

Step 2.5 can be performed in the following way:

2.5 Search for global minima of either $\beta(x)$ or $w(x)$; if at any stage of the search a point x_1 is found such that $f(x_1) \leq f(y_o)$, we accept this point.

Obviously, there is no loss of generality assuming that x_1 is not a local minimum of f(x).

Finally, we state two regularity conditions, which have to be used in the next sections. We shall assume that these conditions are satisfied for any local minimum point $y_o \in D$ of f(x).

(R1) There exist two non-negative numbers M, v such that $x \in D$ implies $|x^T \tilde{g}(x)| \leq M(x^T x)^v \tilde{f}(x)$.

(R2) There exists a positive number K such that $x \in D$ implies:

$$\text{Max}\{|x^T \text{ grad } w|, |\text{grad } w|\} \leq K\, w(x).$$

Here $\tilde{g}(x) = g(x+y_o)$ is obviously the gradient of $\tilde{f}(x) = f(x+y_o)-f(y_o)$, while w(x) is given by 2.5.4.

3. A SEQUENCE OF UNCONSTRAINED LOCAL MINIMISATIONS

In this section we illustrate a technique for performing step 2.5, by finding a global minimum of the function $\beta(x)$, β: $R^n - \{0\} \longrightarrow R$, defined by 2.5.3.

We assume that regularity condition (R.1) holds and, by 2.2, that there exist positive constants ρ_1, l and L such that

3.1. $z^T z \leq \rho_1$ implies $z \in \tilde{D}$

3.1.2 $l x^T x \leq x^T \tilde{G}(z) x \leq L x^T x$, for $z^T z \leq \rho_1$

where G is the Hessian matrix of f.

The gradient of $\beta(x)$ is:

3.2 $$\text{grad}\,\beta = 2\beta\left\{\frac{\tilde{g}(x)}{\tilde{f}(x)} - 3\,\frac{\tilde{G}_o x}{x^T \tilde{G}_o x}\right\}.$$

Now let h, K be two arbitrary positive numbers and let $v > 0$ be such that (R1) is satisfied; then we introduce the auxiliary function:

3.3 $$p(h,K,x) = \beta^h(x)\, \exp\left[-\frac{(x^T x)}{K}\right].$$

3.4 Theorem

For any positive h,K the function p(h,K,x) defined by 3.3 has the following properties:

3.4.1 p is continuous in $R^n \setminus \{0\}$

3.4.2 $\lim\limits_{x^T x \to 0} p(h,K,x) = +\infty$

3.4.3 p is continuously differentiable in the set $\{ x \in R^n : \tilde{f} \neq 0 \}$ and the gradient is:

$$\text{grad } p(h,K,x) = p(h,K,x)\left\{ h \frac{\text{grad}\,\beta}{\beta} - 2\frac{v}{K}(x^T x)^{v-1} x \right\} =$$
$$= 2p(h,K,x)\left\{ h \frac{\tilde{g}(x)}{\tilde{f}(x)} - \left[3h \frac{\tilde{G}_o}{x^T \tilde{G}_o x} + \frac{v(x^T x)^{v-1}}{K} I \right] x \right\}.$$

3.4.4 $p(h,K,x) > 0$ for any $x \in R^n - \{0\}$ and

$p(h,K,x) = 0$ implies $\tilde{f}(x) = 0$, $x \in D$

PROOF of 3.4.2 Since $\tilde{f}(x) = \tilde{f}(0) + x^T \tilde{g}(0) + \frac{1}{2} x^T \tilde{G}(\ x)x$, $0 < \lambda < 1$, 3.1.2 implies that $\tilde{f}(x) = \frac{1}{2} x^T \tilde{G}(\ x)x \geq l\, x^T x$ for $x^T x \leq \rho_1$, while $\beta(x) \geq \frac{l^2}{4} \geq \frac{x^T x}{(x^T \tilde{G}_o x)^3} \geq \frac{l^2}{4L^3}(x^T x)^{-1}$. It follows that

$$p(h,K,x) \geq \left[\frac{l^2}{4l^3} (x^T x)^{-1} \right]^h \exp\left[-\frac{(x^T x)^v}{K} \right] \rightarrow +\infty \text{ as } x^T x \rightarrow 0.$$

We shall now approximate the global minima of $\beta(x)$ by means of a sequence of stationery points of $p(x,h,K)$. We can always assume that at any stage of the research we have $\tilde{f}(x) > 0$, otherwise step 2.5 is performed without any further investigation.

Assume then that for some $h,K > 0$, $x \in R^n$ is a stationary point of $p(h,K,x)$ such that $\tilde{f}(x) > 0$. Then by 3.4.3 grad p (h,K,x) is continuous in the neighbourhood of this point and the following equality must hold:

3.5 $$\tilde{g}(x) = \frac{\tilde{f}(x)}{h} \left\{ \frac{3h\tilde{G}_o}{x^T \tilde{G}_o x} + \frac{v(x^T x)^{v-1}}{K} I \right\} x$$

Multiplying by x^T we get:

3.6 $$x^T \tilde{g}(x) = \frac{\tilde{f}(x)}{h} \left[3h + \frac{v}{K}(x^T x)^v \right] = \tilde{f}(x) \left[3 + \frac{v}{hK}(x^T x)^v \right].$$

We are now ready to state the main theorems.

3.7 Notation

For any $h,K > 0$, we denote by $S(h,K)$ the set of stationary points of $p(h,K,x)$ such that $\tilde{f}(x) \neq 0$ at these points.

We can observe that, since there exist points $y \neq 0$ such that $\tilde{f}(y) = 0$ and since $p(h,K,x) \rightarrow 0$ as $x^T x \rightarrow +\infty$, then $S(h,K)$ mut be nonempty.

3.8 Theorem

Let $S = \{ x \in S(h,K) \text{ for some } h,K > 0 : \tilde{f}(x) > 0 \}$.

Then there exists a $\rho_2 > 0$ such that $x \in S$ implies $x^T x \geq \rho_2$.

Proof. Since $\tilde{G}$ is uniformly continuous in D, there exists a $\rho_2 > 0$ such that by 3.1.1 and 3.1.2, $z_1^T z_1 \leq \rho_1$

$z_2^T z_2 \leq \rho_1$, $(z_1 - z_2)^T (z_1 - z_2) \leq \rho_2$ imply $\left[\tilde{G}(z_1) - \tilde{G}(z_2)\right] \leq \frac{1}{2}$,

where ρ_1 and l are as in 3.1.1 - 3.1.2.

We can always assume $\rho_2 \leq \rho_1$.

Assume now that $x \in S$ is such that $0 < x^T x \leq \rho_2 \leq \rho_1$.

Then we get from 3.6 and a Taylor expansion:

$x^T \tilde{g}(x) = \tilde{f}(x) + \frac{1}{2} x^T \tilde{G}(\lambda_1 x)x = 3\tilde{f}(x) + v(hK)^{-1}(x^T x)^v \tilde{f}(x)$, and since we assume $\tilde{f}(x) > 0$ this implies

$$\tilde{f}(x) + \frac{1}{2} x^T \tilde{G}(\lambda_1 x)\, x > 3\tilde{f}(x)$$

$$2\tilde{f}(x) < \frac{1}{2} x^T \tilde{G}(\lambda_1 x)x.$$

By another Taylor expansion we get

$$2\tilde{f}(x) = x^T \tilde{G}(\lambda_2 x)x$$

and hence

$$x^T G(\lambda_1 x) > 2x^T G(\lambda_2 x)x$$

for some $0 < \lambda_1, \lambda_2 < 1$. On the other hand $x^T x \leq \rho_2$ implies

$$\lambda_1^2 x^T x \leq \rho_1, \ \lambda_2^2 x^T x \leq \rho_1 \text{ and } (\lambda_1 x - \lambda_2 x)^T (\lambda_1 x - \lambda_2 x) < \rho_2$$

which gives a contradiction:

$$\frac{1}{2} \frac{x^T\left[\tilde{G}(\lambda_1 x) - \tilde{G}(\lambda_2 x)\right] x}{x^T x} > \frac{x^T G(\lambda_2 x)x}{x^T x} \geq 1.$$

Q.E.D.

3.9 Theorem

There exists $\delta > 0$ such that, for any $h, K > 0$ for which $hK < \delta$, $x \in S(h,K)$, $\tilde{f}(x) > 0$ imply $x \in \tilde{D}$.

PROOF. We prove that any $\delta \leq \frac{v}{M}$ satisfies the statement.

Indeed assume by contradiction that for some $hK < \frac{v}{M}$ there exists $x \in S(h,K)$ such that $\tilde{f}(x) > 0$, $x \notin \tilde{D}$. Then by (R1) and 3.6 we get:

$M(x^T x)^v \tilde{f}(x) \geq \left| x^T \tilde{g}(x) \right| = \tilde{f}(x)\left[3 + \frac{v}{hK}(x^T x)^v\right] > \tilde{f}(x)\, M(x^T x)^v$, a contradiction.

Q.E.D.

3.10 Theorem

Let $\{h_n\}$ and $\{K_n\}$ be any two positive sequences such that $h_n K_n \to 0$ and let $\{x_n\}$ be such that $\tilde{f}(x_n) > 0$, $x_n \in S(h_n, K_n)$.

Then every limit point y' of $\{x_n\}$ is such that $y' \in \tilde{D}$, $y'^T y' \geq \rho_2$ and $\tilde{f}(y') = 0$.

PROOF. We observe that from 3.8 and 3.9, for large n it must be $x_n \in \tilde{D}$, $x_n^T x_n \geq \rho_2$, which implies that limit points y' of $\{x_n\}$ exist and are such that $y' \in \tilde{D}$, $y'^T y' \geq \rho_2$. Assume now, by contradiction, that there exists a subsequence $\{x_{n_1}\}$ such that $x_{n_1} \rightarrow y'$, $\tilde{f}(y') > 0$.

In a neighbourhood of y', $p(h_{n_1}, K_{n_1}, x)$ is continuously differentiable and 3.6 implies:

$$\left| x_{n_1}^T \tilde{g}(x_{n_1}) \right| = \tilde{f}(x_{n_1}) \left[3 + \frac{v}{h_{n_1} K_{n_1}} (x_{n_1}^T x_{n_1})^v \right] \geq \tilde{f}(x_{n_1}) \left[3 + \frac{v \rho_2}{h_{n_1} K_{n_1}} \right] \rightarrow +\infty$$

as $h_{n_1} K_{n_1} \rightarrow 0$, a contradiction.

Q.E.D.

3.11 Remark

By the proofs of the previous theorems if $h_n K_n < \delta \leq \frac{v}{M}$ and if $x_n \in S(h_n, K_n)$ is such that $\tilde{f}(x_n) > 0$, it follows that

$$0 < \tilde{f}(x_n) < \frac{\delta}{V} \frac{|\tilde{g}(x_n)|}{|x_n|} .$$

3.12 Conclusion

The previous theorems show that if we are able to find any sequence of stationary points of a family of functions p(h,K,x) for $hK \rightarrow 0$, then we can approach a point x_1 that solves step 2.5 In the following figure the behaviour of the functions f, $\tilde{f}$, β and p is illustrated.

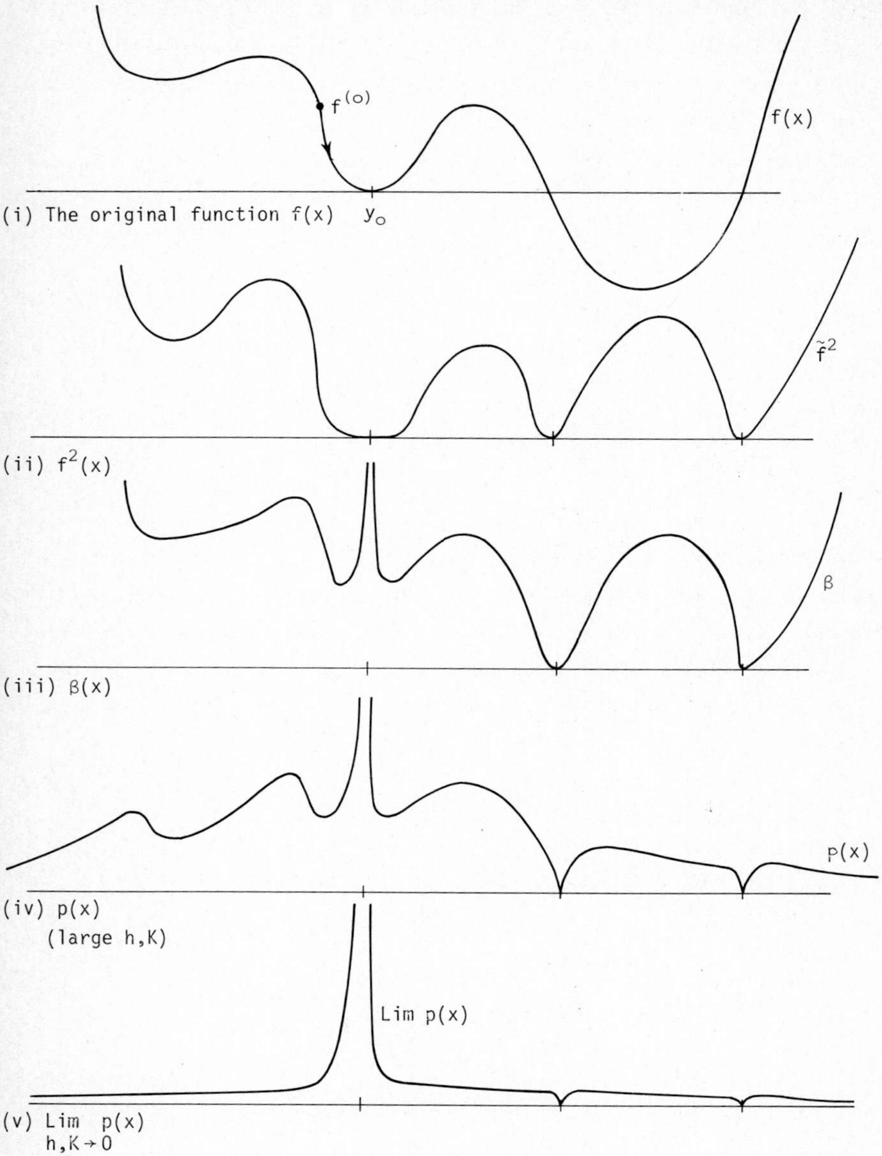

The functions f, $\tilde{f}$, β, p defined in the text.

4. A TRAJCTORY APPROACH.

In this section we illustrate a technique for performing step 2.5, by finding global minima of the function w(x) defined by 2.5.4.

We observe that if f(x) satisfies the assumptions of section 2, then w(x) is continuously differentiable as many times we need in $\{x \in R^n : \tilde{f}(x) \neq 0\}$.

We also assume that (R2) holds for w(x).

For simplicity of notation, g(x), G(x), $G^{-1}(x)$ and H(x) will denote in this section the gradient, the Hessian, its inverse and adjoint of the function w(x) (and not of f(x) as in the previous sections).

We state now the main property.

4.1. Theorem.

Assume that $\lim_{|x| \to +\infty} w(x) = C > 0$ and that (R2) holds.

Then if x(t) is a solution of:

4.1.1 $$g(x(t)) = tw[x(t))] \frac{a(t)}{a(t)^T x(t) - R_1}, \quad R_1 \in R$$

such that $x(t_o) = x_o$, $|a(t)| = |a_o|$, for any $t \geq t_o > 0$, x(t) is bounded and $\lim_{t \to +\infty} w[(x(t))] = 0$

Proof. If for some sequence $t_n \to +\infty$ we have $|a(t_n)^T x(t_n)| \to 0$, then $w[x(t_n)] \to 0$, otherwise for some n (R2) would be violated; since by assumption $\lim_{|x| \to +\infty} w(x) > 0$, the theorem must hold.

If for some sequence $t_n \to +\infty$ we have $|a(t_n)^T x(t_n)| \geq \varepsilon_1 > 0$ then multiplying both sides of 4.1.1 by $x(t_n)^T$, we get again $w[x(t_n)] \to 0$ and, as before, $x(t_n)$ must be bounded.

Q.E.D.

The main reason for which equation 4.1. is suggested for solving the global minimum problem is that if we consider the "penalized function"

4.2 $$p(x) = \frac{[(w(x))]^{1/t}}{R_1 - a^T x}, \quad a^T x < R_1, \quad t > 0, \quad a \in R^n,$$

the equation that gives stationary points of p(x) such that $w(x) \neq 0$ is of the type 4.1.1.

On the other hand, if the global minimum value of w(x) is 0 (that is, if there exist points x_1 such that $f(x_1) = f(y_o)$, $x_1 \neq y_o$) and if w(x) is dominated at infinity by a suitable polynomial, there must exist saddle points of p (possibly at the infinity) for any sufficiently large value of t, while if the global minimum value of w(x) is positive (that is, if y_o is the global minimum of f(x)), for large t such saddle points do not exist. We try now to find a solution of 4.1.1 by solving a suitable ordinary differential equation.

We start with the equation

4.3 $$g(x) = tw(x)\frac{a_o}{a_o^T x - R_1} \qquad t \geq t_o > 2,\ R_1 > 0,\ a_o \in R^n$$

If 4.3 is to be satisfied at some initial point x_o such that $g_o \neq 0$ for $t = t_o$, then

4.4 $$a_o = \frac{R_1 g_o}{g_o^T x_o - t_o w_o}, \qquad t_o f_o > g_o^T x$$

while, for any (x,t) satisfying 4.3, we have

4.5 $$a_o = \frac{R_1 g(x)}{g(x)^T x - tw(x)}$$

Substituting in 4.1.1 we get

4.6 $$g(x(t)) = t\frac{w[x(t)]}{g_o^T x(t) - g_o^T x_o + t_o w_o} g_o = t\, l(x(t)) g_o$$

where

4.7 $$l(x) = \frac{w(x)}{g_o^T x - g_o^T x_o + t_o w_o}$$

We observe that $\frac{\partial l}{\partial x} = 0$ implies $g(x) = l(x) g_o$, which is impossible if x satisfies 4.6 for $t \geq t_o > 2$.

We now derive from 4.6 an ordinary, non-autonomous differential equation such that 4.6 is a solution starting at (x_o, t_o) for $t = t_o$:

4.8 $$\dot{g} = \frac{dg}{dt} = t\,\dot{l}\,g_o + l\,g_o = \left[t\left(\frac{\partial l}{\partial x}\right)^T \dot{x} + 1\right] g_o$$

$$= \left[t\,\frac{1}{w}\,(lt - l)g_o^T x + 1\right] g_o = \frac{1}{wt}\left[(t-1)g^T x + w\right] g.$$

4.9 $$G\dot{x} = \frac{1}{wt}\left[(t-1)g^T\dot{x} + w\right] g$$

If at some $x \in R^n$, G is nonsingular, we can multiply both sides of 4.9 by H, the adjoint matrix of the hessian G, and we get:

4.10 $$\dot{x} \det G = \frac{1}{tw}\left[(t-1)g^T x + w\right] Hg\ .$$

Since $\det G \neq 0$, x must have the form:

4.11 $\dot{x} = q\,Hg$

which implies

4.12 $$q = \frac{w}{\det G\; tw-(t-1)\; g^T Hg}$$

For $q \in R$, it must be $wt \det G \neq (t-1)\, g^T Hg$, which for $\det G \neq 0$ becomes:

4.13 $$g^T G^{-1} g \neq \frac{wt}{t-1}$$

Equations 4.11 and 4.12 can still be used when $\det G = 0$, provided $g^T Hg \neq 0$.

Equation 4.11 cannot be used, on the other hand, when we reach a point $x(t)$ such that either:

4.14 $$\det G \neq 0 \qquad g^T G^{-1} g = t\,\frac{w}{t-1}$$

or:

4.15 $$\det G = 0 \qquad g^T Hg = 0$$

and at such points the trajectory $x(t)$ will terminate.

In the former case, consider the real function:

4.16 $$s(x) = \frac{g^T g}{(g^T x-wt)^2}\,,\ t \text{ fixed.}$$

Consider now the gradient of $s(x)$:

4.17 $$\frac{\partial s(x)}{\partial x} = 2\,\frac{(g^T x-wt)Gg-g^T g[Gx+(1-t)g]}{(g^T x - wt)^3}$$

and the total time derivative of $s(x)$ along the solutions of (4.11, 4.12):

4.18 $$\dot{s}(x) = \left(\frac{\partial s(x)}{\partial x}^T\right)\dot{x} = 2s(x)\,\frac{w}{wt-g^T x} > 0$$

On the other hand, if we reach a point such that $\frac{\partial s}{\partial x} = 0$, by multiplying both sides of 4.17 by $g^T G^{-1}$ we find that $g^T G^{-1} g = \frac{t}{t-1}\,w$ and hence equations (4.11, 4.12) do not hold at this point.

Stationary points of $s(x)$ reached by solutions of (4.11, 4.12) cannot actually be minima; if they are maxima, it is impossible to carry on with equations of type (4.11,4.12) with initial conditions in a sufficiently small neighbourhood of the final point. But if $x(t)$ satisfies 4.14, but it is not a maximum of $s(x)$, we can proceed in the following way.

In a small neighbourhood of $x(t)$ we consider the set of points.

4.19 $$S = \{x(t) + \Delta x:\ s(x(t)+\Delta x) = s(x(t))\}$$

From our assumptions it follows that $S \neq \emptyset$.

Then we compute a vector $q \in R^n$ such that

4.20
$$q_j = \sum_{i,k} \frac{\partial G^{-1}_{ik}}{\partial x_j} g_i g_k$$

Since to the first order in Δx we have

$$g(x+\Delta x)\ G^{-1}(x+\Delta x)\ g(x+\Delta x) - \frac{tf(x+\Delta x)}{t-1}$$

$$\simeq g^T G^{-1} g + 2g^T \Delta x + g^T \Delta (G^{-1}) g - \frac{tw}{t-1} - \frac{tg^T \Delta x}{t-1} = \frac{t-2}{t-1} g^T \Delta x + g^T \Delta (G^{-1}) g$$

where $\left[\Delta(G^{-1})\right]_{ik} = \left(\frac{\partial G^{-1}_{ik}}{\partial x}\right)^T \Delta x$,

and since $g^T \Delta (G^{-1}) g = \sum_j \Delta x_j \sum_{ik} \frac{\partial G^{-1}_{ik}}{\partial x_j} g_i g_k = q^T \Delta x$,

we may conclude that for every sufficiently small x such that $x+\Delta x \in S$ and $\left[\frac{t-2}{t-1} g + q\right]^T \Delta x \neq 0$, we can continue the integration of equations (14.11, 4.12) from the initial point $x + \Delta x$ at the initial time t.

A similar policy can be used also if 4.15 is satisfied at the point x(t), provided that at this point s(x)has not a maximum, q is computed from the adjoint matrix H instead from G^{-1}, and $x + \Delta x \in S$, $\left[\frac{t-2}{t-1} g+q\right]^T \Delta x \neq 0$ hold.

We observe that whenever we continue the integration of (4.11, 4.12) following the strategy we have just described, we change the vector a_o in 4.3 to a new vector a_1, having the same norm but different direction.

In conclusion, if we can carry on this procedure for any $t \geq t_o$, we find a solution x(t) of 4.1.1 which, by theorem 4.1, will converge, as $t \to +\infty$ to a point where w(x) vanishes, and then step 2.5 will have been performed. Clearly, if the global minimum of w(x) is greater than zero, that is if y_o found in step 2.4 was the global minimum of f(x), then it will not be possible to find a solution of 4.1.1 for t geater than some maximum value. Unfortunately when the trajectory reaches a maximum of the function s(x) and hence cannot be continued it does not imply that y_o is the global minimum of f(x).

4.21 CONCLUSION

Two approaches have been described which attempted to solve the global optimisation problem by descending from a local minimum.Such methods would only have to locate a small percentage of the local minima of the function. Unfortunately while both methods perform reasonably satisfactory on a few test functions, neither can be guaranteed to find the global minimum in every situation.

It must be remarked, however, that even if we were to consider a more general form of "penalized function" than 4.2, namely

4.2.1.1 $$p(x) = [w(x)]^{1/t} \eta(x)$$

where $\eta(x)$ is any continuously differentiable function, we still cannot ensure at the same time that the level sets of p(x) are bounded (which implies that we can always find a local minimum) and that a sequence of stationary points of p(x) tends to a global minimum of w(x) as $t \to +\infty$.

References

1. Branin F.H. (1972). I.B.M.J. Res. Develop. pp 504-522.
2. Goldstein & Price (1971). Maths of Computation Vol. 25 No. 115 pp 569-575.

TOWARDS GLOBAL OPTIMISATION 2
L.C.W. Dixon and G.P. Szegö (eds.)

THE APPLICATION OF A GLOBAL OPTIMIZATION METHOD TO SOME TECHNOLOGICAL PROBLEMS (*)

F.ARCHETTI,ISTITUTO DI MATEMATICA,
UNIVERSITA' DEGLI STUDI DI MILANO,
MILANO, ITALIA
F.FRONTINI,C.I.S.E. SEGRATE, ITALIA

In this paper the authors are concerned with the computational problems which arise in the optimization of complex technological systems. The attention is focused on systems whose "cost function" has many local extrema. Two such problems are considered: the design of optical filters and the economic optimization of a dry cooling tower in a thermal power plant. Some numerical results obtained by a stochastic global optimization method are reported.

INTRODUCTION

Technological design has grown up with a relatively simple mathematical methodology; except in some specific areas, e.g. structural engineering, where advanced mathematical methods are now widely used, the designer relies upon a set of mostly heuristic rules, which enables him to evaluate the performance (or the cost) of a given technological system in different situations. This traditional approach has deeply influenced the applications of automatic computers in this field: the same heuristic procedures are now programmed in order to simulate, more completely than before, the behaviour of the system; the possibilities of automatic computing in this field are thus left mostly unused while the ever increasing system complexity, as well as the high energy cost, requires a more precise definition of performance and methods specifically designed in order to optimize it.
The new environmental awareness also clearly requires a new approach:

(*) This work has been partially supported by "Gruppo Nazionale per l' Informatica Matematica" of the Italian National Research Council.

no longer can the environment be considered as a passive substrate of the plant but the technological parameters must be expanded to include the environmental ones.

New comprehensive methodologies must therefore be worked out that can fully exploit the far reaching possibilities of automatic computing in order to solve those problems and result in better performance and a deeper understanding of the behaviour of the system (Hersom (1975)).

Mathematical models have been widely suggested as a powerful tool to deal with such problems and a wide literature exists about this subject.

In this paper the authors are concerned with some computational problems which arise in the minimization of a "cost function" (maximization of a performance index) to which the optimization of the system is often reduced. A formal definition of "cost function" is now given.

Let R^N be the space of design (control) parameters and $D^1=(d_1^1,d_2^1,\dots,d_N^1)$ a "better design" than $D^2=(d_1^2,d_2^2,\dots,d_N^2)$.

Any function $f:R^N\to R$ is a cost function consistent with the system iff $f(D^1)<f(D^2)$. Clearly that function of the design parameters expressing the actual cost of the system is a consistent "cost function" provided all environmental "costs" are included.

Thus the mathematical formulation of the problem is

$$\min f(x), \qquad x\varepsilon\Omega$$

subject to the constraints $x\varepsilon\Omega$, where Ω, the feasible set, is given on technological grounds. The features of $f(x)$ have a marked influence on the computational problems inherent in its minimization.

Most research, as yet performed in the area of numerical optimization, has been directed towards functions enjoying both smoothness and convexity assumptions. As a result of this effort, effective and reliable numerical routines are widely available in this area.

Technological problems pose new and intriguing questions: in many cases $f(x)$ cannot be shown to enjoy the regularity assumptions (e.g. $f(x)\varepsilon C^1$ or $f(x)\varepsilon C^2$) required by most effective methods; sometimes $f(x)$ can be known only on the basis of noisy measurements $f(x,w)$, where w is a noise term of unknown distribution.

Some relevant theoretical results have been obtained about these problems but the numerical weaponry for solving them is still rather poor and only recently some algorithms have been tentatively suggested (Dixon (1975),(1976)).

In many actual problems convexity is lacking or difficult to prove: this may happen by the very nature of the problem (multimodality as in the optical filter design problem of sect. 1) or due to insufficient

a priori information.
In both cases the usual techniques are likely to miss the global minimum: an often used but crude approach, is to single out the region where the global minimum is supposed to lie: this operation, which is accomplished by handling auxiliary constraints, requires some knowledge of the system and a good dose of ingenuity.
Thus an effective method for multiextremal optimization could result in a drastic widening of the mathematical approach to design problems.

Multiextremal optimization problems, or global optimization problems, as this situation has come to be termed in the literature, have been gaining increasing attention in the last years and many methods have been suggested (for a systematic survey see Dixon et al. (1974) and Dixon et al. (1975)). A stochastic method has been recently proposed (Archetti (1975) and Archetti-Betrò (1976)): its structure and numerical performance are given in De Biase-Frontini (1977); in this paper the authors are concerned with some technological applications.

1. TECHNOLOGICAL PROBLEMS

OPTICAL FILTER DESIGN

In many problems of applied optics a multilayer filter is to be designed in order to minimize the reflected energy over a frequency band $[\Lambda_1, \Lambda_2]$. This is achieved by depositing on a substrate m layers, of thickness d_j, $j=1,2,\dots,m$, of materials of refractive index R_j, $j=1,2,\dots,m$. Our results have been derived for a particular laser application: given m, R_j and the order in which the materials are deposited, the optimal thickness vector $d=(d_1,\dots,d_m)$ is to be found which ensures the "best" response on $[\Lambda_1, \Lambda_2]$.
Let the wave interval $[\Lambda_1, \Lambda_2]$ be partitioned into k equal subintervals $[\lambda_{j-1}, \lambda_j]$, $j=1,2,\dots,k$, with $\lambda_o=\Lambda_1$, $\lambda_k=\Lambda_2$.
The classical relations are used to evaluate the reflected energy E: for each layer and each subinterval $[\lambda_{j-1}, \lambda_j]$ the phase difference is

$$\Phi_p^j = 2 \cdot R_p \cdot d_p \cdot \frac{\lambda_o}{\lambda_j} \qquad p=1,2,\dots,m \quad j=1,2,\dots,k. \tag{1.1}$$

By introducing the matrix

$$M_j = \prod_{p=1}^{m} \begin{bmatrix} \cos \Phi_p^j & \dfrac{i \sin \Phi_p^j}{R_p} \\ i\, R_p \sin\Phi_p^j & \cos \Phi_p^j \end{bmatrix} \tag{1.2}$$

we may evaluate the absorbed energy in $[\lambda_{j-1}, \lambda_j]$, which is given by

$$(1.3) \qquad T_j(d) = \left[\frac{2R_o}{R_o(M^j_{11} + R_s M^j_{12}) \quad (M^j_{21} + R_s M^j_{22})} \right]^2$$

where R_o and R_s are respectively the refractive indices of the air and the substrate. As $0<T_j(d)<1$, $R_j(d)=1-T_j(d)$ is the reflected energy in $[\lambda_{j-1}, \lambda_j]$.

Therefore the reflected energy over $[\Lambda_1, \Lambda_2]$ is :

$$(1.4) \qquad F(d) = \sum_{j=1}^{k} R_j$$

as a measure of the reflected energy also the following functions, convexly related to (1.4), could be used:

$$(1.5) \qquad F(d) = \sum_{j=1}^{k} R_j^2$$

$$(1.6) \qquad F(d) = \max_j |R_j| \ .$$

The design problem may thus be reduced to the minimization of F(d): (1.4) is the most often used: it yields quite a flat response; (1.5) yields the design with minimum area under the response design curve; (1.6) gives the design with the least maximum reflectance.

The minimization process could result in physically infeasible results: e.g. layers with thickness less than $0.05 \cdot \lambda_o$, which is very difficult to measure.

The optimization problem is thus to be completed with suitable constraints: the optimal thickness vector d^* is determined by the conditions:

$$F(d^*) = \min F(d) \quad | \quad d_p \geq \bar{d}_p, \quad p=1,2,\ldots,m$$

where $\bar{d}_p$ are determined on technological grounds: $\bar{d}_p \geq 0.05 \cdot \lambda_o$.

The main computational difficulty is that many local minima exist due to sinusoidal terms in the expression of the reflected energy.

For small values of m (3, 4) auxiliary constraints can be introduced in order to bound the region where the global minimum is likely to lie, and a local routine may subsequently be applied (Mc Keown - Nag (1976)).

As some filters may have up to 30 and even more layers, the usefulness of a general tool for multiextremal high dimensionality problems is stressed by this kind of applications: the stochastic method described in De Biase-Frontini (1977) has been successfully applied to actual filters containing up to 19 layers and the numerical results are reported in section 2.

Another interesting application is the economic optimization of a natural convection dry cooling tower in a thermal power plant (Archetti - Frontini (1975)) : dry cooling towers are independent of large water resources and can therefore result in a lower cost of energy transmission and reduced environmental impact.
An environmental variable x_8 is defined as the difference between the average temperature of the environment and that of the condenser.
The pressure at the exchanger and therefore its size and the cost of the tower are thus linked to the location of the plants.

The table below shows the dependence between the different costs and the pressure at the condenser.

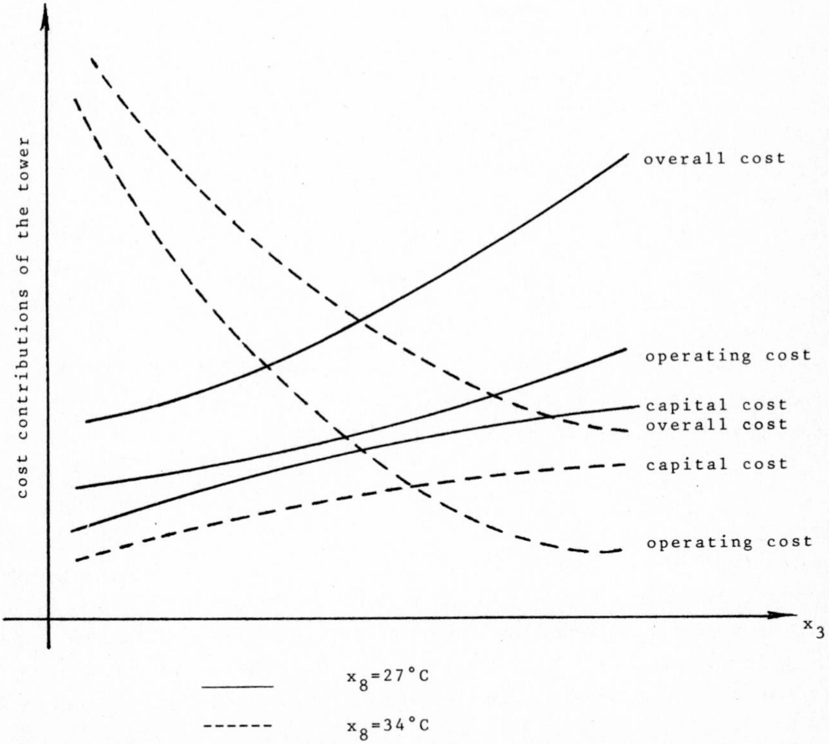

The above picture does not claim any quantitative value: it is quite simplified with respect to the real situation.

The control variables considered in the optimization are:

x_1=	height of exchanger's column [m]	$10 \leq x_1 \leq 50$
x_2=	output diameter [m]	$90 \leq x_2 \leq 150$
x_3=	condenser pressure [at]	$0.03 \leq x_3 \leq 0.2$
x_4=	water rate per column [Kg/s]	$15 \leq x_4 \leq 50$
x_5=	air specific rate [Kg/s·m^2]	$0.5 \leq x_5 \leq 5$
x_6=	width of exchanger's column [m]	$1 \leq x_6 \leq 5$
x_7=	thickness of exchanger's column [m]	$0.05 \leq x_7 \leq 0.3$
x_8=	initial temperature difference [C^o]	$5 \leq x_8 \leq 35$

The function to be minimized is the cost contribution to the tower (keeping account of capital as well as of operating cost) to a kWh produced by the plant.
More complex models are now being studied with additional environmental variables such as wind and humidity which are represented by their yearly or seasonal distribution functions.

The variable x_8 is known to introduce multimodalities into the objective function: in Cavanna-Folli (1974) x_8 is parametrized with a step less than 1/10 C^o; a local minimization routine is started for any of the resulting subproblems and the best result obtained is subsequently picked up.
In Cavanna-Folli (1974) this process has been performed for any site P_i, i=1,2,...,N, which has been previously selected as a would be location for the plant and the best value obtained in this process is eventually considered.
This work has required a lot of problem preparation and computer programming; moreover the final result may be dependent on the step used in the parametrization of x_8.

2. COMPUTATIONAL RESULTS

The structure of the algorithm and its numerical performance in academic examples are reported in De Biase-Frontini (1977).
We only say that it performs a sequential process by samples $\Theta_q^i = \{f(x_j),\ j=1,2,...,q\}$, where x_j is a random variable uniformly distributed in the search domain. After each sample a stochastic

estimate β_i of f^* is derived. Once some complex termination criteria has been satisfied, say at the υ-th sample, $\beta^* = \beta_\upsilon$ is accepted as an approximation to f^*. This result is subsequently improved and validated using a local routine, which gives (x_c, f_c), the final estimate to (x^*, f^*). Here we are mainly concerned with the computational results for the technological problems outlined in section 1.

Optical filter design; four filters have been considered:

A) 3 layer anti-reflection filter with $R_{1,2,3} = (3, 2.3, 1.5)$ and $[\Lambda_1, \Lambda_2] = [7., 14.]\mu$.
B) 3 layer anti-reflection filter with $R_{1,2,3} = (1.35, 2.2, 1.7)$ and $[\Lambda_1, \Lambda_2] = [0.4, 0.6]\mu$.
C) 19 layer reflecting filter with $R_{2i-1,2i} = (1.38, 2.2)$, $i=1,2,\ldots,9$, $R_{19} = 2.2$ and $[\Lambda_1, \Lambda_2] = [0.570, 0.630]\mu$.
D) 3 layer anti-reflection filter (Musset-Thelen (1970)) with $R_{1,2,3} = (1.38, 2.1, 1.6)$ and $[\Lambda_1, \Lambda_2] = [0.4, 0.8]\mu$.

The optimal thickness vectors are respectively:

A) $d_1 = 0.22\ \lambda_o$, $d_2 = 0.15\ \lambda_o$, $d_3 = 0.25\ \lambda_o$, $\lambda_o = 8\mu$.
B) $d_1 = 0.48\ \lambda_o$, $d_2 = 0.24\ \lambda_o$, $d_3 = 0.24\ \lambda_o$, $\lambda_o = 0.5\mu$.
C) $d_i = 0.25\ \lambda_o$, for $i=1,2,\ldots,19$ $\lambda_o = 0.6\mu$.
D) $d_1 = 0.14\ \lambda_o$, $d_2 = 0.24\ \lambda_o$, $d_3 = 0.16\ \lambda_o$, $\lambda_o = 0.55\mu$.

As far as A is concerned (fig. 1) the results are the same as those reported by Mc Keown-Nag (1976); as far as D is concerned the response curve is drawn in figure 2: the comparison between the response curve of the filter obtained with the optimization process and that obtained by traditional techniques shows a marked improvement for the former, mainly in the frequency band [0.52,0.75] .

As far as C is concerned we remark the proposed method can effectively handle a many layer filter. For this filter we list below some details of the computational results. Samples θ_q^i of size q=15 were performed.

i	15	20	25	30	35	40	45
β_i	-4.8007	-4.8160	-4.8160	-4.8540	-4.9194	-4.9286	-4.9333

Figure 1.

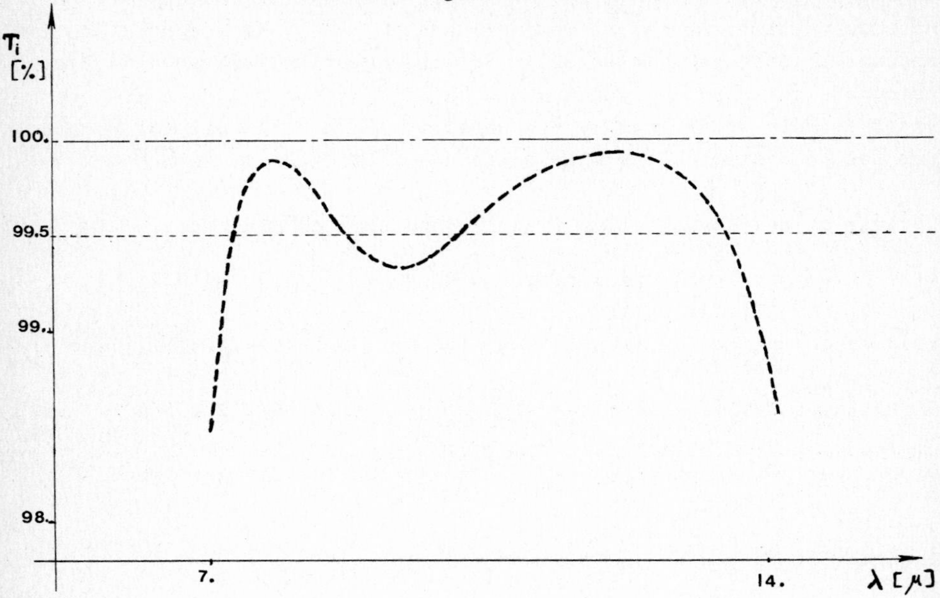

Figure 2.

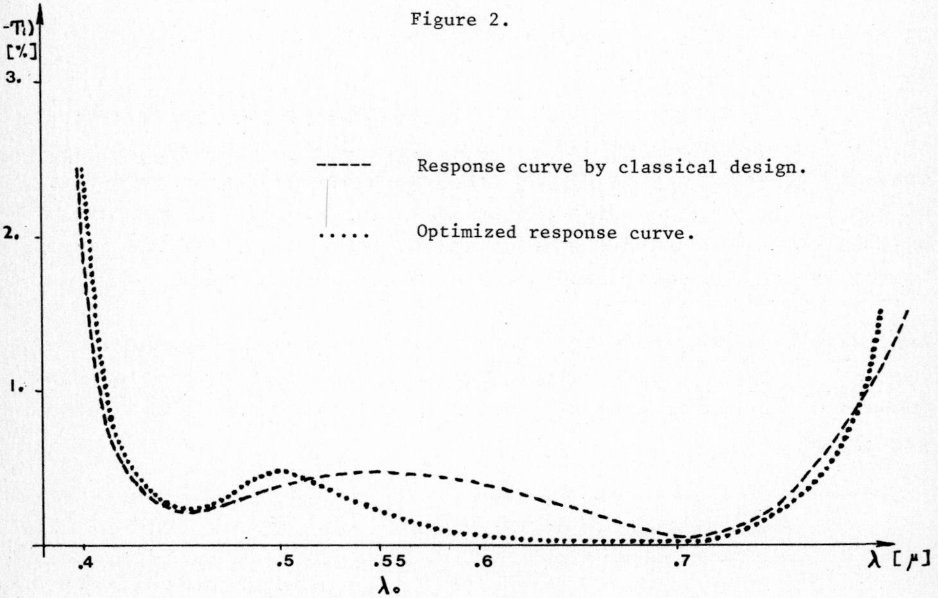

The termination criteria have been satisfied for $\upsilon = 48$, yielding $\beta^* = \beta_\upsilon = -4.9314$. The local routine improved the result yielding $f_c = -4.99$ and the thickness vector listed above.

The tower we examined has exchanger of the Forgò type, of fixed column width: so that $x_7 = 0.15$ m; the other control variables are allowed to vary between the bounds given in section 1.
We obtained the following results, using the stochastic method discussed in De Biase-Frontini (1977):

$x_1 = 43.2$ m	$x_2 = 114.6$ m	$x_3 = 0.093$ at	$x_4 = 32.1$ Kg/s
$x_5 = 2.52$ Kg/s·m^2	$x_6 = 1.80$ m	$x_7 = 0.15$ m	$x_8 = 28.2$ C^o
$f_c = 0.499$ mills\$/kWh		$\beta^* = 0.508$ mills\$/kWh	

This value is lower that those obtained by Cavanna-Folli (1974) which are between 0.54 and 0.70 mills\$/kWh, depending on the different step values used in the parametrization of x_8.

The global method requires 399 function evaluations of which 89 are spent in the local search routine in the last stage of the algorithm. The number of function evaluations required to scan the search region for the approximation of f* is markedly low: this is a most important feature as a function evaluation, when the cost function is given in algorithmic form, is by far the most expensive part of the whole optimization process.
Thus the global optimization approach has some important advantages: it requires a shorter overall computer time, it allows a more comprehensive attitude towards the model of the plant without any additional parametrization and it is therefore a promising tool for optimizing complex systems.

ACKNOWLEDGEMENTS

The authors are grateful to L.C.W. Dixon, for valuable help in the writing of this paper.

REFERENCES

ARCHETTI F. (1975). A sampling technique in global optimization; in: Towards Global Optimization. Dixon L.C.W.&Szego G.P. eds.North Holland.

ARCHETTI F. and BETRO' B. (1976). Recursive stochastic evaluation of the level set measure in global optimization problems; Quaderni del

Dipartimento di Ricerca Operativa e Scienze Statistiche, A-21, Università di Pisa.

CAVANNA A. and FOLLI A. (1974). Le torri di raffreddamento: parte 3$^{\underline{a}}$, ottimizzazione tecnico-economica delle torri di raffreddamento a secco; C.I.S.E.-TORRI 2/1.

DE BIASE L. and FRONTINI F. (1977); A stochastic method for global optimization: its structure and numerical performance; this volume.

DIXON L.C.W. (1975). On the convergence of the variable metric method with numerical derivatives and the effect of noise in the function evaluation; in: Optimization and Operations Research. Springer Verlag.

DIXON L.C.W. (1976). The optimization of industrial processes; N.O.C. Tech. Rep. N° 73.

DIXON L.C.W., GOMULKA J. and SZEGO G.P. (1975). Towards Global Optimization; in: Towards Global Optimization. Dixon L.C.W. and Szego G.P. eds. North-Holland.

DIXON L.C.W., GOMULKA J. and HERSOM S.E. (1975). Reflections on the global optimization problem; N.O.C. Tech. Rep. N° 64

HERSOM S.E. (1975). The practice of optimization; in: Towards Global Optimization. Dixon L.C.W. and Szego G.P. eds. North Holland.

Mc KEOWN J.J. and NAG A. (1976). An application of optimization techniques to the design of an optimal filter; in: Optimization in Action. Dixon L.C.W. ed. Academic Press.

MUSSET A. and THELEN A. (1970). Multilayer anti-reflection coatings; in Progress in Optics. Wolf E. ed. North-Holland.

PART TWO

LOCAL UNCONSTRAINED OPTIMISATION

TOWARDS GLOBAL OPTIMISATION 2
L.C.W. Dixon and G.P. Szegö (eds.)

QUASI-NEWTON METHODS FOR NONLINEAR UNCONSTRAINED MINIMIZATION: A REVIEW

E. Spedicato, CISE, Segrate (Italy)

INTRODUCTION

Quasi-Newton methods for nonlinear unconstrained minimization have been developed as an alternative to the Newton-Raphson method to solve the following problem:

find a local minimizer x^+ of $F=F(x)$, $x \in R^n$, $F \in C^2$
given an estimate x_o of x^+

Such methods use the following iterative process

$$x' = x - \lambda Hg \quad (1)$$

where x is the actual estimate of the minimizer x^+, x' is an improved estimate, g is the gradient of F, H is an estimate of the inverse of the hessian matrix G of F and λ is a scalar, called stepsize or linear search parameter, chosen so that $F(x') \leq F(x)$ or ideally, for perfect line searches, so that $F(x') = \min_{\lambda} F(x- \lambda Hg)$. In methods that will be considered the matrix H, often called the variance matrix, is initially assigned and then is updated using current information as follows

$$H' = H+D(p,y,a) \quad (2)$$

where a is a vector of parameters and

$$p = x'-x$$
$$y = g'-g \quad (3)$$

The correction (2) is made such that the following relation, known as the Quasi-Newton condition, is satisfied

$$H'y = p \quad (4)$$

Note that relation (4) is true when H' is the inverse matrix of G and g is linear and is satisfied at first order when g is general. The rationale to develop Quasi-Newton methods lies in the attempt to eliminate defects of the Newton-Raphson iteration while keeping its fast convergence (this is usually quadratic for the Newton-Raphson method and superlinear for the Quasi-Newton methods). The defects avoided by Quasi-Newton methods are the following:

- they don't need to exactly evaluate the second derivatives, buil

ding an approximation to them
- they do not require the solution of a linear system at every iteration, allowing the determination of the search vector s=Hg by a matrix-vector multiplication or by updating the factors of H'^{-1} in $0(n^2)$ multiplications
- they can control singularity and often ill-conditioning by a suitable choice of parameter a.

The first paper on Quasi-Newton methods was written in 1959 by Davidon and the framework in which research has subsequently developed was set by Broyden in 1965 in a paper devoted to solving systems of nonlinear equations. In the last ten years much research has been done concerning the various theoretical and practical aspects of Quasi-Newton methods, as shown by the about 150 papers listed in the bibliography.

GENERATION OF FORMULAS FOR THE VARIANCE MATRIX

The Quasi-Newton condition (4) constitutes a set of n linear algebraic equations for the n^2 unknown components of the variance. Clearly other conditions have to be added to define H' uniquely. Classes of formulas for the variance have been obtained in the literature following essentially two approaches. The first is based on a functional model. One considers a strictly convex quadratic function

$$Q = x^T Ax + b^T x + c \tag{5}$$

with A positive definite ($A > 0$). Q can be considered as a first order approximation of F around x^+ if $A = G(x^+)$. In the hope that good behaviour on a quadratic function implies fast convergence near the minimizer of a general function one requires that when F=Q and perfect line searches are performed then the search vectors are A-conjugate, i.e. they satisfy

$$s_k^T As_j = 0 \qquad k \neq j \tag{6}$$

because A-conjugacy implies determination of the minimizer with exact line searches in at most n iterations (quadratic termination). The hopeful assumption at the basis of this approach has been theoretically substantiated in 1975 by the results of Dixon and of Stoer discussed below. The first Quasi-Newton method of Davidon, refined by Fletcher and Powell in 1963 and known as the DFP method, has the quadratic termination property. This method is a member of a general class, proposed by Broyden in 1967, all of whose members possess quadratic termination if not singular, which reads

$$H' = H + \frac{\sigma + \tau\theta}{\sigma^2} pp^T + \frac{\theta - 1}{\tau} Hyy^T H - \frac{\theta}{\sigma} (py^T H + Hyp^T) \tag{7}$$

where θ is a free scalar parameter and

$$\sigma = p^T y$$
$$\tau = y^T Hy \tag{8}$$

in (7) $\theta = 0$ corresponds to the DFP method and $\theta = 1$ to a method, proposed independently in 1970 by Broyden, Fletcher and Shanno, known as the BFS method. A more general three parameter class, containing Broyden class as a special case, was proposed by Huang in 1970; in 1971 Adachi derived the most general formulas which generate A-conjugate search vectors. Adachi's formulas are quite complex and subsequent research has been concerned essentially with Huang's class.
A second approach is the variational approach proposed by Greenstadt in 1970. One requires that the correction D minimizes a generalized Frobenius norm, namely

$$\| D \|^2 = Tr(W^T DWD) \tag{9}$$

with $W > 0$, subject to the symmetry condition $(H=H^T \rightarrow H'=H'^T)$ and the Quasi-Newton condition. The result is the following formula

$$H' = H + \frac{(Hy-p)^T y}{(d^T y)^2} dd^T - \frac{(Hy-p)d^T + d(Hy-p)^T}{d^T y} \tag{10}$$

$$d = W^{-1} y$$

The class of matrices given by (10) has a nonzero intersection with Broyden's class, for instance the BFS method belongs to both classes, however in general its members do not possess the property of quadratic termination.
A mixed approach was proposed by Tewarson in 1970. He minimizes again a generalized Frobenius norm, subject not to the symmetry condition, which is dropped, but to the following hierarchy of Quasi-Newton conditions, which implies for exact line searches A-conjugacy of the search vectors.

$$H_k^T y_j = p_j \qquad j < k \tag{11}$$

The solution has a simple form when W satisfies the relations

$$y_k^T W^{-1} y_j = 0 \qquad j < k \tag{12}$$

in which case it reads

$$H' = H - \frac{d(H^T y - p)^T}{d^T y} \qquad d = W^{-1} y \tag{13}$$

choice $W^{-1} = H + \theta A^{-1}$ satisfies (12) and in this case a subclass of Huang class, also considered independently by Broyden and Johnson in 1972, is obtained.
In 1968 Bard observed that if F is multiplied by a scalar α (linear scaling) then the matrices of Broyden's class and the sequence $\{x_k\}$ are altered unless H is suitably scaled. This observation led Oren in 1972 to modify Broyden's class in order that the sequence $\{x_k\}$ is independent of the linear scaling. This results in a class of formulas with a new parameter γ, namely

$$H' = \gamma H + \frac{\sigma + \gamma\theta\tau}{\sigma^2} pp^T + \frac{\gamma(\theta-1)}{\tau} Hyy^T H - \frac{\gamma\theta}{\sigma}(Hyp^T + py^T H) \tag{14}$$

Invariancy is obtained taking γ in a suitable interval; finite termination holds for all nonsingular updates. The algorithms better known and most efficient belong to Oren's class and we shall concentrate our discussion on it.

STABILITY QUESTIONS IN OREN'S CLASS

A very important requirement is that variances are nonsingular. Indeed, even if iteration (1) can be performed for singular H, once H_k becomes singular, then all $H_{k'}$ are singular for $k' > k$, because columns of the correction D are linear combination of columns of H; therefore all search vectors $s_{k'}$ lie in the range of H_k and this fact clearly can make convergence impossible. The determinant of the variance is easily expressed analytically and the following conditions for nonsingularity are obtained

$$\gamma \neq 0 \qquad \theta \neq \frac{-\sigma^2}{\varepsilon\tau - \sigma^2} \qquad \varepsilon = p^T H^{-1} p \tag{15}$$

Another computationally useful condition is that the variances are positive definite, as this insures the vector $-s$ to point in a descent direction. Necessary and sufficient conditions for maintenance of positive definiteness, derived by Spedicato in 1971 for the equivalent symmetric subclass of Huang class, are the following

$$\sigma > 0 \qquad \gamma > 0 \qquad \theta > \frac{-\sigma^2}{\varepsilon\tau - \sigma^2} \tag{16}$$

Rounding errors can make the variance lose positive definiteness even if relations (16) are satisfied; this however happens rarely, according to results of Spedicato in 1972. Positive definiteness can be checked factorizing the variance or its inverse in the form

$$H^{-1} = LDL^T \tag{17}$$

where L is triangular and D is diagonal, and then checking that the elements of D are positive; note however that if H is not positive definite the factorization may not exist. Procedures to update the factors have been proposed by Gill and Murray in 1972, Goldfarb in 1973, Fletcher and Powell in 1973. The iteration (1) in this approach is substituted by the following iteration

$$LDL^T s=g \tag{18}$$
$$x'=x - \lambda s$$

Control over rounding errors and possible tendency of the variance to quasi-singularity can be done to some extent choosing the free parameters to reduce the condition number. Exact condition number in Frobenius norm has been derived and minimized by Spedicato in 1975; however the formulas are complex and prone to numerical instability. Strict bounds to the condition number in the natural norm have been obtained by Brodlie, Gourlay and Greenstadt in 1973 and by Spedicato in 1975, in the framework of canonical forms for matrices of rank two. The first authors use a factorization of H while the second author uses the Raleigh quotients; formulas equivalent to those of Spedicato have also been obtained by Davidon in 1975 with a different approach, and further discussed by Schnabel in 1976. The algorithms of Oren's class which have the property that the Raleigh bound is minimized are given by the subclass where γ and θ are related as follows

$$\theta = \frac{\sigma(\varepsilon - \gamma\sigma)}{\gamma(\varepsilon\tau - \sigma^2)} \tag{19}$$

Many authors have considered how to choose optimally the free parameters. Dixon in 1972 has established that algorithms of Broyden's class with exact line searches generate a sequence $\{x_k\}$ which is independent of $\{\theta_k\}$. A similar result was also proved by Adachi in 1973 for his general class. In practice exact line searches cannot be performed and different choices of the parameters result in different efficiencies.

Numerical evidence has been reported that the BFS method is the best in Broyden's calss. Oren and Spedicato in 1976 have proposed the following selection of parameters (algorithm OS)

$$\gamma = \sqrt{\frac{\varepsilon}{\tau}} \qquad \theta = \frac{1}{1 + \sqrt{\dfrac{\varepsilon\tau}{\sigma^2}}} \tag{20}$$

Algorithm OS has the following properties:

- the variance is optimally conditioned with respect to the Raleigh bound in the natural norm
- sequence $\{x_k\}$ is not affected by a linear scaling
- the variance is self-dual, in the sense of duality defined by Fletcher in 1970
- the single-step rate of convergence, defined in the next section, is optimal.

Numerical experiments of Spedicato and of Brodlie in 1975 show that for small n, say $n < 20$, the BFS is preferable, while for large n the OS performs better. This relates probably to the tendency, observed sperimentally by McCormick and Pearson in 1969, of methods of Broyden's class to concentrate reduction of function value at the end of n iterations cycles, while methods of Oren's class, due to their optimal single-step rate of convergence, tend to spread uniformly the function reduction; also for large problems the conditioning of H becomes important and this favours the OS method.

CONVERGENCE QUESTIONS IN OREN'S CLASS

Algorithms of Oren's class have quadratic termination for exact line searches; for non exact line searches Dixon has discussed in 1973 modifications which maintain termination on quadratic functions. Termination can be extended to a class of nonquadratic functions, which are obtainable from a quadratic function Q by a transformation of "nonlinear scaling", say

$$F = \varphi(Q) \qquad \frac{\partial\varphi}{\partial F} > 0 \quad \text{at} \quad x \neq x^+ \tag{21}$$

This transformation does not affect the contour lines of Q and therefore to obtain termination it is sufficient that the sequence $\{H_k\}$ does not depend on φ. This can be attained defining vector y in (3) as follows

$$y = g' - \frac{(\partial\varphi/\partial F)_k}{(\partial\varphi/\partial F)_{k+1}} g = g' - \psi g \tag{22}$$

This definition was given by Spedicato in 1974, generalizing a special case considered by Goldfarb in 1972. It extends to Quasi-Newton methods the nonquadratic model introduced by Fried in 1971 and by Davison and Wong in 1973 for conjugate gradient methods. The formula to calculate ψ has been derived by Spedicato in 1976 and reads for the case of λ exactly determined

$$\psi = \lim_{\hat{\lambda}\to\lambda} \frac{g(x+\hat{\lambda}Hg)^T Hg}{(\lambda-\hat{\lambda})g^T Hg} \tag{23}$$

The introduction of the parameter ψ, given by (23) where the limit is approximated taking for instance $\hat{\lambda} = 0.9\lambda$, is useful when line searches are accurate; otherwise a heuristic determination like that proposed by Spedicato in 1974 may be preferable.

For exact line searches and quadratic functions the single-step reduction in function value can be expressed as follows

$$F' - F^+ \leq Z(H)\ (F - F^+) \tag{24}$$

with Z a function of the condition number of $\sqrt{A}\, H \sqrt{A}$. From the quadratic termination it follows that $Z(H_n) = 0$, but for $k < n$ it is possible that $Z(H_k) > 1$ and that $F' \cong F$. This phenomenon relates to the behaviour observed by McCormick and Pearson and, as noted by Oren in 1972, becomes more serious with increased inaccuracy in the line search; it may cause a Quasi-Newton iteration to perform worse than a steepest descent iteration. A technique to reduce monotonically $Z(H)$ was developed by Oren and it results in choosing the free parameters in the following interval

$$0 \leq \theta \leq 1 \qquad \frac{\sigma}{\tau} \leq \gamma \leq \frac{\varepsilon}{\sigma} \tag{25}$$

Quadratic termination implies a fast rate of convergence for general functions in the following sense: if some additional mild conditions are satisfied, then a method with quadratic termination possesses n-quadratic rate of convergence on general functions, namely it holds that

$$\lim_{k\to\infty} \frac{\| x_{k+n} - x^+ \|}{\| x_k - x^+ \|^2} = c \qquad c < \infty \tag{26}$$

This result, due to work of Dixon and of Stoer in 1975, is valid for Oren's class with the restriction (24) and for Broyden's class with $0 \leq \theta \leq 1$, for perfect line searches or asimptotically perfect line searches. A similar result was proved in 1972 by McCormick and Ritter, but without establishing the relation with quadratic termination.

Local convergence results for Broyden's class (starting point sufficiently close to x^+) have been proved first in 1973 by Broyden, Dennis and Moré for exact line searches. Non exact line searches have been considered by Burmeister (1973), Schuller and Stoer (1973), Dennis and Moré (1974), Rheinboldt and Vandergraft (1974) and Stoer (1975); extensions to Oren's class under restrictions (25) are due to

Baptist (1975). For locally convergent iterations the rate of convergence is shown to be superlinear if the hessian is nonsingular at x^+. Knowledge about the global convergence properties for arbitrary starting point is still incomplete. A conjecture of Powell (1972) that Quasi-Newton methods converge to a stationary point of F if $F \in C^2$ has only be proved for n=2. Under essentially the assumption of convexity Powell proved in 1971 global convergence for the DFP method for exact line searches, and hence, from Dixon's theorem, of Broyden's class. In 1975 Powell has extended the result to the case of approximate line searches for the BFS method.

The convergence of the sequence of variances is similarly not well understood. Convergence of the DFP matrix to the true inverse hessian has been studied by Ribière in 1970. A counterexample of Powell shows that such a convergence is not to be expected in general.

Finally, Broyden, Dennis and Moré have proved that superlinear rate of convergence implies that the norm and the direction of the search vector must tend to those of the Newton-Raphson iteration and that consequently the stepsize λ_k must tend to one.

QUASI-NEWTON METHODS FOR LEAST SQUARES

When F is a sum of squares, $F = \sum_{i=1}^{m} r_i^2 = r^T r$, a modification of the Newton-Raphson iteration which is frequently used is the Gauss-Newton iteration. This can be derived observing that the hessian of a sum of squares can be written as follows

$$G = 2(J^T J + \sum_{i=1}^{m} r_i R_i) \qquad (27)$$

where G is the hessian of F, J is the jacobian of r and R_i is the hessian of r_i. In the Gauss-Newton method the term $\sum_{i=1}^{m} r_i R_i$ is considered small and is neglected.

Quasi-Newton techniques can be used in different ways to improve the Gauss-Newton method:

- instead of evaluating J at every iteration, it can be estimated by Quasi-Newton updates, as done by Powell in subroutine VA05A of the AERE library
- instead of neglecting the term $\sum_{i=1}^{m} r_i R_i$, it can be estimated evaluating the single hessian R_i by Quasi-Newton updates, as done by Brown and Dennis in 1971. While Gauss-Newton methods converge locally only if $F(x^+)$ is sufficiently small and at only a linear

rate of convergence, unless $F(x^+)=0$, the modified algorithm of Brown and Dennis converges locally independently of the value of $F(x^+)$ and at a superlinear rate of convergence

- let H be an approximation to G and write $H=J^TJ+D$; D can be determined in a variety of ways requiring that H' satisfies the Quasi-Newton equation. This approach, considered by Dennis in 1974 and developed by Biggs in 1975, requires less memory occupation than the previous method and is very efficient.

CONCLUSIONS

The theory and computational practice developed in the fifteen years since Davidon's fundamental paper constitute now a large body of results; see for instance the excellent review of the state of art in 1974 by Dennis and Moré. Some research areas are still open, for instance:

- global convergence for continuously differentiable but nonconvex functions
- methods which take into account the structure of the problem, in particular the sparsity of the hessian

Finally codes of high quality should be made available through hand books like the Rheinsch and Wilkinson monograph on linear algebra and treatises should be written to give a detailed and unified vision of this field of numerical analysis.

BIBLIOGRAPHY

A tentatively complete bibliography on Quasi-Newton methods is given below, extending to 1976. About 150 papers are listed, due to more that 50 researchers. A few papers do not deal directly with Quasi-Newton methods but are strictly related to them. The lack of quotations of Soviet authors is not meant to indicate absence of research in Soviet Union in this field, but only our ignorance of their literature.

The following abbreviations are used for commonly quoted Journals:

CJ	The Computer Journal
MC	Mathematics of Computation
MP	Mathematical Programming
MS	Management Science
JACM	Journal of the Association for Computing Machinery
JIMA	Journal of the Institute of Mathematics and its

	Applications
JOTA	Journal of Optimization Theory and Applications
SIAM JAM	SIAM Journal of Applied Mathematics
SIAM JNA	SIAM Journal of Numerical Analysis
ZAMM	Zeitschrifts von Angewandte Mathematik und Meckanik

Adachi N., On variable metric algorithm, JOTA 7, 391-409, 1973.

- On the uniqueness of search directions in variable metric algorithm, JOTA 11, 590-604, 1973.

Bard Y., On a numerical instability of Davidon-like methods, MC 22, 665-666, 1968.

- Comparison of gradient methods for the solution of non linear parameter estimation problems, SIAM JNA 7, 157-185, 1970.

Biggs M.C., Minimization algorithms making use of non quadratic properties of the objective function, JIMA 8, 315-327, 1971.

- A note on minimization algorithms which make use of non quadratic properties of the objective function, JIMA 12, 337-338, 1973.

- Some recent matrix updating method for minimizing sums of squared terms, Report NOC 67, Hatfield, 1975.

Brodlie K.W.,Gourlay A.R., Greenstadt J., Rank-one and rank-two corrections to positive definite matrices expressed in product form, JIMA 11, 73-82, 1973.

Brodlie K.W., Some topics in uncontrained minimization, PhD dissertation, University of Dundee, 1973.

- An assessment of two approaches to variable metric method, Working Paper, University of Leicester, 1976.

Broyden C.G., A class of methods for solving nonlinear simultaneous equations, MC 19, 577-593, 1965.

- Quasi Newton methods and their applications to function minimization, MC 21, 368-381, 1967.

- A new method of solving nonlinear simultaneous equations, CJ 10, 406-410, 1969.

- A new double-rank minimization algorithm, AMS Notices 16, 670,1969.

- The convergence of single-rank Quasi-Newton method, MC 24, 365-382, 1970.

- The convergence of a class of double-rank minimization algorithms, I and II, JIMA 6, 222-231, 1970.

- The convergence of an algorithm for solving sparse nonlinear simultaneous equations, MC 25, 285-294, 1971.

Broyden C.G., Johnson P., A class of rank-one optimization algorithms, in Numerical methods for nonlinear optimization, F.A. Lootsma ed.,

Academic Press, 1972.

Broyden C.G., Dennis J.E., Moré J.J., On the local and superlinear convergence of Quasi-Newton methods, JIMA 12, 223-245, 1973.

Broyden C.G., Quasi-Newton or modification methods, in Numerical solution of systems of nonlinear algebraic equations, G.D. Byrne, C.A. Hall eds., Academic Press, 1974.

Brown K.M., Dennis J.E., New computational algorithms for minimizing a sum of squares of nonlinear functions, Report 71-6, Yale University, 1971.

Burmeister W., Die konvergenzordnung des Fletcher-Powell Algorithmus, ZAMM 53, 696-699, 1973.

Callier F.M., Toint P.L., On quadratic function minimization: the n-step termination property can imply the conjugateness of the search directions, Internal report, Université de Namur, 1975.

Davidon W.C., Variable metric method for minimization, Report ANL-5990, Argonne, 1959.

- Variance algorithm for minimization, CJ 10, 406-410, 1967.

- Optimally conditioned optimization algorithm without line searches, MP 9, 1-30, 1975.

- New least squares algorithms, JOTA 18, 187-197, 1976.

Davison E.J., Wong P., A robust conjugate gradient algorithm that minimizes L-functions, Automatica 11, 297-308, 1975.

Dennis J.E., On the convergence of Broyden's method for nonlinear simultaneous equations, Report 69-46, Cornell University, 1969.

- On the convergence of Newton-like method, in Numerical methods for nonlinear algebraic equation, P. Rabinowitz, Gordon and Breach, 1970.

- On the convergence of Broyden's method for nonlinear simultaneous equations, MC 25, 559-567, 1971.

- Toward a unified convergence theory of Newton-like methods, in Nonlinear Functional Analysis and Applications, Rall, Academic Press, 1971.

- On some methods based on Broyden's secant approximation to the hessian, in Numerical methods for nonlinear optimization, F.A. Lootsma ed., Academic Press

- Some computational techniques for the nonlinear least squares problems, in Numerical solution of systems of nonlinear algebraic equations, G.D.Byrne, C.A. Hall eds., Academic Press, 1974

Dennis J.E., Moré J.J., A characterization of superlinear convergence and its application to Quasi-Newton methods, MC 28, 549-560,1974.

- Quasi-Newton methods: motivation and theory, Report TR 74-217, Cornell University, 1974.

Dixon L.C.W., Variable metric algorithms: Necessary and sufficient conditions for identical behaviour on nonquadratic function, Journal of Optimisation Theory & Applications Vol. 10 No. 1, pp 34 - 40 1972.

- Quasi-Newton algorithms generate identical points, MP 2, 383-387, 1972.

- The choice of step lenght, a crucial factor in the performance of variable metric algorithm, in Numerical methods for nonlinear optimization, F.A. Lootsma ed., Academic Press, 1972.

- Conjugate directions without linear searches, JIMA 11, 317-328, 1973.

- Nonlinear optimization; a survey of the state of the art, in Software for numerical mathematics, D.J. Evans ed., Academic Press, 1974.

- On quadratic termination and second order convergence: two properties of unconstrained optimization algorithm, in Towards global optimization, L.C.W. Dixon and G.P. Szegö ed., North Holland, 1975.

Fletcher R., A new approach to variable metric algorithms, CJ 13, 317-322, 1970.

- An efficient globally convergent algorithm for unconstrained and linearly constrained optimization problems, Report TP 431, AERE, 1970.

- Quasi-Newton methods for the solution of optimization problems, Report TP 432, AERE, 1971.

- A FORTRAN subroutine for minimization by Quasi-Newton method, Report R7 1125, AERE, 1972.

Fletcher R., Powell M.J.D., A rapidly convergent descent method for minimization, CJ 6, 163-168, 1963.

- On the modification of LDL^T factorizations, Report TP 419, AERE, 1973.

Fletcher R., Reeves C.M., Function minimization by conjugate gradients, CJ 6, 149-153, 1963.

Fried I., N-step conjugate gradient minimization scheme for nonquadratic function minimization, AIAAJ 9, 2286-2287, 1971.

Gay D., Representing symmetric rank-two updates, Report 124, National Bureau of Economic Research, 1976.

Gill P.E., Murray W., Quasi-Newton methods for unconstrained minimization, JIMA 9, 91-108, 1972.

- Newton-type method for unconstrained and linearly constrained opti

mization, MP 7, 311-350, 1972.
- Safeguarded steplength algorithm for optimization using descent methods, Report NAC 37, NPL, 1974.

Gill P.E., Murray W., Pitfield R.A., The implementation of two revised Quasi-Newton methods for unconstrained optimization, Report NAC 11, NPL, 1972.

Gill P.E., Murray W., Saunders M.A., Methods for computing and modifying the LDV factors of a matrix, Report NAC 56, NPL, 1974.

Goldfarb D., Extension of Davidon variable metric method to maximization under linear inequality and equality constraints, SIAM JAM 17, 739-764, 1969.
- A family of variable metric methods derived by variational means, MC 24, 23-26, 1970.
- Sufficient conditions for the convergence of variable metric algorithm, in Optimization, R. Fletcher ed., Academic Press, 1969.
- Variable metric and conjugate direction method in unconstrained optimization: recent developments, ACM proceeding, 1972.
- Factorized variable metric method for unconstrained optimization, IBM Report RC 4415, New York, 1973.

Greenstadt J., Variations on variable metric method, MC 24, 1-18, 1970.
- Product form for variable metric corrections, Working Paper, IBM, Palo Alto, 1971.
- A Quasi-Newton method with no derivatives, MC 26, 145-166, 1972.
- Variational derivation of Broyden like formulas, Working Paper, IBM, Palo Alto, 1975.
- Revision of a Quasi-Newton method with no derivatives, Working Paper, IBM, Palo Alto, 1975.

Huang H.Y., Unified approach to quadratically convergent algorithms for function minimization, JOTA 5, 405-423, 1970.

Huang H.Y., Levy A.V., Numerical experiments on quadratically convergent algorithms for function minimization, JOTA 6, 269-282, 1970.

Hoshino S., A formulation of variable metric method, JIMA 10, 394-403, 1972.

Jones D.S., The variable metric algorithm for nondefinite quadratic function, JIMA 12, 63-71, 1973.

Lenard M.L., Practical convergence conditions for unconstrained optimization, MP 4, 309-323, 1973.
- Practical convergence conditions for the DFP method, Report TR 1356, University of Wisconsin, Madison, 1973.

Matthews A., Davies D., A comparison of modified Newton methods for unconstrained optimization CJ 14, 293-294, 1971.

McCormick G., The variable reduction method for nonlinear programming, MS 17, 146-160, 1970.

- On the convergence and rate of convergence of the reset variable metric method, Report 1012, University of Wisconsin, Madison, 1973.

McCormick G., Pearson J.D., Variable metric method and unconstrained optimization, in Optimization, R. Fletcher ed., Academic Press,1969.

McCormick G., Ritter K., Method of conjugate direction versus Quasi-Newton method, MP 3, 101-116, 1972.

McKeown J.J., Specialized versus general-purpose algorithm for minimizing functions that are sums of squares, Report 50, NOC, 1973.

Mei H.H.W., On the conditioning of DFP and BFGS updates for unconstrained optimization Report 75-265, Cornell University, 1975.

Meyer R.R., The validity of a family of optimization mehtod, SIAM JC 8, 41-54, 1970.

- On the convergence of algorithms with restart, Report 225, University of Wisconsin, Madison, 1974.

Moré J.J., Tragenstein J., On the global convergence of Broyden's method, Report 74-216, Cornell University, 1974.

Murray W., The relationship between the approximate hessian matrices generated by a class of Quasi-Newton methods, Report NAC 12, NPL, 1972.

Oren S.S., Selfscaling variable metric algorithm for unconstrained minimization, PhD dissertation, Stanford University, 1972.

- Selfscaling variable metric algorithm without line searches for unconstrained minimization, MC 27, 873-885, 1973.

- On the selection of parameters in selfscaling variable metric algorithm, MP 7, 351-367, 1974.

- On Quasi-Newton and pseudo Quasi-Newton algorithm, JOTA 20, 155-170, 1976.

Oren S.S., Luenberger D., The selfscaling variable metric algorithm, Proceeding 5th Haway Conference System Science, 1972.

- Selfscaling variable metric algorithm I: Criteria and Sufficient conditions for scaling a class of algorithm, MS 20, 845-862,1974.

- Selfscaling variable metric algorithm II: Implementation and experiments, MS 20, 863-874, 1974.

Oren S.S., Spedicato E., Optimal conditioning of sefscaling variable metric algorithm, MP 10, 70-90, 1976.

Pearson J., On variable metric method of minimization, CJ 12, 171-178, 1969.

Powell M.J.D., A theorem on rank-one modifications to a matrix and its inverse, CJ 12, 288-290, 1969.

- A survey of numerical methods for unconstrained optimization, SIAM Review 12, 79-97, 1970.
- A new algorithm for unconstrained optimization, Report TP 393, AERE, 1970.
- Rank-one methods for unconstrained optimization, in Integer and nonlinear programming, J. Abadie ed. North Holland, 1970.
- A FORTRAN subroutine requiring first derivatives of the objective function, Report R-6469, AERE, 1970.
- Recent advances in unconstrained optimization, MP 1, 26-57, 1971.
- On the convergence of the variable metric algorithm, JIMA 7, 21-36, 1971.
- Problems related to unconstrained minimization, Report TP 439, AERE, 1971.
- On search directions of minimization algorithm, Report TP 492, AERE, 1972.
- Some properties of the variable metric method, in Numerical methods for nonlinear optimization, F.A. Lootsma ed., Academic Press, 1972.
- Quadratic termination properties of a class of double rank minimization algorithms, Report TP 471, AERE, 1972.
- Some theorems on quadratic termination properties of minimization algorithms, Report TP 472, AERE, 1972.
- Convergence properties of a class of minimization algorithms, Report CSS 8, AERE, 1974.
- A view of minimization algorithms that do not require derivatives, Report CSS 9, AERE, 1974.
- A view on unconstrained optimization, Report CSS 14, AERE, 1975.
- Some global convergence properties of a variable metric algorithm for minimization without exact line searches, Report CSS 15, AERE, 1975.

Rheinboldt W.C., Vandergraft J.S., On the local convergence of update method, SIAM JNA 11, 1069-1085, 1974.

Ribiere G.E., Sur la methode de DFP pour la minimisation de functions, MS 16, 572-592, 1970.

Rosen E.M., A review of Quasi-Newton method in nonlinear equation solving and unconstrained optimization, Proceedings 21th National Conf. ACM, Thomson Book, 1966.

Schnabel R.B., Optimal conditioning in the convex class of rank two updates, Report TR 76-288, Cornell University, 1976.

Schubert L.K., Modification of Quasi-Newton method for nonlinear equations with a sparse jacobian, MC 24, 27-30, 1970.

Schuller G., On the order of convergence of certain Quasi-Newton me thods, Numerische Mathematik 23, 181-192, 1974.

Schuller G., Stoer J., Uber die Konvergenzordnung geweisser Rang-2 Verfahren zur Minimierung von Funktionen, Int. Series of Num. Math. 23, Birkhauser, Basel, 1974.

Shanno D.F., Conditioning of Quasi-Newton method for function minimization, MC 24, 647-656, 1970.

- Inexact step length in Quasi-Newton method, Report WP 74.09, University of Toronto, 1974.

- Effective comparison of unconstrained optimization techniques, Working paper, University of Mississippi, 1976.

Shanno D.F., Kettler P.C., Optimal conditioning of Quasi-Newton methods MC 24, 657-664, 1970.

Shanno D.F., Phua K.H., Selfscaling variable metric algorithm, an alternative view, Working paper, University of Mississippi, 1975.

Spedicato E., Problems related to the solution of the Quasi-Newton equation, Report CISE-N-147, Milano, 1971.

- Conditioning in Huang's two parameter class for the conjugate gradient method, Report CISE-N-154, Milano, 1972.

- Computational performance of Huang's symmetric update for the conjugate gradient method, Calcolo 10, 1-27, 1973.

- Stability of Huang's update for the conjugate gradient method, JOTA 11, 469-479, 1973.

- A bound on the condition number of rank-two updates and applications to the variable metric method, Calcolo 12, 185-200, 1975.

- On condition numbers of matrices in rank-two optimization algorithm, in Towards global optimization, L.C.W. Dixon and G.P.Szego ed., North Holland, 1975.

- Recent developments in the variable metric method for unconstrained minimization, in Towards global optimization, L.C.W. Dixon and G.P. Szego ed., North Holland, 1975.

- A variable metric method derived from invariancy to nonlinear scaling, JOTA 20, 315-329, 1976.

- Computational experience with Quasi-Newton algorithms for minimization problems of moderately large size, Report CISE-N-175, Milano, 1975.

- A note on the determination of the scaling parameter in a class of Quasi-Newton methods for unconstrained minimization, Working Paper, Stanford University, 1976.

- On some classes of Quasi-Newton methods for systems of nonlinear algebraic equations, Working paper, Stanford University, 1976.

- Metodi Quasi-Newtoniani per equazioni algebriche e minimizzazione non lineare: sviluppi e prospettive, Bollettino Unione Matematica Italiana, 14-A, 545-567, 1976.

Stewart G.W., A modification of Davidon minimization method to accept difference approximations of derivatives, JACM 14, 72-83, 1967.

Stoer J., On the convergence of some minimization algorithm, IFIP Conference, Stockholm, 1974.

- On the convergence rate of imperfect minimization algorithm in Broyden's class, MP 9, 313-335, 1975.

- On the relation between quadratic termination and convergence properties of minimization algorithms, Report SU 326, P 30-42, Stanford University, 1975.

- Tewarson R.P., On the use of generalized inverses in function minimization, Computing 6, 241-248, 1970.

Thomas S.W., Sequential estimation techniques for Quasi-Newton algorithm, Report TR, 75-227, Cornell University, 1975.

Williamson W.E., Square root variable metric method for function minimization, AIAA J 13, 107-109, 1975.

Wolfe M.A., A Quasi-Newton method with memory for unconstrained function minimization, JIMA 15, 85-94, 1975.

Wolfe P., Another variable metric methods, Working paper, IBM, New York, 1967.

- Convergence conditions for ascent methods, SIAM Review 11, 226-235, 1967.

- Convergence conditions for ascent methods II: some corrections, SIAM Review 13, 185-186, 1971.

Zeleznik F.J., Quasi-Newton method for nonlinear equations, JACM 15, 265-271, 1968.

TOWARDS GLOBAL OPTIMISATION 2
L.C.W. Dixon and G.P. Szegö (eds.)

COMPUTATIONAL EXPERIENCE WITH QUASI-NEWTON ALGORITHMS FOR MINIMIZATION PROBLEMS OF MODERATELY LARGE SIZE

E. Spedicato, CISE, Segrate (MI), Italy

Three Quasi-Newton algorithms, the BFS method due to Broyden, Fletcher and Shanno, the OS method due to Oren and Spedicato, and Davidon's new (1976) method, are compared on a set of problems of moderately large size (dimension n of variables up to 80). Results show that the BFS method may be preferable for small n ($n < 20$) but that for large n it tends to be inferior to both the OS and Davidon's methods, which possess a better control of rounding errors. Also they show that scaling the initial matrix is beneficial.

INTRODUCTION

Quasi-Newton methods are among the best techniques for the unconstrained minimization of a nonlinear function $F=F(x)$ whose gradient $g=g(x)$ is available. These methods are based upon the iteration

$$x_{k+1}=x_k - \lambda_k H_k g_k \qquad k=0,1,2,.... \tag{1}$$

where H_k is a square matrix characterizing the method, λ_k is a scalar chosen usually so that $F(x_{k+1}) < F(x_k)$ and x_o, H_o are given; if H_k is theoretically guaranteed to be positive definite an equivalent but numerically more preferable version of iteration (1) is the following

$$\begin{aligned} L_k^T D_k L_k s_k &= g_k \\ x_{k+1} &= x_k - \lambda_k s_k \end{aligned} \qquad k=0,1,2,.... \tag{2}$$

where D_k and L_k are respectively a diagonal and triangular matrix such that $H_k^{-1} = L_k^T D_k L_k$.

There are different ways to specify the matrix H_k. A common feature of quasi-Newton algorithms is that they satisfy an equation of the form

$$H_{k+1}(g_{k+1} - \psi_k g_k) = \rho_k p_k \tag{3}$$

where ψ_k, ρ_k are scalars; for $\psi_k=1$, $\rho_k=1$ relation (3) is known as the quasi-Newton condition.

Most recent research has analyzed algorithms belonging to a class characterized by the following update of the matrix H_k

$$H_{k+1} = \gamma_k H_k + \frac{\sigma + \gamma_k \theta_k \tau}{\sigma^2} p_k p_k^T + \gamma_k \frac{\theta_k - 1}{\tau} H_k y_k y_k^T H_k - \gamma_k \frac{\theta_k}{\sigma} (p_k y_k^T H_k + H_k y_k p_k^T \tag{4}$$

where γ_k and θ_k are free scalar parameters, vectors p_k and y_k are given by relations

$$p_k = x_{k+1} - x_k \quad , \qquad y_k = g_{k+1} - g_k \tag{5}$$

and the scalars σ, τ, ε (this to be used later) are given by the inner products

$$\sigma = p_k^T y_k \quad , \qquad \tau = y_k^T H_k y_k \qquad \varepsilon = \lambda_k^2 g_k^T H_k g_k \tag{6}$$

This class, introduced by Oren [1], satisfies relation (3) with $\psi_k = 1$, $\rho_k = 1$. For $\gamma_k = 1$ it reduces to a class considered by Broyden [2]; it is shown that, under a suitable mapping of parameters and a mild condition on the search parameter λ_k, the algorithms using Oren's update are equivalent to the algorithms using the symmetric Huang [3] update, which satisfies equation (3) with $\psi_k = 1$ and ρ_k arbitrary. There is experimental evidence that the choice of the parameters affects the efficiency of the algorithms and the question of the optimal selection of the parameters is therefore important. For Broyden's class, numerical results of the literature (Broyden [2], Shanno [4], Gill and Murray [5], Powell [6], Dixon [7], Spedicato [8]) have suggested that the algorithm with $\theta_k = 1$, known as BFS or Broyden-Fletcher-Shanno method, is preferable.

Another algorithm of Broyden's class is also considered here. It is the algorithm recently proposed by Davidon [9], where parameter θ_k is chosen according to the following switch

$$\begin{aligned} &\text{if} \quad \sigma \leq \frac{2\varepsilon\tau}{\varepsilon + \tau} \qquad \theta_k = \frac{\sigma(\varepsilon - \sigma)}{\varepsilon\tau - \sigma^2} \\ &\text{if} \quad \sigma > \frac{2\varepsilon\tau}{\varepsilon + \tau} \qquad \theta_k = \frac{\sigma}{\sigma - \tau} \end{aligned} \tag{7}$$

This switch follows requiring that the condition number (in natural norm) of the matrix $H_k^{-1} H_{k+1}$ is minimized. This algorithm, named the DS method, will be considered in the basic iteration form (1), though the use of search vectors different than $s_k = H_k g_k$ has been proposed by Davidon.

For the general class (4) results of Oren and Luenberger [10] on invariancy to linear scaling of the function and optimal single-step rate of convergence (on quadratic functions and exact line searches) have suggested the choice of parameters in the following intervals

$$0 \leq \theta_k \leq 1 \qquad \frac{\sigma}{\tau} \leq \gamma_k \leq \frac{\varepsilon}{\sigma} \tag{8}$$

Further results of Oren and Spedicato [11] have suggested the following choice

$$\theta_k = \frac{1}{1+\sqrt{\frac{\varepsilon\tau}{\sigma^2}}} \qquad \gamma_k = \sqrt{\frac{\varepsilon}{\tau}} \tag{9}$$

This choice, giving the OS method, follows requiring that a strict bound to the ratio of the condition numbers (in natural norm) of H_{k+1} to H_k is minimized and that the algorithm is also self-dual in the sense of Fletcher [12]. If self-duality is dropped, then a one-parameter class of algorithms with minimal bound to the condition number can be considered, where γ_k and θ_k are related as follows (subject to restrictions (8))

$$\theta_k = \frac{\sigma(\varepsilon - \sigma\gamma_k)}{\gamma_k(\varepsilon\tau - \sigma^2)} \tag{10}$$

Note that for $\gamma_k = 1$ one of the values given by Davidon's switch is obtained; the second value of Davidon's switch corresponds to a value of θ_k larger than one, for which a different relation between γ_k and θ_k than (10) applies (canonical form of the rank-two correction of positive type instead than of mixed type). It can be shown that Davidon's and Oren and Spedicato bounds are equivalent.
In the following we introduce a set of problems with variable n to test the relative efficiency of the BFS, OS and DS algorithms.

A SET OF TEST PROBLEMS WITH VARIABLE N

Most test problems encountered in the literature have few variables ($n < 10$) and are inadequate for a comparison over a significant range of n. We considered two types of problems where n can be fixed at will. The first, described under section (a), consists of problems which are naturally dependent on n; the second, described under section (b), consists of problem which are artificially generated from literature problems in small number of variables. They possess an intrinsic symmetry and they could naturally be solved in their original space. It can be shown that, with infinite word length and a mild condition on λ_k, the variable metric method, started with a suitably symmetric initial point, generates a path to the minimum which is a symmetric extension of the path in the original space. In practice however these conditions are violated, the symmetry is lost and the path becomes dependent on n.

a) Natural n-dimensional problems

- Hilbert quadratic

$$F = x^T Ax, \qquad A_{ij} = \frac{1}{i+j-1}$$

$n = 10,40,80$

$x_o = (4,4/2,4/3,\ldots,4/n)$

This problem was considered by Oren [14]; A is the Hilbert matrix which is very ill-conditioned; in practice however rounding errors on the representation of A_{ij} probably improve the actual condition number. The minimum is at the nul vector.

- Power of quadratic convex function

$$F = (x^T Ax+c)^r, \qquad A_{ij} = i\,\delta_{i,j},$$

$n = 10,20,40,60,80$

$x_o = (1,1,\ldots,1)$

This problem with $c = 0$ and $r = 2$ was considered by Oren [14] and with $c = 0$ and a variety of values of r by Spedicato [15]. We tested this function in three cases: $c = 0$, $r = 2$ giving a problem with singular hessian in the minimum $x = 0$; $c = 0$, $r = 1/2$, giving a nonconvex problem with the minimizer $x = 0$ as a cuspidal point; with $c = 1$, $r = 1/2$, giving a well-behaved problem.

- Trigonometric function

$$F = \sum_{i=1}^{n} \left[n+i - \sum_{j=1}^{n} (A_{ij}\sin(x_j) + B_{ij}\cos(x_j)) \right]^2$$

$$A_{ij} = \delta_{i,j}, \qquad B_{ij} = i\delta_{i,j} + 1$$

$n = 20,40,60,80$

$x_o = (1/n,1/n,\ldots\ldots.1/n)$

This function is a special case of the trigonometric function with random coefficients known in the literature. The expected minimum is at the nul vector, but other minima exist. The indices to x are component indices.

- Mancino function

$$F = \sum_{i=1}^{n} f_i^2$$

$$f_i = \sum_{\substack{j=1 \\ j\neq i}}^{n} \left[\sqrt{x_j^2+i/j}\,(\sin^{\alpha} \log \sqrt{x_j^2+i/j} + \cos^{\alpha} \log \sqrt{x_j^2+i/j}) \right]$$

$$+ n x_i \beta + (i-n/2)^{\gamma}, \qquad \alpha=5, \quad \beta=14, \quad \gamma=3$$

$n=5,10,20,30$

$$x_o=(af_1(0),\ldots\ldots.af_n(0)) \qquad a=-\frac{n}{\left[\beta^2 n^2-(\alpha+1)^2(n-1)^2\right]}$$

Gheri and Mancino considered the problem of solving the nonlinear system $f_i=0$ and showed that for $\gamma>0$, $\beta>\alpha>1$ the solution was unique and a good estimate was that given[16]. The location of the zero varies with n. The calculation of this function is time consuming.

- Extended Rosenbrock function

$$F=\sum_{i=1}^{n/2} \left[100(x_{2i}-x_{2i-1}^2)^2+(1-x_{21-1})^2\right]$$

$n=2,8,20,40,80$

$x_o=(-1.2,1,-1.2,1,\ldots\ldots,-1.2,1)$

- Extended Powell function

$$F=\sum_{i=1}^{n/4} \left[(x_{4i-3}+10x_{4i-2})^2+5(x_{4i-2}-x_{4i})^2+(x_{4i-2}-2x_{4i-1})^4 +10(x_{4i-3}-x_{4i})^4\right]$$

$n=4,20,40,60$

$x_o=(3,-1,0,3,3,-1,0,3.\ldots\ldots\ldots.3,-1,0,3)$

The extended Rosenbrock function was considered by Oren[14] and by Dixon[13]. Oren found that the number of iterations was increasing with n, while Dixon, quoting results of Whinfield, found that the number of iterations was about independent of n; this discrepancy can be explained by different word length and line search precision.

NUMERICAL EXPERIMENTS

The code used to test the three methods was AERE subroutine VA10A, with suitable changes in the part relating to the update of H_k. The code sets $H_o=I$, updates the inverse hessian and selects λ_k by a moderately accurate line search (requiring about two function and gradient evaluations per iteration). The use of a moderately accurate line search is convenient when dealing with large problems, because there is experimental evidence, see for instance Spedicato[8] and Dixon[13], that the number of iterations, and consequently the overhead time (which is $O(n^2)$) increases when the accuracy of the line search is relaxed. This observation is important in our case, where six out of eight functions require only $O(n)$ multiplications to be calculated. Updating the inverse hessian may be marginally less stable than updating a factorization of the hessian. However in no case loss of positive definiteness was indicated by testing the sign of $g_k^T H_k g_k$. The input parameters were set as follows: maximum number of iterations 50+2n; required precision on all components of x, 0.001. Calculations were performed in single and double precision on an IBM 370/125. The total CPU time was about 30 hours, some of the largest problems requiring more than one hour each.

In Tables 1-3 comparisons are made of relative performances. The first number gives final function value, the second the number of iterations, the third the number of function evaluations (the cost of the gradient assumed equal to the cost of the function). Table 1 relates to double precision calculations. We observe that for many functions there is a striking similarity of results. They are virtually identic for the Hilbert, trigonometric and well-behaved power functions and marginally different for the Rosenbrock,Powell and Mancino functions. Significant differences are only seen on the ill-behaved power function, where the OS method is clearly the best, followed by the DS method. For exact line searches and no rounding errors the identity of behaviour of methods of Broyden's class is a well known result of Dixon. The present results lead us to conjecture* that Dixon's theorem is valid also for Oren's class.Table 2 relates to calculations in single precision. Here the picture is more varied. Performance on Hilbert, trigonometric and well-behaved power function is still virtually identic and similar to the case of double precision. On the other functions the performance varies, with the OS showing best, followed by the DS; there is evidence moreover that differences tend to grow with n. These results show that rounding errors have important effects and that, especially for large n, it is convenient to keep a low condition number of the hessian.

In Table 3 we consider the three functions which gave most dissimilar results in single precision and we explore the effect of scaling the initial matrix. We set $H_o=D$, where D is a diagonal matrix whose elements are the inverse of the (positive) diagonal elements of the hessian of the function, calculated analytically. Clearly in this way the hessian is scaled. Results show that the scaled BFS method, here called BFS^+, is radically improved. It becomes competitive with the OS method on Rosenbrock function but is still inferior on the power functions. On the other side, only marginal improvement is observed on the OS^+ method. These results suggest that most of the deterioration of the condition number occurs, when $H_o=I$, at the first iteration and therefore can be easily removed; however on some classes of problems, to which the power function belongs, scaling at every iteration may prove beneficial.

Editorial note: The conjecture has been proved incorrect. - L.C.W.D.

CONCLUSIONS

The experiments described show that, when rounding errors are not important, the BFS,OS and DS methods perform similarly. When rounding errors are significant methods which control the condition number growth are preferable, especially for large n. It is suggested to pay particular attention to scaling the initial matrix. This can be done by using the OS method or more simply initially scaling the BFS method. This can be obtained with the method described or just taking $H_o = \frac{\sigma_0}{\tau_0} I$, in agreement with relation (10). This scaled BFS algorithm has also been considered with good results in recent experiences by Brodlie[17] and Shanno and Phua[18].

REFERENCES

Brodlie D.K.W. (1977). An assessment of two approaches to variable metric methods, Mathematical Programming, 12,344-355

Broyden C.G. (1967). Quasi-Newton methods and their application to function minimization, Mathematics of Computation,21,368-381

Davidon W.C. (1975). Optimally conditioned optimization algorithms without line searches, Mathematical Programming 9, 1-30

Dixon L.C.W. (1972). The choice of step length, a crucial factor in the performance of variable metric methods, F.Lootsma ed., Numerical methods for nonlinear optimization, 149-170,Academic Press

Dixon L.C.W. (1974). Nonlinear optimization: a survey, D.J.Evans ed., Software for numerical mathematics, 193-218, Academic Press

Fletcher R. (1970). A new approach to variable metric algorithms, Computer Journal, 13, 317-322

Gheri,G., Mancino,O. (1971). A significant example to test methods for solving systems of nonlinear equations, Calcolo, 8, 107-114.

Gill, P.E., Murray, W. (1972). Quasi-Newton methods for unconstrained optimization, Journal of the Institute of Mathematics and its Appli cation, 9, 91-108.

Huang, H.Y. (1970). Unified approach to quadratically convergent algorithms for function minimization, Journal of Optimization Theory and Applications, 5, 405-423.

Oren, S.S. (1972). Self-scaling variable metric algorithms for unconstrained minimization, Ph. D. Dissertation, Stanford University.

Oren, S.S. (1974). On the selection of parameters in self-scaling variable metric algorithms, Mathematical Programming, 7, 351-367.

Oren, S.S., Luenberger, D.G. (1974). Self-scaling variable metric (SSVM) algorithms - I: Criteria and sufficient conditions for scaling a class of algorithms, Management Science, 20, 845-862.

Oren, S.S., Spedicato, E. (1976). Optimal conditioning of self-scaling variable metric algorithm, Mathematical Programming, 10, 70-90.

Powell, M.J.D. (1970). Recent advances in unconstrained optimization, Report n. T.P.430, AERE, Harwell.

Shanno, D.F. (1970). Conditioning of Quasi-Newton methods for function minimization, Mathematics of Computation, 24, 647-656.

Shanno, D.F., Phua, K.H. (1976). Matrix conditioning and nonlinear optimization, Working paper, University of Arizona (Tucson) and Nanyang University (Singapore).

Spedicato, E. (1973). Computational performance of Huang's symmetric update for the conjugate gradient method, Calcolo, 10, 1-27.

Spedicato, E. (1976). A variable metric method for function minimization derived from invariancy to nonlinear scaling, Journal of Optimization Theory and Application, 20, 315-329.

AKNOWLEDGMENT

The author thanks W. Murray, of NPL, M.J.D. Powell, of AERE, and L.C.W. Dixon, of NOC, for discussion on the subject.

Table 1 - Results on IBM 370/125, double precision

Function	n	BFS	OS	DS
Rosenbrock	2	6E-9,26,102	5E-8,31,104	3E-9,31,122
	8	2E-4,44,184	6E-4,50,196	7E-4,49,194
	20	9E-5,85,366	4E-5,89,354	4E-5,90,358
	40	4E-4,112,452	4E-4,120,458	4E-4,125,460
Powell	4	6E-17,34,115	4E-6,19,74	6E-16,32,119
	20	2E-5,21,80	2E-5,21,78	5E-5,21,74
	40	4E-7,41,154	1E-8,41,154	8E-11,41,152
	60	1E-8,61,230	1E-10,61,204	4E-7,61,202
Hilbert	10	3E-14,13	1E-13,13	3E-14,13
	40	1E-11,43	8E-10,43	1E-11,43
	60	8E-12,63	5E-11,63	9E-12,63
Power, c=0 r=2	10	1E-6,37	5E-17,38	2E-9,41
	20	2E-5,43	4E-18,41	5E-8,53
	40	8E-5,70	8E-14,45	3E-4,61
	60	3E-5,101	4E-14,61	6E-4,88
Power, c=0 r=1/2	10	2E-3,35	1E-3,14	2E-3,24
	20	6E-3,55	3E-4,23	2E-3,39
	40	4E-3,100	4E-4,42	6E-3,59
	60	8E-3,128	3E-4,62	2E-2,63
Power, c=1 r=1/2	20	1,22,84	1,22,72	1,22,82
	40	1.,42,164	1.,42,126	1.,42,142
	60	1.,62,244	1.,62,178	1.,62,184
Trigonometric	10	2E-8,11	2E-10,11	8E-9,11
	20	5E-9,21	5E-8,21	3E-8,21
	40	5E-9,41	3E-8,41	1E-8,41
Mancino	5	6E-5,6,20	5E-7,6,14	6E-5,6,20
	10	2E-3,15,56	3E-4,11,24	3E-6,11,38
	20	1E-1,21,80	1E-3,21,46	1E-5,21,72

Table 2 - Results on IBM 370/125, simple precision

Function	n	BFS	OS	DS
Rosenbrock	2	3E-9,26,104	2E-9,40,162	3E-9,31,120
	8	4E-3,41,163	9E-4,66,266	5E-4,50,198
	20	9E-4,75,318	8E-3,47,188	4E-5,77,315
	40	2E-3,130,578	2E-2,59,236	9E-5,128,530
	80	7E-4,199,892	3E-7,86,320	
Powell	4	5E-14,32,124	4E-6,18,68	1E-12,31,120
	20	2E-6,39,156	2E-5,21,80	2E-7,51,198
	40	4E-8,74,296	2E-9,41,154	7E-8,62,238
	60	5E-7,75,306	2E-10,61,204	2E-6,64,248
Hilbert	10	4E-11,13,36	6E-10,13,38	1E-10,13,34
	40	5E-9,43,96	4E-10,43,102	4E-11,42,73
	80	1E-10,83,180	4E-10,83,188	2E-10,81,183
Power, c=0	10	1E-6,36,142	3E-17,38,150	2E-9,41,160
r=2	20	1E-5,43,176	3E-18,41,162	5E-8,53,212
	40	3E-5,69,288	2E-14,45,188	4E-4,62,258
	60	8E-5,94,400	2E-14,61,218	5E-4,89,380
	80	2E-4,129,576	1E-14,81,264	
Power, c=0	10	2E-3,35,154	1E-3,14,52	2E-3,24,92
r=1/2	20	3E-3,59,288	3E-4,23,82	2E-3,39,150
	40	6E-3,99,536	3E-4,42,130	5E-3,59,224
	60	8E-3,124,666	4E-4,62,176	2E-2,62,234
	80	1E-2,146,820	3E-4,82,230	
Power, c=1	5	1.,9,32	1.,8,28	1.,8,28
r=1/2	20	1.,22,72	1.,22,72	1.,22,28
	40	1.,42,164	1.,42,124	1.,42,144
	60	1.,62,244	1.,62,180	1.,62,188
Trigonometric	20	7E-10,21,54	1E-8,21,60	4E-9,21,54
	40	1E-8,41,98	2E-8,41,98	6E-9,41,92
	60	3E-9,63,148	2E-8,63,152	1E-8,64,152
	80	4E-8,86,224	4E-8,83,202	3E-8,83,199
Mancino	5	8E-5,6,22	5E-7,6,16	2E-6,6,18
	10	7E-5,15,58	3E-4,11,26	1E-6,11,26
	20	8E-4,22,86	7E-4,21,48	5E-5,23,50
	30	4E-1,36,144	5E-4,31,68	3E-2,33,72

Table 3 - Scaled versus unscaled initial matrix (IBM 370/125 simple precision)

Function	n	BFS	BFS$^+$	OS	OS$^+$
Rosenbrock	2	3E-9,26,104	1E-10,29,116	2E-9,40,162	2E-11,38,152
	8	4E-3,41,163	3E-8,30,118	9E-4,66,266	7E-6,55,220
	20	9E-4,75,318	8E-7,30,120	8E-3,47,188	1E-5,72,288
	40	2E-3,130,578	1E-5,41,144	2E-2,59,236	1E-3,59,231
	80	7E-4,199,892	4E-5,86,338	3E-7,86,320	1E-5,82,320
Power c=0, r=2	5	2E-8,32,126	1E-8,24,96	8E-23,35,138	4E-16,30,120
	20	1E-5,43,176	1E-11,33,126	3E-18,41,162	8E-14,52,104
	40	3E-5,69,288	6E-7,41,120	2E-14,45,188	3E-13,41,132
	60	8E-5,94,400	2E-7,61,162	2E-14,61,218	2E-12,61,170
	80	2E-4,129,576	6E-8,81,204	1E-14,81,264	6E-8,81,210
Power c=0, r=1/2	5	2E-3,22,90	2E-3,12,56	5E-4,13,48	4E-3,8,34
	20	3E-3,59,288	4E-3,36,204	3E-4,23,82	8E-4,22,64
	40	6E-3,99,536	5E-3,56,346	3E-4,42,130	2E-4,42,110
	60	8E-3,124,666	1E-2,71,454	4E-4,62,176	1E-3,62,140
	80	1E-2,146,820	7E-3,85,536	3E-4,82,230	2E-3,82,182

TOWARDS GLOBAL OPTIMISATION 2
L.C.W. Dixon and G.P. Szegö (eds.)

A BRIEF NOTE ON MODIFIED QUASI-NEWTON METHODS

E. G. H. Crouch
Department of Mathematics and Statistics
Teesside Polytechnic
Middlesbrough
Cleveland

MQN (Modified Quasi-Newton) methods are defined as differing from Quasi-Newton methods in one essential feature, namely that a different method of generating directions is employed such that, in the case of a quadratic function, the directions are conjugate with respect to the Hessian matrix without the use of exact minimisation. It is essestial to test that the direction of search is downhill since otherwise MQN methods may fail to minimise a quadratic form because a null direction-vector may be generated.

1. INTRODUCTION

Dixon (1973) described an optimisation algorithm similar to Quasi-Newton algorithms but with a different method of generating the directions used.

We shall use the term QN (Quasi-Newton) methods to refer to methods of the Davidon-Fletcher-Powell type (Davidon, 1959; Fletcher and Powell, 1963) which generate a direction p from the formula

$$p = - Hg \tag{1}$$

where H is an approximation to the inverse Hessian, g is the gradient at the current point x; we shall not distinguish variants such as that of Gill and Murray (1972) which uses

$$Bp = - g \tag{2}$$

where B is an approximation to the Hessian, stored in factorised form, instead of (1). Either (1) or equivalently (2) will be referred to as the QN formula and a direction so generated as a QN direction.

Dixon's algorithm retains the use of an updating formula, but instead of using the QN formula directions are generated by a method which, when minimising a quadratic function, gives directions conjugate to the Hessian matrix without needing to use perfect line searches.

We shall use the term MQN (Modified QN) method to refer to algorithms which possess this property and which use an updating formula of the one-parameter family discussed by Broyden (1970) and Fletcher (1970) to generate the inverse Hessian (or equivalently the Hessian itself).

A new MQN method is defined in section 2: that it is an MQN method as defined above follows from section 3. For brevity we shall generally assume use of the inverse Hessian, H, and omit reference to variants of the algorithm using the approximate Hessian B.

Only in section 4 do we specifically consider the use of B.

2. THE NEW ALGORITHM: DEFINITIONS

A typical iteration of the new MQN method is defined as follows:

Iteration i

Initially a current point x_{i-1}, the gradient g_{i-1}, a direction of search p_{i-1} and an approximation H_{i-1} to the inverse Hessian are given.

Stage 1.

It is checked that the direction p_{i-1} is downhill and satisfies

$$p_{i-1}^T g_{i-1} + 10^{-9} < 0.$$

Otherwise the direction is rejected and the QN direction given by (1) or (2) used instead. If this too is rejected then H_{i-1} is reset to I. A new point

$$x_i = x_{i-1} + d_i = x_{i-1} + t_i p_{i-1} \tag{3}$$

is generated (using $t_i = 1$ or any suitable value).

Stage 2.

Calculate g_i, the gradient at x_i, and

$$y_i = g_i - g_{i-1}. \tag{4}$$

Stage 3.

A new direction is generated from

$$p_i = k_i u_i = k_i \left[\frac{d_i}{d_i^T y_i} - \frac{H_{i-1} y_i}{y_i^T H_{i-1} y_i} \right] \tag{5}$$

for some value of the scalar quantity k_i.

Stage 4.

If $d_i^T y_i$ is greater than 10^{-12}, H_{i-1} is updated to give H_i. Otherwise H_{i-1} is retained.

(Note that the next iteration can retain x_{i-1} as current point, rejecting x_i given by (3), as in Dixon's method).

We shall refer to an iteration as defined above as an MQN iteration. If the QN formula is used instead of (5) and stages 3 and 4 are interchanged we shall refer to the iteration as a QN iteration or if t is chosen to give a perfect line search as a PQN iteration. Then a QN or PQN algorithm consists exclusively of QN or PQN iterations and an MQN algorithm is defined as consisting predominantly but not necessarily exclusively of MQN iterations. In particular the check carried out in stage 1 of the algorithm to ensure that the direction is downhill may force the use of the QN direction.

Except where stated otherwise it is to be assumed that p_o is the QN direction.

3. SOME PROPERTIES OF THE MQN METHOD

It is well known that the properties of the PQN method, when used to minimise a quadratic function, arise from a certain inductive property.

Let P_i denote the proposition that (6a) and (6b)

$$H_i y_j = d_j, \quad 0 < j \leqslant i \tag{6a}$$

$$p_i^T y_j = 0, \quad 0 < j \leqslant i \tag{6b}$$

are both true for a particular algorithm when the function is quadratic. Then iteration i is inductive if P_{i-1} implies P_i. The iterations of the PQN method are inductive in this sense and therefore the method generates conjugate directions and H_n is exact (Fletcher and Powell, 1963).

It is easily shown that MQN iterations are inductive. The proof is simpler than for the PQN method because it does not depend on previous steps having satisfied a condition for a perfect line search of the form

$$0 = p_{j-1}^T g_i = d_j^T g_i, \quad 0 < j \leqslant i \ . \tag{7}$$

Moreover, for both the MQN and PQN methods the inductive property holds for arbitrary p_o, although the only practical choice in the MQN method seems to be the QN direction. With other choices of p_o the method can fail with what in section 5 is called prior termination. (That p_o may be chosen arbitrarily is shown in an appendix).

We now state three properties of the new algorithm as described above.

Property 1. For a quadratic function the new method generates n conjugage directions without perfect linear searches and also generates the inverse Hessian, p_o being chosen arbitrarily.

Property 2. The new method and the QN method with standard p_o (that is, the QN direction) are equivalent (in a sense defined below) if and only if perfect line searches are used.

Property 3. For a quadratic function the new method, using arbitrary steps, gives the same directions and the same matrices as the PQN method with standard p_o.

By equivalent in property 2 we mean that if iteration i of two algorithms start with the same x_{i-1}, H_{i-1} and p_{i-1} then they both give the same x_i, H_i and p_i, assuming that both choose t so as to give the same stepsize $||d_i||$ and both use the same updating formula. Thus two algorithms whose iterations are equivalent are able to compute the same points x and matrices H.

Property 1 follows from the inductive property referred to above (or alternatively is a consequence of property 3 in the case when p_o is the QN direction).

To prove property 2 we consider the directions generated: under the conditions assumed in the above definition of "equivalent", if the directions in the two algorithms are the same then the x and H will be the same. To give the same p_i it is necessary that

$$k_i u_i = - H_i g_i \tag{8}$$

where

$$u_i = \frac{d_i}{d_i^T y_i} - \frac{H_{i-1} y_i}{y_i^T H_{i-1} y_i} . \tag{9}$$

The usual updating formula for H is

$$H_i = H_{i-1} + \frac{d_i d_i^T}{d_i^T y_i} - \frac{H_{i-1} y_i y_i^T H_{i-1}}{y_i^T H_{i-1} y_i} + \phi (y_i^T H_{i-1} y_i) u_i u_i^T \tag{10}$$

with $\phi = 1$ and other formulae of the Broyden-Fletcher one-parameter family take values of ϕ in the range 0 to 1.

If we assume that (1) holds so that, from (3)

$$- H_{i-1} g_{i-1} = p_{i-1} = d_i / t_i ,$$

then it follows from (10), (4) and (9) that

$$- H_i g_i = - H_{i-1} g_{i-1} - \left[\frac{d_i^T g_i}{d_i^T y_i}\right] d_i + \left[\frac{y_i^T H_{i-1} g_{i-1}}{y_i^T H_{i-1} y_i}\right] H_{i-1} y_i + c u_i$$

$$= k u_i - \left[\frac{d_i^T g_i}{d_i^T y_i}\right] d_i \tag{11}$$

where c and k are scalars. For (11) to agree with (8) we require perfect line searches.

Thus we have established property 2.

It follows from property 2 that the new standard method gives the same directions

$\bar{p}_i$ as the PQN algorithm, apart from a scalar factor depending on the choice of k. Since the updating formula (10) is, for a quadratic function, independent of k and t, as shown by Dixon (1973), property 3 follows at once from property 2. (The choice of k is discussed in section 4.)

Property 3 establishes that the new method is an MQN method as defined in section 1.

An important feature of MQN methods follows from property 3. For a quadratic function the MQN p_n is the same as for the PQN method, namely the zero vector, and to obtain the exact minimum the (n+1)th step should be in the QN direction with t = 1, so that

$$d_{n+1} = p_n = - H_n g_n.$$

(This is discussed further in section 5).

4. THE CHOICE OF k

A reexamination of (8) and (11) indicates the value of k_i to be used in (5) to give the same p_i as the PQN method. Using (10) with $\phi = 1$ this value is

$$k_i = - \frac{d_i^T g_{i-1}}{d_i^T y_i} \cdot y_i^T H_{i-1} y_i \tag{12}$$

and this gives p_i independently of the choice of t_i in (3).

If the Davidon-Fletcher-Powell formula (with $\phi = 0$) is used to update H, or if a formula analogous to (10) with $\phi = 1$ is used to update B, then we require

$$k_i = p_{i-1}^T y_i. \tag{13}$$

In one sense, it is unimportant which updating formula is used, at least when the linear searches are exact (Dixon, 1971) but computational considerations sometimes suggest that taking $\phi = 1$ is preferable to using the Davidon-Fletcher-Powell formula. Thus Fletcher (1970) shows that it gives greater assurance that the matrices will remain positive-definite in the presence of rounding errors and Greenstadt (1970) and Goldfarb (1970) show that a norm of $H_i - H_{i-1}$ (more exactly $||A^{\frac{1}{2}}(H_i - H_{i-1})A^{\frac{1}{2}}||$ where A is the exact Hessian) is minimal when using (10) with $\phi = 1$.

5. DOWNHILL DIRECTIONS AND PRIOR TERMINATION

The QN method has two attractive features which MQN methods do not necessarily possess. These are that the QN method always gives downhill directions and cannot generate a zero direction vector, p, unless the gradient is zero. These properties follow immediately from (1), because the matrices generated by the updating formula can be expected to be positive-definite, particularly if safeguards against rounding errors such as those proposed by Gill and Murray (1972) are taken.

We have no immediate assurance that MQN methods possess either of these features. It follows from property 3 that the x_i given by MQN and PQN methods differ by a linear combination of d_i's and hence for a quadratic function the gradients g_i differ only by a linear combination of y_i's with $j \leqslant i$. Because

$$p_i^T y_j = 0, \quad j \leqslant i$$

$p_i^T g_i$ will be the same for the MQN and PQN (if k is chosen appropriately). This gives some assurance of obtaining downhill directions with MQN methods for a quadratic function, neglecting the effect of rounding errors.

Nevertheless, whereas the PQN method terminates with g_n zero for a quadratic function, the MQN method does not, and to calculate the minimum an additional QN step is necessary. Without this refinement, the MQN method can terminate with p zero and g nonzero; we call this prior termination.

The absence of these two properties may slow down convergence of MQN methods. In practice the test for search directions to be downhill was found sufficient to avoid prior termination and to ensure that a final QN step was employed when minimising a quadratic function. (An alternative method would be to force the use of the QN direction every nth iteration.)

6. NUMERICAL RESULTS

Using four test functions ROS2, EXP4, POW4 and TEST10 given by Dixon (1973) it was found that the MQN and QN algorithms gave similar performance, as indicated by the table below in which the effective function evaluations required are given (where the n components of the gradient vector are regarded as equivalent to n function evaluations). In each case the calculations continued until the function value was reduced to 10^{-4}.

The use of a DFP-type formula for updating B was found to be superior in general to the use of the BFGS-type.

The best results for the QN method were obtained without interpolation in the search, using values of t taken as 1, 0.1, 0.01 as suggested in Fletcher (1970). For the MQN method best results were obtained using an interpolation based on Dixon (1973) which makes use of the gradient.

Table of Results

Problem	EFE (QN method)	EFE (MQN method)
ROS2	120	131
EXP4	122	173
POW4	124	137
TEST10	778	1007
Total	1144	1448

The assistance of Roy Duncan is gratefully acknowledged in programming and obtaining these numerical results.

APPENDIX: ON ARBITRARY CHOICE OF p_o AND H_o

The inductive property holds for PQN and MQN methods when the function minimised is quadratic. This is if

$$H_{i-1}y_j = d_j, \quad 0 < j < i \tag{14}$$

and

$$p_{i-1}^T y_j = 0, \quad 0 < j < i \tag{15}$$

then

$$H_i y_j = d_j, \quad 0 < j \leq i \tag{16}$$

and

$$p_i^T y_j = 0, \quad 0 < j \leq i. \tag{17}$$

Since

$$y_r = Ad_r$$

for a quadratic function with Hessian matrix A, it follows from (15) that

$$d_j^T y_i = d_i^T y_j = p_{i-1}^T y_j = 0, \quad 0 < j < i. \tag{18}$$

The proof of (16) is the same for MQN and PQN methods, namely: for the case $j = i$ (16) follows at once from the updating formula (10), and for $j < i$ equation (16) can be proved using (14), (15) and (18). In particular we establish (19) below as a special case ($j = i = 1$).

The proof of (17) depends on the formula used to generate the direction of search, either the QN formula or (5) and (9). For the MQN method, from (9),

$$u_i^T y_i = 0$$

and (18) follows in the case $i = j$. For the case $j < i$ (17) follows from the use of (9) with (14) and (18).

To initiate the inductive proof we need

$$H_1 y_1 = d_1, \tag{19}$$

$$p_1^T y_1 = d_2^T y_1 = 0, \qquad (20)$$

of which (19) was established above.

For the PQN method

$$y_1^T p_1 = - y_1^T H_1 g_1 = - d_1^T g_1,$$

using (19), for a perfect line search. This proves (20).

For the MQN method

$$y_1^T p_1 = k y_1^T u_1 = 0, \quad \text{from (5) and (9).}$$

Again this proves (20).

For the PQN and MQN methods both p_o and H_o are arbitrary, in the sense that the above proofs make no assumptions about either. (Other considerations may be important, such as that H_o should be positive- definite.)

REFERENCES

Broyden, C. G. (1970). The Convergence of a Class of Double-Rank Minimisation Algorithms, JIMA, 6, 76-90 and 222-231.

Davidon, W. C. (1959). Variable Metric Method for Minimisation, ANL-5990, Argonne Nat Labs Report.

Dixon, L. C. W. (1971). Variable Metric Algorithms: Necessary and Sufficient Conditions for Identical Behaviour on Non-Quadratic Functions, TR 26, NOC. Hatfield Polytechnic.

Dixon, L. C. W. (1973). Conjugate Directions without Linear Searches, JIMA, 11, 317-328.

Fletcher, R. (1970). A new Approach to Variable Metric Algorithms, Computer Journal, 13, 317-322.

Fletcher, R. and Powell, M. J. D. (1963). A rapidly Convergent Descent Method for Minimisation, Computer Journal, 6, 163-168.

Gill, P. E. and Murray, W. (1972). Quasi-Newton Methods for Unconstrained Optimisation, JIMA, 9, 91-108.

Goldfarb, D. (1970). A family of Variable Metric Methods derived by Variational Means, Math Comp, 24, 23-26.

Greenstadt, J. (1970). Variations on Variable Metric Methods, Math Comp, 24, 1-22.

TOWARDS GLOBAL OPTIMISATION 2
L.C.W. Dixon and G.P. Szegö (eds.)

NECESSARY AND SUFFICIENT CONDITIONS FOR THE CONVERGENCE OF AN ALGORITHM IN UNCONSTRAINED MINIMIZATION

M.GAVIANO
ISTITUTO MATEMATICO
UNIVERSITA' DI CAGLIARI, ITALY

ABSTRACT: This paper contains the main results established in [1], [2] and [3] concerning the convergence of an algorithm for unconstrained minimization. Specifically, it is shown that if for an algorithm α the condition (i) $\sum|f(z_{i+1})-f(z_i)|/\|\nabla f(z_i)\|^2=+\infty$ holds, then α constructs a sequence which among its limit-points has at least one stationary point. Further, it is shown that for any algorithm α which converges to a minimium point, (i) must hold, while if α converges to a saddle point (i) may not hold.
Finally, for uniformely convex functions, it is shown that to obtain convergence to the unique minimum, the condition (ii) $\sum\rho_i^2=+\infty$, $\rho_i=-h_i^T\nabla f(z_i)/\|h_i\|\,\|\nabla f(z_i)\|$ must hold, while if f(x) is convex but not uniformely convex, (ii) does not need to hold.

1. INTRODUCTION

In this paper we shall be concerned with the problem of locating a minimum of a twice continuously differentiable function $f(z)$, $z \in R^n$.
We shall refer to an algorithm that could solve such a problem as a mechanism and consider those mechanisms which generate a sequence $z_0, z_1, z_2, \ldots$ with $f(z_0) \geq f(z_1) \geq f(z_2) \geq \ldots$; 'A' will denote the set of all these algorithms.
Moreover, for any $\alpha \in A$, we shall assume that at the i-th iteration, during which z_{i+1} is computed, the two following problems are solved

(i) a search direction h_i is determined;

(ii) a step size $\lambda_i > 0$ along h_i is determined

and z_{i+1} is given by $z_{i+1} = z_i + \lambda_i h_i$.
Given any $\alpha \in A$, the aim of our investigation will be that of finding necessary or sufficient conditions on the search directions h_i and on the relative step sizes λ_i so that α is a convergent algorithm.
This problem has been treated by many authors in the literature. The results given

in [1], [2] and [3] will be the basis of this paper. We may also refer to the results given by Lenard [4], Ortega and Rheinboldt [5], Polak [6], Zangwill [7] and Wolfe [8], [9].

2. CONVERGENCE TO A STATIONARY POINT

In this section we shall give some conditions under which an algorithm $\alpha \varepsilon A$ constructs a sequence $\{z_i\}_{i=0}^{\infty}$ which among its limit-points has at least one stationary point. Our conditions as it will be seen later, may be considered a generalization of the ones given by Wolfe in [8].

Before stating our convergence theorem, we recall that for a twice continuously differentiable function we may express the Taylor formula for second-order expansions as

$$(2.1)\quad f(z_{i+1}) = f(z_i) - \lambda_i \rho_i \|\nabla f(z_i)\|\,\|h_i\| + \lambda_i^2 \int_0^1 (1-t) h_i^T \nabla^2 f(z_i + t\lambda_i h_i) h_i \, dt$$

with

$$\rho_i = -h_i^T \nabla f(z_i) / \|h_i\|\,\|\nabla f(z_i)\|^{(1)}, \quad \nabla^2 f(z) = d^2 f(z)/\,dz^2 .$$

Defining ν_i and μ_i as

$$\lambda_i = \nu_i \rho_i \|\nabla f(z_i)\| / \|h_i\| ,$$

$$\mu_i = \|h_i\|^2 \Big/ \int_0^1 (1-t) h_i^T \nabla^2 f(z_i + t\lambda_i h_i) h_i \, dt,$$

we may write from (2.1)

$$(2.2)\qquad f(z_{i+1}) = f(z_i) - \nu_i \rho_i^2 \|\nabla f(z_i)\|^2 + \nu_i^2 \rho_i^2 \|\nabla f(z_i)\|^2 / \mu_i .$$

We shall often make use of (2.2).

From (2.2) we derive

$$(2.3)\qquad [f(z_{i+1}) - f(z_i)] / \|\nabla f(z_i)\|^2 = -\nu_i \rho_i^2 (1 - \nu_i/\mu_i).$$

The quantity $-\nu_i \rho_i^2 (1-\nu_i/\mu_i)$ or $[f(z_{i+1}) - f(z_i)]/\|\nabla f(z_i)\|^2$ will play a major role in our convergence theorems. We are now ready to prove

2.4 Theorem. Let $f(z)$ be twice continuously differentiable and let $\alpha \varepsilon A$ be an algorithm which constructs a sequence $\{z_i\}_{i=0}^{\infty}$. Assume that the level set $L[f(z_0)]$ is bounded and that

$$(2.5)\qquad \sum_{i=0}^{\infty} \nu_i \rho_i^2 (1-\nu_i/\mu_i) = +\infty ,$$

then

$$(2.6)\qquad \liminf_{i\to\infty} \|\nabla f(z_i)\| = 0.$$

---

(1) We shall assume throughout the paper $h_i^T \nabla f(z_i) \le 0$.

__Proof.__ We shall prove the theorem by contradiction. Let (2.6) be false. That implies the existence of a constant $k_1>0$ and an index i_1 such that for $i>i_1$

$$\|\nabla f(z_i)\| \geq k_1. \tag{2.7}$$

Further, there exists a $z^*\varepsilon R^n$ such that for $i=1,2,\ldots$

$$f(z^*) < \delta \leq f(z_i) \tag{2.8}$$

with $\delta = \lim f(z_i)$.

Indeed, if we could not find any z^* satisfying (2.8), then, necessarily each limit-point of the sequence $\{z_i\}$ is a minimum point and by the assumptions on $f(z)$ a stationary point.

By (2.7) and (2.8), we may easily see that there exist a constant $k_2>0$ and an index $i^*\geq i_1$ such that for $i>i^*$

$$\|\nabla f(z_i)\|^2 \geq k_2[f(z_i) - f(z^*)]. \tag{2.9}$$

From (2.2) we have

$$f(z_{i+1}) - f(z^*) = f(z_i) - f(z^*) - \nu_i\rho_i^2 \|\nabla f(z_i)\|^2 (1-\nu_i/\mu_i). \tag{2.10}$$

Combining (2.10) and (2.9) and taking for convenience $\omega_i = \nu_i\rho_i^2(1-\nu_i/\mu_i)$, we may write for $i>i^*$

$$f(z_{i+1}) - f(z^*) \leq f(z_i) - f(z^*) - k_2\omega_i[f(z_i) - f(z^*)], \tag{2.11}$$

that is,

$$f(z_{i+1}) - f(z^*) \leq [f(z_i) - f(z^*)](1-k_2\omega_i). \tag{2.12}$$

Finally, since for $i\geq i^*$

$$(1-k_2\omega_i) > 0, \tag{2.13}$$

from (2.12) we obtain for $i\geq i^*$

$$f(z_{i+1}) - f(z^*) \leq [f(z_{i^*}) - f(z^*)] \prod_{j=i^*}^{i} (1-k_2\omega_j). \tag{2.14}$$

By hypothesis we have

$$\sum_{j=i^*}^{\infty} k_2\omega_j = +\infty \tag{2.15}$$

from which we get

$$\sum_{j=i^*}^{\infty} -\log(1-k_2\omega_j) = +\infty; \tag{2.16}$$

that yields

$$\lim_{i\to\infty} -\log \prod_{j=i^*}^{i} (1-k_2\omega_j) = +\infty. \tag{2.17}$$

From (2.17) we may easily derive that

$$\lim_{i\to\infty} \prod_{j=i^*}^{i} (1-k_2\omega_j) = 0. \tag{2.18}$$

Recalling (2.14), by (2.18), we have

(2.19) $$\lim_{i\to\infty} f(z_{i+1}) - f(z^*) = 0.$$

Clearly, (2.19) contradicts (2.8).
As a consequence the theorem is established.

We now compare the convergence conditions given in 2.4 with the ones given by Wolfe. For convenience we summarize Wolfe's conditions as follows

2.20 Proposition. Let $f(z)$ be a twice continuously differentiable function, $\alpha \varepsilon A$ and the level set $L[f(z_0)]$ be bounded. Assume that α constructs a sequence $z_0, z_1, z_2, \ldots$ such that

(i) $$-h_i^T \nabla f(z_i) \geq \rho \|h_i\| \, \|\nabla f(z_i)\|, \qquad \rho \varepsilon (0,1)$$

(ii) the sequence $\{z_i\}$ contains a subsequence of serious steps.

Then, α either locates a stationary point in a finite number of iterations or any limit-point of $\{z_i\}$ is a stationary point.

Wolfe defines a 'serious' step as follows: first he distinguishes five different properties that a step size may have

(2.21)-(i) $f(z_{i+1}) - f(z_i) \leq -\eta$;

-(ii) $\|z_{i+1}-z_i\| \geq \eta \min\{1, -h_i^T \nabla f(z_i)/\|h_i\|\}$;

-(iii) $-h_i^T \nabla f(z_i^*)/\|h_i\| \leq -(1-\eta) h_i^T \nabla f(z_i)/\|h_i\|$ for some $z_i^* \varepsilon [z_i, z_{i+1}]$ such that $f(z)$ is non increasing from z_i^* to z_i;

-(iv) $f(z_{i+1}) - f(z_i) \leq -\eta h_i^T \nabla f(z_i) \|z_{i+1}-z_i\| / \|h_i\|$;

-(v) with z_i^* as in (iii), $f(z_{i+1}) \leq f(z_i^*)$,

where the quantity η is fixed, $0<\eta<1$, then a step size is called serious if one of these four sets of conditions is satisfied: (i) alone; (ii) and (iv); (iv) and (iii); or (iii) and (v).

The main difference between our approach for obtaining convergence conditions and Wolfe's one is that we consider conditions on the whole optimization process while Wolfe considers conditions on each iteration. Moreover, any algorithm which satisfies the assumptions of 2.20 satisfies those of 2.4. Indeed we could show that if for an algorithm 2.20-(i) and 2.20-(ii) hold, then

(2.22) $$|f(z_{i+1}) - f(z_i)| / \|\nabla f(z_i)\|^2 \geq k_1 > 0$$

for $i=0,1,2,\ldots$ and with k_1 constant.
For instance, if for a step size 2.21-(ii) and 2.21-(iv) hold, then

(2.23) $$\begin{aligned} f(z_{i+1}) - f(z_i) &\leq -\eta \|z_{i+1}-z_i\| \, \|\nabla f(z_i)\| \rho_i \\ &\leq -\eta^2 \rho_i \|\nabla f(z_i)\| \min\{1, \rho_i \|\nabla f(z_i)\|\}. \end{aligned}$$

From (2.23) if $\rho_i \|\nabla f(z_i)\| = \min\{1, \rho_i \|\nabla f(z_i)\|\}$ we can derive

$$|f(z_{i+1}) - f(z_i)| / \|\nabla f(z_i)\|^2 \geq \eta^2 \rho^2, \tag{2.24}$$

that is (2.22).

If $\min\{1, \rho_i \|\nabla f(z_i)\|\} = 1$ then

$$|f(z_{i+1}) - f(z_i)| / \|\nabla f(z_i)\| \geq \eta^2 \rho. \tag{2.25}$$

Since $\lim\sup \|\nabla f(z_i)\| < +\infty$, clearly (2.25) also implies (2.22).

The convergence conditions proposed by Ortega and Rheinboldt, by Polak and by Zangwill are different from the ones given in 2.4.

3. CONVERGENCE TO A MINIMUM POINT

In order to evaluate the convergence conditions given in 2.4, we shall consider the following question in this section: Does (2.5) need to hold for an algorithm which constructs a sequence converging to a minimum point?

To answer such a question, we need to prove the following lemma first

3.1 Lemma. Let z^* be any minimum point of a twice continuously differentiable function $f(z)$. Then for any sequence $\{z_i\}$ with $z_i \to z^*$ as $i \to \infty$ there exists a positive constant k such that for $i=1,2,\ldots$

$$\|\nabla f(z_i)\|^2 / [f(z_i) - f(z^*)] \leq k. \tag{3.2}$$

Proof. Assume the lemma to be false. Then there exists a sequence $\{x_i\} \subset \{z_i\}$ such that

$$\lim_{i\to\infty} \|\nabla f(x_i)\|^2 / [f(x_i) - f(z^*)] = +\infty. \tag{3.3}$$

Consider the sequence $\{\bar{x}_i\}$ defined as follows

$$\bar{x}_i = x_i - k_1 \nabla f(x_i) \tag{3.4}$$

where k_1 is a positive constant satisfying the relation

$$k_1 < 2/M \tag{3.5}$$

where $M = \sup_{z \in B(z^*)} \|\nabla^2 f(z)\|$ with $B(z^*)$ any neighborhood of z^* containing $\{z_i\}$.

We have

$$\begin{aligned} f(\bar{x}_i) - f(x_i) &= -k_1 \|\nabla f(x_i)\|^2 + k_1^2 \int_0^1 (1-t) \nabla f(x_i)^T \nabla^2 f(\cdot) \nabla f(x_i) dt \\ &\leq -k_1 \|\nabla f(x_i)\|^2 + \tfrac{1}{2} k_1^2 M \|\nabla f(x_i)\|^2 \\ &= -k_1 (1 - \tfrac{1}{2} k_1 M) \|\nabla f(x_i)\|^2. \end{aligned} \tag{3.6}$$

From (3.6) and (3.3) we may write

$$\lim_{i\to\infty} \sup [f(x_i) - f(\bar{x}_i)] / [f(x_i) - f(z^*)] \geq \tag{3.7}$$

$$\geq \lim_{i\to\infty} \sup k_1(1-\tfrac{1}{2}k_1M)\, \|\nabla f(x_i)\|^2/[f(x_i) - f(z^*)]$$

$$= +\infty.$$

On the other hand, since z* is a local minimum, there exists an index $\bar{i}$ such that for $i>\bar{i}$

(3.8) $$f(\bar{x}_i) \geq f(z^*)$$

which gives

(3.9) $$f(x_i) - f(\bar{x}_i) \leq f(x_i) - f(z^*).$$

Clearly (3.9) and (3.7) yield a contradiction.
As a consequence the proof is completed.

We are now able to prove

<u>3.10 Theorem.</u> Let f(z) be twice continuously differentiable, then an algorithm $\alpha\varepsilon A$ constructs a sequence $\{z_i\}_{i=0}^{\infty}$ converging to a minimum z* only if

(3.11) $$\sum_{i=1}^{\infty} \nu_i\rho_i^2\,(1-\nu_i/\mu_i) = +\infty.$$

<u>Proof.</u> From (2.2) and making use of (3.2), we may write

(3.12) $$\begin{aligned} f(z_{i+1}) - f(z^*) &= f(z_i) - f(z^*) - \nu_i\rho_i^2\, \|\nabla f(z_i)\|^2(1-\nu_i/\mu_i) \\ &= [f(z_i) - f(z^*)][1 - \nu_i\rho_i^2(1-\nu_i/\mu_i)\|\nabla f(z_i)\|^2/ \\ &\qquad /(f(z_i) - f(z^*))] \\ &\geq [f(z_i) - f(z^*)][1 - k\nu_i\rho_i^2(1-\nu_i/\mu_i)]. \end{aligned}$$

Assume now that (3.11) does not hold; that is

(3.13) $$\sum_{i=1}^{\infty} \omega_i = k_1 < \infty,$$

with $\omega_i = \nu_i\rho_i^2(1-\nu_i/\mu_i)$.
Necessarily, we have

(3.14) $$\omega_i \to 0 \quad \text{as} \quad i\to\infty$$

and there exists a suitable index i* such that for i>i*

(3.15) $$1 - k\omega_i > 0.$$

Hence, we may write for i>i*

(3.16) $$f(z_{i+1}) - f(z^*) \geq [f(z_{i*}) - f(z^*)] \prod_{j=i*}^{i} (1 - k\omega_j).$$

Further (3.13) implies

(3.17) $$\prod_{i=i*}^{\infty} (1 - k\omega_i) = k_2 > 0.$$

By (3.16) and (3.17) we can say that $\{z_i\}$ does not converge to z*. Obviously such a result contradicts our hypothesis; hence the theorem is proved.

As a consequence of 3.10, we may answer the initial question of this section positively.

4. CONVERGENCE TO A SADDLE POINT

In this section we shall consider the same question as the previous section, but for an algorithm converging to a saddle point. We shall answer the question negatively. Indeed we will give an example which shows that an algorithm may construct a sequence converging to a saddle point without (3.11) being satisfied. Consider the following function

$$f(z) = x^2 - y^2 \tag{4.1}$$

which has a saddle point at $z^*=(0,0)$.

Consider now an algorithm α which operates as in Table 4.2

Iteration	x	y
1	1+1	1
2	$1/2 + 1/2^2$	1/2
⋮	⋮	⋮
i	$1/i + 1/i^2$	1/i
1+i	$1/(1+i) + 1/(1+i)^2$	1/(1+i)

Table 4.2

We have

$$f(z_{i+1}) = 1/(1+i)^4 + 2/(1+i)^3 < 1/i^4 + 2/i^3 = f(z_i). \tag{4.3}$$

Hence α is a descent algorithm.

We now compute the quantity

$$\sum_{i=1}^{\infty} [f(z_i) - f(z_{i+1})]/ \|\nabla f(z_i)\|^2. \tag{4.4}$$

From Table 4.2 we compute

$$\begin{aligned} f(z_{i+1}) - f(z_i) &= 1/(1+i)^4 + 2/(1+i)^3 - 1/i^4 - 2/i^3 \\ \|\nabla f(z_i)\|^2 &= (2/i + 2/i^2)^2 + 4/i^2. \end{aligned} \tag{4.5}$$

Hence, we get

$$[f(z_i) - f(z_{i+1})]/ \|\nabla f(z_i)\|^2 = [1/i^4 + 2/i^3 - 1/(1+i)^4 - 2/(1+i)^3]/ /[8/i^2 + 4/i^4 + 8/i^3] \tag{4.6}$$

and after simple computations

$$\lim_{i\to\infty} i^2[f(z_i) - f(z_{i+1})]/\|\nabla f(z_i)\|^2 = 3/4. \tag{4.7}$$

Finally, since $\sum 1/i^2 = k < +\infty$, we may write

$$\sum_{i=1}^{\infty} [f(z_i) - f(z_{i+1})]/\|\nabla f(z_i)\|^2 = \bar{k} < +\infty . \tag{4.8}$$

Therefore, we may conclude this section by stating that for an algorithm which constructs a sequence converging to a saddle point, the relation

$$\sum_{i=1}^{\infty} [f(z_{i+1}) - f(z_i)]/\|\nabla f(z_i)\|^2 = +\infty \tag{4.9}$$

does not need to hold.

5. CONVERGENCE FOR UNIFORMELY CONVEX FUNCTIONS

In this section, we shall be concerned with convergence conditions for algorithms which minimize uniformely convex functions. An immediate result may be obtained combining Theorems 2.4 and 3.10.

We have

5.1 Theorem. Let $f(z)$ be twice continuously differentiable and uniformely convex. Then, an algorithm $\alpha\varepsilon A$ either converges in a finite number of iterations to the unique minimum z^* of $f(z)$ or converges to z^* if and only if

$$\sum_{i=1}^{\infty} \upsilon_i\rho_i^2(1-\upsilon_i/\mu_i) = +\infty . \tag{5.2}$$

The sufficient convergence condition of 5.1 may be seen as a generalization of a result due to Zoutendijk and reported in page 48 of ref.[6]. Zoutendijk under the same assumptions on $f(z)$ as 5.1 proved that an algorithm for which $\sum\rho_i^2 = +\infty$ and for which the step sizes λ_i are chosen such that $f(z)$ is exactly minimized along h_i, constructs a sequence $\{z_i\}$ converging to the unique minimum.

From 5.1 we now derive a necessary convergence condition on the search directions h_i, no matter how the step sizes λ_i have been chosen.

5.3 Theorem. Let the assumptions of 5.1 hold. Then, an algorithm $\alpha\varepsilon A$ either converges in a finite number of iterations or converges only if

$$\sum_{i=1}^{\infty} \rho_i^2 = +\infty . \tag{5.4}$$

Proof. From (2.2) we can write

$$f(z_{i+1}) \geq f(z_i) - \upsilon_i\rho_i^2 \|\nabla f(z_i)\|^2 + \tfrac{1}{2}m\upsilon_i^2\rho_i^2 \|\nabla f(z_i)\|^2 \tag{5.5}$$

with m a positive constant such that

$$m \leq \min\{\|\nabla^2 f(z)\| \mid z\varepsilon R^n\}.$$

From (5.5) we derive

(5.6) $$-\upsilon_i\rho_i^2 \|\nabla f(z_i)\|^2 + \tfrac{1}{2}m\upsilon_i^2\rho_i^2 \|\nabla f(z_i)\|^2 \leq 0$$

and

(5.7) $$\upsilon_i \leq 2/m.$$

Further, we obtain by the definition of υ_i, μ_i and since f(z) is uniformely convex,

(5.8) $$0 \leq \upsilon_i/\mu_i \leq 1$$

and

(5.9) $$0 \leq 1-\upsilon_i/\mu_i \leq 1.$$

We now recall by 5.1 that

(5.10) $$\sum_{i=1}^{\infty} \upsilon_i\rho_i^2(1-\upsilon_i/\mu_i) = +\infty$$

must hold for a converging algorithm. Clearly, (5.9) and (5.7) combined with (5.10) imply

(5.11) $$\sum_{i=1}^{\infty} \rho_i^2 = +\infty .$$

The proof is therefore completed.

Results similar to those of 5.3 have been given by Lenard [4] and Wolfe [8]. In Lenard's paper (5.4) is proved to be a necessary convergence condition under the hypothesis that $(1-\theta_i^2) \geq c$, with c constant and θ_i defined by

(5.12) $$\theta_i = h_i^T\nabla f(z_{i+1})/h_i^T\nabla f(z_i);$$

moreover, θ_i is assumed positive.
The latter condition is equivalent to requiring that the approximation $z_i+\lambda_i h_i$ to the minimum along h_i lies between the starting point and the location of the exact minimum.
In Wolfe's paper, (5.12) is proved to be a necessary convergence condition assuming that $\{z_i\}$ is a Curry sequence; that is, it is assumed that λ_i gives the exact minimum along h_i.
We conclude this section by giving an example which shows that, when the function to be minimized is not uniformely convex, then, the necessary convergence condition of 5.1, $\sum\rho_i^2 = +\infty$, does not need to be satisfied.
Consider the problem of minimizing the following function

(5.13) $$f(z) = x^2 + y^4.$$

Clearly $z^*=(0,0)$ is the unique minimum of f(z); moreover f(z) is convex but not uniformely convex.

Consider now an algorithm α which starting from $z_1=(1,1)$ constructs the sequence in Table 5.14

Iteration	x	y
1	1	1
2	1/4	1/2
3	1/9	1/3
⋮	⋮	⋮
i	$1/i^2$	1/i
i+1	$1/(i+1)^2$	1/(i+1)

Table 5.14

We have

(5.15) $$\nabla f(z_i) = \begin{bmatrix} 2/i^2 \\ 4/i^3 \end{bmatrix}, \quad h_i = \begin{bmatrix} 1/(i+1)^2 - 1/i^2 \\ 1/(i+1) - 1/i \end{bmatrix}.$$

Since

(5.16) $$f(1/(i+1)^2, 1/(i+1)) = 2/(i+1)^4 < 2/i^4 = f(1/i^2, 1/i),$$

α is a descent algorithm; moreover, since $z_i \to (0,0)$ as $i\to\infty$, α is convergent. Compute

(5.17) $$\begin{aligned}\rho_i &= - h_i^T \nabla f(z_i)/\,\|h_i\|\,\|\nabla f(z_i)\| \\ &= - [2(1/(i+1) - 1/i)(1/(i+1) + 1/i)/i^2 + 4(1/(i+1) - 1/i)/i^3]/ \\ &\quad / [4/i^4 + 16/i^6]^{\frac{1}{2}}[(1/(i+1) - 1/i)^2(1/(i+1) + 1/i)^2 + (1/(i+1) - 1/i)^2]^{\frac{1}{2}} \\ &= [1/(i+1) + 3/i]/[1 + 4/i^2]^{\frac{1}{2}}[(1/(i+1) + 1/i)^2 + 1]^{\frac{1}{2}}.\end{aligned}$$

We can now easily verify that

(5.18) $$\lim_{i\to\infty} \rho_i^2/(1/i^2) = 16.$$

Because $\sum 1/i^2 = k < \infty$, we get by (5.18)

(5.19) $$\sum_{i=1}^{\infty} \rho_i^2 = \bar{k} < \infty .$$

Consequently, though α is a convergent algorithm, $\sum\rho_i^2 = +\infty$ does not hold.

ACKNOWLEDGMENTS

The author wishes to thank L.W.C.Dixon, for many helpful discussions and valuable suggestions.

This research was carried out while the author was supported by the CNR, at the Hatfield Polytechnic, N.O.C. .

REFERENCES

[1] M.Gaviano: "Some new results on the convergence of an algorithm in unconstrained minimisation", Bollettino U.M.I. (5), 13-A, (1976), 647-650.

[2] M.Gaviano, G.D'Ambra: "Weak conditions for the convergence to a stationary point of an algorithm in unconstrained minimisation", Rendiconti del Seminario della Facolta' di Scienze dell'Universita' di Cagliari (1977) (to appear).

[3] M.Gaviano: "Remarks on the convergence of algorithms in unconstrained minimisation", N.O.C. Technical Report No. 87, Hatfield Polytechnic (1977).

[4] M.Lenard: "Practical convergence condition for unconstrained Optimisations", Math. Programming 4 (1973), 309-323.

[5] J.Ortega, W.Rheinboldt: "Iterative solution of nonlinear equations is Several variables", Academic Press 1970.

[6] E.Polak: "Computational methods in optimisation", Academic Press 1971.

[7] W.Zangwill: "Nonlinear Programming", Prentice Hall 1969.

[8] P.Wolfe: "Convergence conditions for ascent methods", Siam Review 11 (1969), 226-235.

[9] P.Wolfe: "Convergence conditions for ascent methods II: Some corrections", Siam Review 13 (1971), 185-188.

TOWARDS GLOBAL OPTIMISATION 2
L.C.W. Dixon and G.P. Szegö (eds.)

SOME SAFEGUARDS FOR DESCENT MINIMIZATION ALGORITHMS TO AVOID NUMERICAL NON CONVERGENCE

By G. Resta, University of Genova, ITALY
and C. Sutti, University of Parma, ITALY

ABSTRACT

In this paper we present some safeguards which may be introduced in a descent minimization algorithm. These safeguards guarantee numerical convergence on a very wide class of objective function. This class includes non convex functions which have linear segments on level surfaces. In practice these safeguards are sufficient to stop any such algorithm after a finite number of iterations. The ultimate point will either be an approximate stationary point or a vertex of the objective function.

INTRODUCTION

This paper investigates the convergence of a minimization procedure given by a difference equation $x_{n+1} = x_n + \lambda_n p_n$, where p_n is some unit search direction and λ_n is the step length.

In paper (2) a set of necessary and sufficient conditions is given which guarantee that $\lim_i \lambda_i = 0$.

Some of the conditions concern the objective function, the others the step lengths.

The condition on the step length is that two non negative scalars ε and θ must exist, such that the monodimensional function $\theta(\lambda) = \varphi(x_n + \lambda p_n)$ must satisfy the following inequality 1,1 , for a non-finite number of iterations.

1.1 $$\theta(0) \geq \theta(\lambda) \geq \theta(\lambda_n) \qquad \forall \lambda \in [0, \varepsilon\lambda_n]$$

The line search can however be more arbitrary provided that the step is less than or equal to θ times the last step obtained satisfying the above mentioned condition.

These conditions are not restrictive at all, because they permit the actual point generated by the numerical procedure to skip from a connected component of a level set to another component, if the function to be minimized is not convex.

In Fig. 1 the indications of condition 1.1 in terms of a graph of the mono-dimensional function $\theta(\lambda)$ are shown. Namely the couple $(\lambda,\theta(\lambda))$ is compelled to be contained in the rectangle $\forall\lambda\in[0,\varepsilon\lambda_n]$.

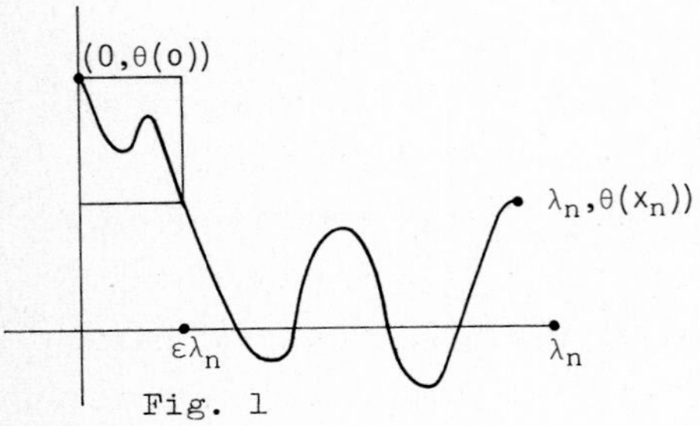

Fig. 1

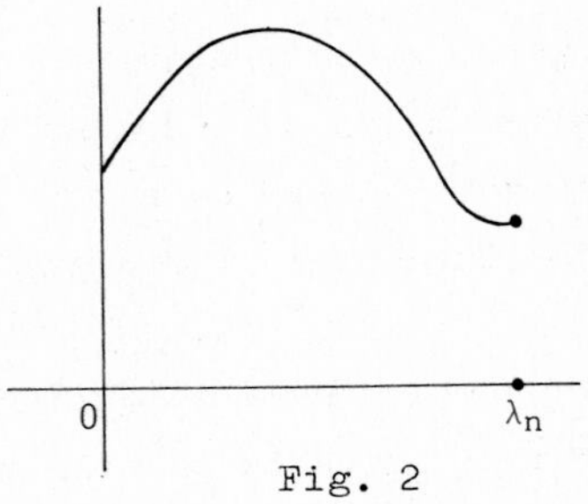

Fig. 2

In Fig. 2 a monodimensional function which does not satisfy condition 1.1 is presented. Such a step is still allowable provided that it is less than or equal to the last one satisfying 1.1.

The class of functions considered in (2) is very large, for instance it contains non-convex and non-differentiable functions, but it does not contain those functions which have linear segments on some of their level sets.

In Fig. 3 some allowable shapes for a level surface $\{x\in\mathbb{R}:\varphi(x) = \theta\}$ for a function considered in (2) are presented:

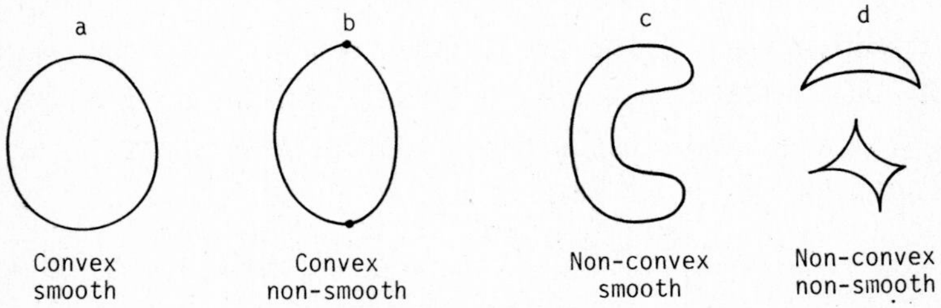

Fig.3

In Fig. 4 a non allowable shape is presented:

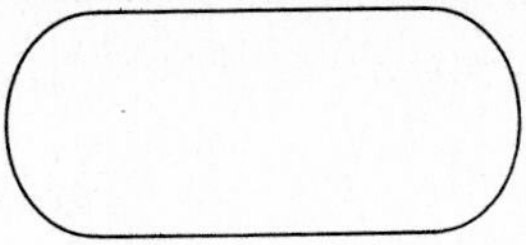

Fig. 4

This paper is devoted to a discussion of the possibility of transferring these properties required of the objective function to be minimized into properties of an algorithm.

In other words we intend to introduce in an algorithm some safeguards and some stopping criteria which ensure that whenever the objective function is continuous the algorithm stops after a finite number of steps, producing in any case a point which is, within a certain accuracy a local minimum.

It is not guaranteed that the ultimate point lies in an ε-neighborhood of a local minimum, but it is warranted that, if the function is differentiable the gradient g^* at the ultimate point is in norm smaller than ε.

This is not a disadvantage of the proof. It is strictly related to tha fact that no condition at all is imposed on the objection function.

It is sufficient to consider the example of a function having no stationary point to see that the result is the best result we can obtain. Consider for instance $\varphi : \mathbb{R} \to \mathbb{R}$ with $\varphi(x) = e^x$. Any algorithm provided with the proposed safeguards would stop at a point in which the derivative of the function is small.

Certainly this result is not surprising because any algorithm provided with a stopping criterion on the norm of the gradient would stop in such a point, however an analogous result is proved for non differentiable functions and for algorithms which do not require the evaluation of the gradient.

Furthermore when no condition is required on the objective function very strange level surfaces are allowed and all these cases are examined in detail and through this examination it becomes evident why and when the proposed safeguards are necessary and useful.

These safeguards are very simple to introduce in a direct search algorithm, but some of them are difficult to introduce in a method using the gradient. We also examine the maximal class of function, on which an algorithm not possessing some of these safeguards, will certainly converge.

From the above introduction it becomes evident that a number of results are presented and it is therefore necessary for them to be ordered in some way.

The order we have choosen for their presentation is the one of progressive enlargement of the class of minimizable objective functions It is also the order in which the number of conditions and safeguards which the algorithm must contain, increases.

The first condition we shall introduce is the following:

(a.5) If at some iteration $i, \lambda_i < \delta$, where δ is a prefixed precision, stop.

This condition guarantees that if the original algorithm generates a sequence such that $\text{minlim}_i \lambda_i = 0$, than the applied algorithm surely stops after a finite number of iterations.

2. DEFINITIONS

Let φ be a continuous function

Let Ω be the class of all the lower bounded functions which satisfy the following condition:

2.1 For any pair of sequences $\{x_n\}_{n\in\mathbb{N}}, \{y_n\}_{n\in\mathbb{N}}$ such that $\varphi(x_n) \geq \varphi(y_n) \geq \varphi(x_{n+1})$ $\forall n \in \mathbb{N}$ and, such that $\text{minlim}_n \|y_n - x_n\| > 0$; a sequence of scalars $\{\mu_n\}_{n\in\mathbb{N}}$ exists, such that $\mu_n \in (0,1)$ $\forall n \in \mathbb{N}$ and $\lim_n \varphi(x_n + \mu_n (y_n - x_n)) \neq \lim_n \varphi(x_n) = \lim_n \varphi(y_n)$.

Condition 2.1 is certainly not satisfied by a function having linear segments on some of its level surfaces.

To clarify this fact consider the following example (Fig. 5):

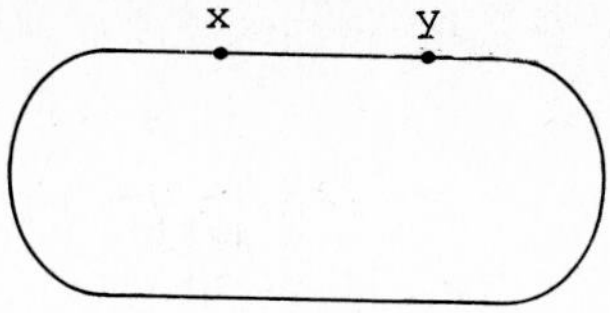

Fig. 5

Consider two sequences $\{x_n\}_{n\in\mathbb{N}}$ and $\{y_n\}_{n\in\mathbb{N}}$, such that respectively $\lim_n x_n = x$ and $\lim_n y_n = y$, and such that $\varphi(x_n) \geq \varphi(y_n) \geq \varphi(x_{n+1})\ \forall n\in\mathbb{N}$, evidently for any sequence of scalars $\{\mu_n\}_{n\in\mathbb{N}}$ such that $\mu_n \in (0,1)$
$\lim_n \varphi(x_n + \mu_n (y_n - x_n)) = \varphi(x) = \lim_n \varphi(x_n) = \varphi(y) = \lim_n \varphi(y_n)$.

Condition 2.1 can however also not be satisfied by functions which do not have any linear segments on any of their level surfaces.

Consider for instance the following example:
Set $\varphi: \mathbb{R}^2 \to \mathbb{R}$ be defined by $\varphi(x) = \cos(\|x\|)$.

The surface of level -1, that is the $\{x\in\mathbb{R}^2 : \varphi(x) = -1\}$ is the union of all the circumferences of radius $\pi + 2K\pi$, $K\in\mathbb{N}$, and evidently does not contain any linear segments.

We can however generate a sequence which does not satisfy condition 2.1 as explained by the following and illustrated in Fig. 6.

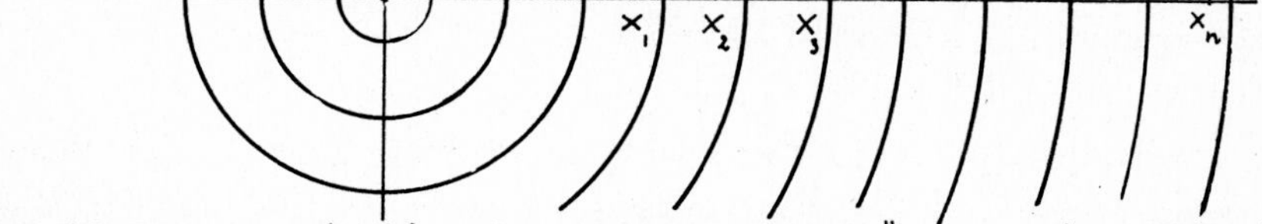

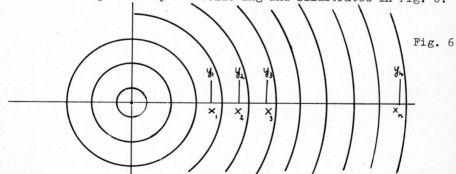

Fig. 6

Provided that $\varphi(x_n) \geq \varphi(y_n) \geq \varphi(x_{n+1})$ and $\|y_n - x_n\| = 1\ \forall n\in\mathbb{N}$, and that $\lim_n \varphi(x_n) = -1$, $\lim_n \|x_n\| = +\infty$, it is certainly true that $\forall \{\mu_n\}_{n\in\mathbb{N}}$, such that $\mu_n \in (0,1)$ $\lim_n \varphi(x_n + \mu_n (y_n - x_n)) = -1$.
In this example though the objective function has no linear segments on level surfaces, it has curved segments where the

radius of curvature tends to infinity.

In the following it will become evident that an algorithm which is not provided with safeguard 1, can perform very badly if applied to a function which does not satisfy 2.1.

Let α be an iterative downhill algorithm, i.e. an algorithm satisfying the following conditions:

(a.1) $x_{n+1} = x_n + \lambda_n p_n$ (where $\lambda_n \in \mathbb{R}^+$ is the step length and $p_n \in \mathbb{R}^m$ is some unit direction)

(a.2) $\varphi(x_{n+1}) \leq \varphi(x_n)$

Let Φ be the set of all the algorithms which satisfy the following conditions:

(a.3) $\exists \{n_k\}_{k\in\mathbb{N}}, \varepsilon \in (0,1]$, such that
$\varphi(x_{n_k}) \geq \varphi(x_{n_k} + \lambda p_{n_k}) \geq \varphi(x_{n_k} + \lambda_{n_k} p_{n_k}), \forall \lambda \in (0, \varepsilon\lambda_{n_k}), \forall k \in \mathbb{N}$.

(a.4) $\exists \eta \in \mathbb{R}^+$ such that if $i \in \mathbb{N}$ and $j = \max\{k$ such that $n_k \leq i\}$, $\lambda_i \leq \eta \lambda_{j'}$.

(a.5) $\exists \delta \in \mathbb{R}^+$ such that if $\lambda_n \leq \delta$, the algorithm stops at the n-th iteration.

In section 4) we will need the following definitions.

Def.$_1$ A triplet (α, φ, x_1) (where α is an algorithm, φ is a function, x_1 a point belonging to $\mathbb{R}^m$) is numerically convergent if the sequence generated by the algorithm α when it is applied to the function φ starting at point x_1 , is such that $\operatorname{minlim}_n \lambda_n = 0$; it is numerically non convergent if the generated sequence is such that $\operatorname{minlim}_n \lambda_n = \psi > 0$.

If φ is a continuous function we give the following definitions:

If $\rho \in \mathbb{R}$ $H(\rho) = \{x \in \mathbb{R}^m : \varphi(x) = \rho\}$

Def.$_2$ A constancy direction for φ with respect to ρ is a direction q such that $\|q\| = 1$, $\exists x \in H(\rho), \exists \mu \in \mathbb{R}^t$:
$\emptyset(\lambda) = x + \lambda q \in H(\rho)$ $\forall \lambda \in [0,\mu]$.

Def.$_3$ $Q(\varphi,\rho)$ is the set of all the constancy directions for φ in respect to ρ , i.e. a subset of the unit-sphere in $\mathbb{R}^m$.

Def.$_4$ A vertex for φ with respect to ρ is a point z such that: $\exists \mu \in \mathbb{R}^+$ and $\exists (q, q_2, \ldots, q_m)$ such that $q_i \in Q(\varphi, \rho)$ the q_i are linearly indipendent vectors and $y(\lambda, i) = z + \lambda q_i \in (\rho)$ $\forall \lambda \in [0, \mu]$, $i=1, \ldots, m$.

In Fig. 7 we illustrate a level surface of a function which has no vertices but is such that the vectors belongings to $Q(\varphi, \rho)$ span all the space.

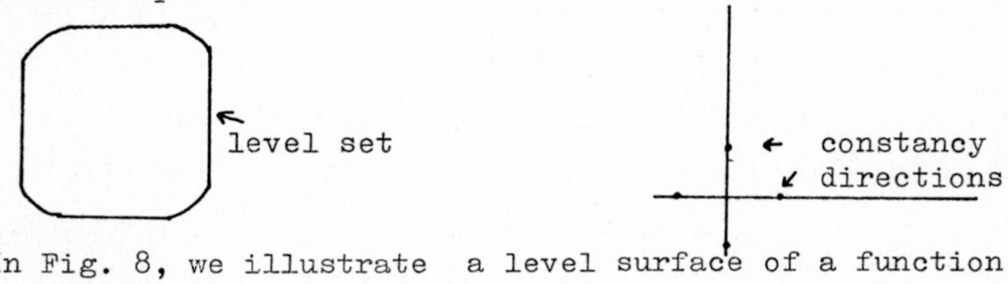

Fig. 7

In Fig. 8, we illustrate a level surface of a function φ having a vertex z.

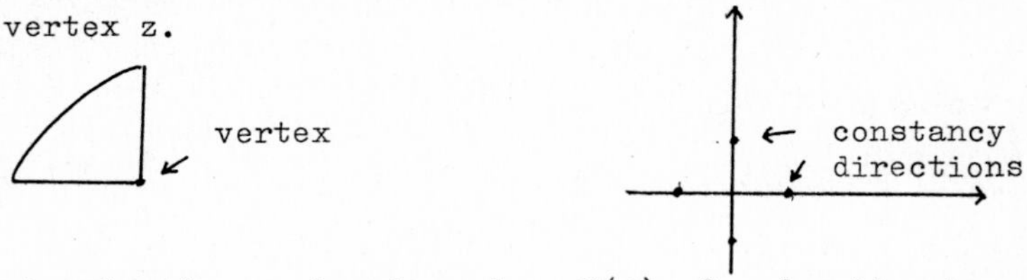

Fig. 8

Obviously if some level surface $H(\rho)$ of a function has an interior part then there are infinitely many vertices.
In fact let $z \in \mathring{H}(\rho)$, then $\exists \mu \in \mathbb{R}^+ = S(z, \mu) C \mathring{H}(\rho)$, i.e. $\forall p \in S(0,1)$, $\varphi(z + \lambda p) = Q(z), \forall \lambda \in (0, \mu)$.

We note that the definitions Def.$_2$, Def.$_3$, Def.$_4$ concern objective functions having linear segments on some of their level surfaces of level ρ, i.e. functions which do not satisfy the following 2.2.

2.2 $\exists \tau \in \mathbb{R}$: $\forall \mu \le \tau$ if $\varphi(x) = \varphi(y) = \mu$, there exists $\lambda \in \mathbb{R}^+$ such that $\varphi(x + \lambda(y - x)) \neq \mu$.
Remember however that we have shown above that a function can satisfy condition 2.2 and not satisfy 2.1, for instance $\varphi(x) = \cos(\|x\|)$.

For functions not satisfying condition (2.1) we give the following definitions:

Def.$_5$ A pseudoconstancy direction for φ in respect to ρ is a direction q such that $\|q\| = 1$, and

$\exists \{x_n\}_{n\in\mathbb{N}}, \{q_n\}_{n\in\mathbb{N}}, \exists \mu \in \mathbb{R}^+$ such that $\lim_n \varphi(x_n + \lambda q_n) = \lim_n \varphi(x_n) = \rho$, $\forall \lambda \in (0,\mu)$.

Def.$_6$ $\underline{Q}$ is the set of the pseudo constancy directions for φ in respect to ρ.

Def.$_7$ A "pseudovertex sequence" for φ with respect to ρ is a sequence $\{x_n\}_{n\in\mathbb{N}}$ such that: $\lim \varphi(x_n) = \rho$, $\exists (q_1, q_2 \ldots q_m)$, $q_i \in \underline{Q}$ for $i = 1,2,\ldots m$, is linearly independent such that:

$\exists \mu \in \mathbb{R}^+ : \lim_n \varphi(x_n + \lambda q_i) = \lim_n \varphi(x_n), \forall \lambda \in (0,\mu)$, $i = 1,2,\ldots m$.

To give an example of $\underline{Q}(\varphi,\rho)$ we can observe that if $\varphi(x) = \cos(\|x\|)$, $\underline{Q}(\varphi,\rho)$ is equal to all the unit-sphere, $\forall \rho \in [-1,1]$. Furthermore we can observe that no pseudovertix sequence existsfor that function and for every $\rho \in [-1,1]$.

3. A TEST ON THE OBJECTIVE FUNCTION

In this paragraph we shall examine the shape of the level surfaces of the objective function to test whether or not the level surfaces contain linear segments.

Linear objective functions obviously have linear segments in their level surfaces. But twice differentiable and no objective functions exist, which contain linear segments on their level surfaces. Examples of such functions are presented in (1).

For twice differentiable functions we introduce the following Theorem 1 that provides a technique to test the presence of linear segments on the level surfaces.

Let $\varphi : \mathbb{R}^m \to \mathbb{R}$ be a twice differentiable function

$g : \mathbb{R}^m \to \mathbb{R}^m$ be its gradient

$G : \mathbb{R}^m \to \mathbb{R}^{m\times m}$ be its Hessian matrix.

Consider furthermore the following conditions:

(3.1) $B = \{x \in A : \det(G(x)) = 0\}$ is a set of isolated points

(3.2) $\forall \theta$ the surface $H(\theta)$ of level θ does not contain linear segments.

THEOREM 1

Condition (3.1) implies (3.2), that is if the set of the points where G(x) is singular consists only of isolated points, the level surfaces of the objective function cannot have linear segments.

Proof: we procede by contradiction assuming (3.2) is not verified. So (3.3) $\exists\, x,y : \varphi(x) = \varphi(y) = \varphi(x + \lambda(y - x)) = \ < \tau,\ \forall\lambda\in(0,1)$
Let $d = y - x \neq 0$. Evidently
(3.4) $\forall\lambda\in[0,1],\ g(x + \lambda d)^T d = 0$. (Because the function is constant along d)
Because of condition (3.1) in the segment joining x and y, there are at most a finite number of points belonging to B.

In this eventuality let us restrict our interval in such a way that it contains no point of B. Then for the Taylor expansion of φ
(3.5) $\varphi(x + \lambda d) = \varphi(x) + \lambda g(x)^T d + \lambda^2/2\ d^T G(x + \mu d)d\ ,\ \mu\in(0,\lambda)$
But for (3.3), (3.4) and (3.5), $d^T G(x + \mu d)d = 0$, then $(x + \mu d)\in B$

Q.E.D.

As we can see from the proof of Theorem 1, the eventual segments along which φ is constant are contained in B.

Therefore, if φ is twice differentiable, in order to see if condition (3.2) is satisfied we can proceed as follows:
i) determine B
ii) if B is a set of isolated points, (3.2) is certainly satisfied.
iii)if B contains some segments we can test if there the function is constant.

We admit that this procedure is not in general a real advance because it implies an a priori analysis of the objective function.

As we have seen in this paragraph it is not very simple to test if a function φ satisfies condiction (2.2); it is therefore obvious that to test if a function φ satisfies condition (2.1) is even more difficult because it should involve the study of the behaviour of function φ at the infinity.

However the class of function Ω is very interesting because in (2) it is proved that whenever we apply an algorithm satisfying (a.1), (a.2), (a.3) and (a.4) to a function φ belonging to Ω starting at a point such that $\varphi(x_1) \leq \tau$, the generated sequence of steps is such that $\lim_n \lambda_n = 0$.

This fact implies that the triplet (α, φ, x_1) is numerically convergent.

We can observe that if $\alpha \in \Phi$ the fact that a triplet (α, φ, x_1) is numerically convergent implies (because of (a.5)) that after a finite number of iterations the algorithm α stops.

From now on we will be mainly interested in the weak property of convergence related to the fact that an algorithm stops after a finite number of iterations.

In the following sections we will transfer the conditions required on the objective function into conditions required on the algorithm so as to obtain more sophisticated algorithms which converge whenever the objective function has a lower bound.

4. NUMERICAL SAFEGUARDS

Let φ be a lower bounded continuous function and $\alpha \in \Phi$.
Suppose we have applied the algorithm α to φ starting at a point $x_1 \in \mathbb{R}^m$ and we have obtained a sequence $\{x_n\}_{n \in \mathbb{N}}$.

Because φ is lower bounded and α satisfies (a.2) the sequence of real numbers $\{\varphi(x_n)\}_{n \in \mathbb{N}}$ is convergent being not increasing and lower bounded.

Let $\rho = \lim_n \varphi(x_n)$ and $\underline{Q}$ the set of the pseudoconstancy directions of φ with respect to ρ.
Now we present theorem 2, concerning functions not belonging to Ω.

THEOREM 2

A necessary condition for numerical non convergence of α is that all the cluster points of the sequence of the search directions $\{p_{n_k}\}_{k\in\mathbb{N}}$ are contained in $\underline{Q}$.

Proof

Suppose the triplet (α,φ,x_1) is numerically non convergent and let p^* a cluster point of the sequence $\{p_{n_k}\}_{k\in\mathbb{N}}$.

This fact implies that there exists a sequence $\{q_j\}_{j\in\mathbb{N}}$ with $q_j=p_{n_{k_j}}$ $\forall j\in\mathbb{N}$ and $\lim_j q_j = p^*$.

Let $\{y_j\}_{j\in\mathbb{N}}$ with $y_j = x_{n_{k_1}}$ $\forall j\in\mathbb{N}$.

Let $\mu=\varepsilon\psi$, where $\varepsilon\in(0,1]^j$ is the real number which appears in condition (a.3) and $\psi = \text{minlim}_k \lambda_{n_k}$, evidently $\mu\in\mathbb{R}^+$.

But this means that p^* is a pseudoconstancy direction for φ with respect to ρ, in fact the sequences $\{y_j\}_{j\in\mathbb{N}}$ and $\{q_j\}_{j\in\mathbb{N}}$ and the positive real number μ are such that they satisfy Def.$_5$, because the fact that the algorithm satisfies condition (a.3) implies that $\forall\lambda\in(0,\mu)$ $\lim \varphi(y_j + \lambda q_j) = \rho$. Q.E.D.

Theorem 2 implies that it is sufficient to introduce conditions on the search directions to obtain convergence for functions which do not satisfy condition (3.2).

We note that no hypothesis on the search directions has been introduced until now.

One reasonable suggestion is to use a sequence of m linearly independent vectors as search directions.

Let us define $\mathcal{Q}$ as the class of all the algorithms belonging to Φ and such that they satisfy the following conditions:

(a.6) There exists a subsequence $\{n_{k_j}\}_{j\in\mathbb{N}}$ of the sequence $\{n_k\}_{k\in\mathbb{N}}$ defined in (a.3), such that $n_{k_j+j'} = n_{k_{j+j'}}$, $j' = 1,2,\dots,m$ and $p_{n_{k_j+j'}}$, $j' = 1,2,\dots,m$ are linearly independent vectors.

Clearly to test if an algorithm satisfies condition (a.6) is not very difficult, while for some algorithms it is nearly impossible to test if they satisfy (a.7).

For instance the univariate search method satisfies condition (a.6).

As a simple consequence of theorem 2 any algorithm belonging to $\mathcal{A}$ generates a finite sequence of points whenever it is applied to a lower bounded function satisfying the following condition:
(4.1) $\forall \rho \in \mathbb{R}$ the set $\underline{Q}$ of the pseudoconstancy directions of φ with respect to ρ does not span all the space $\mathbb{R}^m$.

If we consider again the example in (1), in each case the objective function φ does not satisfy (4.1). So even if the minimization algorithm employed belongs to $\mathcal{A}$, numerical non convergence can happen.

Let us now impose a less restrictive condition on the lower bounded objective function:
(4.2) No pseudovertex sequence exists.

In paper (3) some search paths for functions not satisfying (4.1); but satisfying (4.2), have been examined and it has been shown that it is possible to avoid numerical convergence of the univariate search method by changing the order of the search directions.

As an extension of this result for a more general method we introduce the following condition:
(a.8) $\exists \delta_1 \in \mathbb{R}^+, \delta_2 \in \mathbb{R}^+$ such that if $\varphi(x_{j'}) - \varphi(x_{j'+1}) < \delta_2$, but $\lambda_{j'} > \delta_1$, we reset $\lambda_{j'} = 0$. The (a.8) is an extension of the results in (3) because in practice it generates a change in the order of the search directions.

Obviously we suppose that whenever a step length has been reset to zero by (a.8) we skip step (a.5) because otherwise the procedure would automatically stop.

In order to prevent the algorithm producing a sequence of zero steps it is necessary to introduce a stopping criterion in the algorithm:

(a.9) If $\lambda_{n_{k_j-m+1}} = \lambda_{n_{k_j-m+2}} = \ldots = \lambda_{n_{k_j}} = 0$ (where the sequence $\{n_{k_j}\}_{j\in\mathbb{N}}$ is the one introduced in (a.6)) stop.

Definition

Let $\mathbb{B}$ be the set of all the algorithms belonging to $\mathcal{Q}$ which satisfy (a.8) and (a.9).

Stopping criterion (a.9) prevents the possibility of numerical non convergence as stated in the following Theorem 3.

THEOREM 3

If α is an algorithm belonging to $\mathbb{B}$, φ is a lower bounded continuous function, x_1 is a point belonging to $\mathbb{R}^m$ then the triplet (α, φ, x_1) is numerically convergent.

Proof

By contradiction suppose that $\min\lim_n \lambda_n > 0$, with $\lim_n (\varphi(x_n) - \varphi(x_{n-1})) = 0$.

Then, from (a.8), $\lambda_{j'} = 0$, $j' \geq \overline{J}'$ and, for (a.9), the procedure stops after a finite number of iterations. Q.E.D.

THEOREM 4

The point which is eventually reached, using the safeguards (a.8) and (a.9), is a vertex and the last m spanned directions are constancy vectors.

Of course, if φ is continuously differentiable, all possible vertices are stationary points.

However the most important effect of conditions (a.8), (a.9) is that they compel any algorithm to stop after a finite number of iterations on very wide class of functions, i.e. on any lower bounded continuous function. For functions not bounded below, it is sufficient to add (a.10).

(a.10) If $\varphi(x_n) \leq \xi$ (where ξ is a prescribed tolerance) stop.

REFERENCES

(1) Powell,M.J.D., (1973). On search directions for minimization algorithms. Math. Progr., 4, pp. 193/201

(2) Resta, G., A set of necessary and sufficient conditions for the convergence of minimization algorithms (to appear)

(3) Sutti, C., (1975). Remarks on conjugate directions methods for minimization without derivatives in 'Towards global optimisation' (L.C.W.Dixon and G.P.Szegö eds.) North-Holland, pp. 290/304

TOWARDS GLOBAL OPTIMISATION 2
L.C.W. Dixon and G.P. Szegö (eds.)

NUMERICAL EXPERIENCES ON A MINIMISATION ALGORITHM WITHOUT DERIVATIVES

CARLA SUTTI
MATHEMATICAL INSTITUTE
UNIVERSITY OF PARMA, ITALY

SUMMARY: This paper describes an implementation of the minimisation method without derivatives developed in [2]. The numerical experiences and comparisons with other non-gradient algorithms, which have been performed, and the numerical results obtained are also described.

1. INTRODUCTION

Several minimisation methods that do not require the evaluation of derivatives of the objective function, are known [1]. Such methods are used when it is not possible to find an analytic expression of the derivatives and also when these evaluations would be unduly difficult, inaccurate or time-consuming.
Most non-gradient techniques depend on the properties of the conjugate directions. In order to achieve fast convergence in a neighbourhood of the minimum, such techniques generate search directions which are mutually conjugate with respect to the Hessian matrix of the objective function. The procedures used to construct mutually conjugate directions are different: some methods use onedimensional searches, some others consider matrix eigenvectors [1].
The iterative algorithm [2], whose numerical implementation is presented in this paper, is a mutually conjugate directions method which does not require the evaluations of the gradient. Search vectors are generated using the results of onedimensional searches. At each iteration a number of line searches are performed, in general, this number is less than the dimension of the space of the variables. The algorithm [2] terminates in a finite number of iterations for quadratic objective functions and is convergent for strictly convex functions. Moreover it has been proved that this method also converges on nonconvex functions, if some additional conditions [3] apply.
In this paper the numerical experience that has been obtained with the algorithm [2] is presented. The procedure has been coded in FORTRAN, for the CDC 7600 computer. The program has been tested with several well-known functions, which are commonly used for the comparison of different optimisation techniques.
In section [2] of this paper we describe the main subroutines of the numerical program. In section [3] the details of our particular implementation are

discussed. In section [4] the results of the numerical tests are summarized and compared with results of other nongradient methods.

2. SUBROUTINES DESCRIPTION

The subroutine CNS implements the minimisation of an objective function f(x) of n variables, $x \epsilon R^n$, according to the procedure [2]. The whole minimisation program consists of the main routine calling CNS; the subroutine CNS; the onedimensional minimisation subroutine and the subroutine for the computation of the function values. These last two subroutines are, respectively, called SEARCH and CALFUN. This code easily allows the user to substitute different techniques for the line searches and to change the functions to be tested.

From the calling sequence the following data are transfered to the subroutine CNS:

- the starting point,
- the number of variables,
- the initial search directions (the coordinate axes),
- the requested accuracies on the location and on the value of the minimum,
- the required precision for stopping the onedimensional minimisations,
- an integer to limit the total number of function evaluations,
- the initial step for the line search.

The subroutine CNS then implements the iterative procedure [2] which consists of successive cycles, each containing at most (n-1) iterations. At the k-th iteration of the i-th cycle, $1 \leq k \leq (n-1)$, $i \geq 1$, the subroutine performs the following steps:

- a line search along the n-th direction,
- an arbitrary step along the 1-st direction,
- k line searches along the (n-k+1)-th, (n-k+2)-th,... n-th directions.

If the arbitrary step along the first direction is not one of descent, arbitrary steps along the other directions, 2,3,...,(n-k), are sequentially performed until a descent point is identified. If no descent point is found, a line search along the first direction is performed and if no descent point is found, line searches along the other directions, 2,3,...,(n-k), are sequentially performed.

If the optimum step and reduction of the function are less than the prescribed accuracies, the procedure stops. The first successful direction in these sequence is termed the l-th direction.

Let $x_a(i_k)$ be the point obtained by the first line search along the n-th direction and $x(i_{k+1})$ the final point of the actual iteration, then the direction $[x(i_{k+1})-x_a(i_k)]/|x(i_{k+1})-x_a(i_k)|$, is used as the n-th direction for the next (k+1)-th iteration and the l-th direction is rejected. The new direction is normalised and it is mutually conjugate to the (n-k+1), (n-k+2),...,n-th directions.

In order to compute the function values, CNS calls CALFUN and, in order to perform the onedimensional minimisations CNS calls SEARCH.

The subroutine SEARCH requires the following data:

- the direction d of the line,
- the starting point x on d,
- the precision with which the required minimum must be found,
- the initial step length α.

The minimum position is predicted by constructing a quadratic function defined by three consecutive estimates. Initially $f(x+\alpha d)$ is computed. Then, if $f(x)$ is greater than $f(x+\alpha d)$, $f(x+2\alpha d)$ is calculated; otherwise, if $f(x)$ is less than $f(x+\alpha d)$, $f(x-\alpha d)$ is calculated, and, if $f(x)$ is equal to $f(x+\alpha d)$, $f(x+3\alpha d)$ is computed. In the first case, let $f_a= f(x)$, $f_b= f(x+\alpha d)$, $f_c= f(x+2\alpha d)$ and $a=x$, $b=(x+\alpha d)$, $c=(x+2\alpha d)$; in the second case, let $f_a= f(x+\alpha d)$, $f_b= f(x)$, $f_c= f(x-\alpha d)$ and $a=(x+\alpha d)$, $b=x$, $c=(x-\alpha d)$; in the third case, $f(x)$ and $f(x+3\ d)$ are compared again. In the first and second cases, the turning value v of the quadratic interpolating three points is computed by the following formula:

$$v = \tfrac{1}{2}[(b^2-c^2)f_a+(c^2-a^2)f_b+(a^2-b^2)f_c]/[(b-c)f_a+(c-a)f_b+(a-b)f_c].$$

If v is predicted to be a minimum and it is within the required accuracy, absolute or relative, from the point c, f_v, f_b and f_c are compared; then either v or b or c is chosen as the minimum point, depending upon which of the values f_v, f_b and f_c is the least. A comparison between the minimum function value and $f(x)$ is also performed, so as to avoid the use of a point worse than the initial one. If the stopping test is not satisfied, either a or b or c is replaced by v, depending on which of the values f_a, f_b, and f_c is the highest. Sometimes a different point is rejected in order to have a definite bracket: so the prediction may be repeated. If the point v corresponds to a maximum of the quadratic interpolating function, a new current point and function value are obtained by moving in the direction in which the function decreases.

In order to compute the value of the function for any value of the variables, SEARCH calls CALFUN.

CALFUN codes the sequence of orders to calculate different test functions.

3. PROGRAMMING DETAILS

The crucial points in the implementation of the procedure [2] are the choices of the arbitrary step and of the line search.

The arbitrary step has been fixed less than the previous optimum step and the best tested factor of reduction has been 0.5.

Precaution is taken to prevent reduction to a very small step. Note that this choice of the arbitrary step is consistent with the convergence theory of the algorithm for non convex functions [3].

It has often been observed that a sizable decrease of the function value occurs, without successive onedimensional minimisations being undertaken along spanned directions. It has been considered sufficient in order to proceed further without a line search for the reduction of the function value to be more than a prefixed accuracy this has been chosen smaller than the precision used in the stopping criterion on $f(x)$.

Regarding the line search, it is worthwhile to remark that a quadratic interpolation technique has been chosen, because the method [2] is based on the properties of the quadratic functions and does not involve evaluations of derivatives.

Some care is required in the choice of the initial step length. Infact a bad estimate implies a very high number of function evaluations, even if this choice does not affect the final convergence. The initial step has been set equal to the previous optimum step, provided that this choice is consistent with the required accuracy in the minimum position.

The precision of the stopping criterion of the line search must be chosen so that the onedimensional line searches are sufficiently accurate to give approximate conjugate directions and also sufficiently accurate to terminate the n dimensional search when all n line searches fail.

We have therefore introduced two accuracies, one absolute and one relative, within which the required minimum position is to be found. Since unreasonably high accuracy would require too many function evaluations, we have chosen these accuracies between 10^{-2} and 10^{-3}.

A higher accuracy is requested in the stopping criterion of the n-dimensional minimisation. To stop the procedure we require that both the minimum position and the function value are located within prefixed accuracies. In particular the test on the function values uses a relative precision. We have tested our algorithm with accuracies ranging from 10^{-2} to 10^{-6}.

4. NUMERICAL RESULTS

The program has been run on a CDC 7600 computer in a single precision.

We shall next list the test functions together with the corresponding starting point x^0 and their minimum point (x^*, f^*).

Rosenbrock [4]

$$f(x) = 100(x_2-x_1^2)^2+(1-x_1)^2,$$

$$x^0 = (-1,2,1);$$

$$x^* = (1,1), \quad f^* = 0.$$

Powell [5]

$$f(x) = (x_1+10x_2)^2 + 5(x_3-x_4)^2 + (x_2-2x_3)^4 + 10(x_1-x_4)^4;$$
$$x^0 = (3,-1,0,1);$$
$$x^* = (0,0,0,0); \quad f^* = 0.$$

Wood [6]

$$f(x) = 100(x_2-x_1^2)^2 + (1-x_1)^2 + 90(x_4-x_3)^2 + (1-x_3)^2 + $$
$$+ 10.1[(x_2-1)^2 + (x_4-1)^2] + 19.8(x_2-1)(x_4-1);$$
$$x^0 = (-3,-1,-3,-1);$$
$$x^* = (1,1,1,1); \quad f^* = 0.$$

Sargan [7]

$$f(x) = \sum_{i=1}^{n} x_i^2 + c \sum_{i \neq j} x_i x_j;$$
$$c = 0.4; \quad n = 20;$$
$$x_i^0 = 0.2236068; \quad i = 1,\ldots,20;$$
$$x_i^* = 0; \quad i = 1,\ldots,20; \quad f^* = 0.$$

This is a quadratic example for which the Powell's procedure [7] fails to converge.

Trig (n) [8]

$$f(x) = \sum_{i=1}^{n} [\sum_{j=1}^{n} (A_{ij} \sin x_j + B_{ij} \cos x_j) - E_i]^2;$$
$$n = 2,3,\ldots,20;$$

A_{ij} and B_{ij} are generated to be random integers between −100 and +100; $x_i,\ldots,x_n$ are chosen randomly between $-\pi$ and $+\pi$, for these values the parameters E_i are calculated to be $E_i = \sum_{j=1}^{n} (A_{ij} \sin x_j + B_{ij} \cos x_j)$; the starting values of the variables $x_1^0,\ldots,x_n^0$, are displaced from the values defined above, by increments chosen randomly between − 0.1 and +0.1;

$$x_i^* = 0, \quad i = 1,\ldots,n; \quad f^* = 0.$$

The numerical results are summarized in Table 1

Table 1: Detailed Performances

Function	c	i	n_f	f_{min}
Rosenbrock	12	1	215	$0.1\ 10^{-10}$
Powell	4	3	199	$0.1\ 10^{-8}$
Wood	17	1	828	$0.6\ 10^{-11}$
Sargan	2	1	930	$0.1\ 10^{-12}$
Trig (4)	4	1	162	$0.9\ 10^{-8}$
Trig (6)	4	5	356	$0.2\ 10^{-5}$
Trig (10)	9	4	2083	$0.5\ 10^{-6}$
Trig (20)	5	17	4340	$0.8\ 10^{-1}$

where c denotes the total number of cycles performed, and i the number of the iterations in the last cycle; n_f is the total number of the function evaluations and f_{min} the computed minimum function value.
In order to compare the performance of the implemented algorithm with other non gradient methods, Table 2 and Table 3 are presented.

Table 2: Function Evaluations n_f

Method	Rosenbrock's Function	Powell's Function	Wood's Function
Sutti	215	181	679
Powell	153^R	242^R	817^R
Zangwill	188^R	229^R	218^R
Brent	120^B	219^B	452^B
Brodlie	194^B	217^B	751^B
Stewart	152^B	158^B	2079^B
f^*-f_{min} less than	10^{-11}	10^{-8}	10^{-5}

Table 3: The Trigonometric Function

Function	Powell's Method		Sutti's Method	
	f_{min}	n_f	f_{min}	n_f
Trig (4)	$0.5\ 10^{-6}$	137	$0.9\ 10^{-8}$	162
Trig (6)	$0.9\ 10^{-6}$	349	$0.2\ 10^{-5}$	356
Trig (10)	$0.8\ 10^{-4}$	701	$0.5\ 10^{-6}$	2083
Trig (20)	$0.2\ 10^{+1}$	1872	$0.8\ 10^{-1}$	4340

The results of Table 2 are reported from Rhead [9] (the numbers indexed by R) and from Brodlie [10] (the numbers indexed by B); those in Table 3 from Fletcher [11]. The method [2] appears in general competitive with the best non-gradient algorithms, for problems of small dimensions. But the results for the trigonometric functions are not so satisfying, especially in the high dimensions. However, let us note that, for the case Trig (20), it is possible to get a better accuracy; infact the result presented has been obtained by assuming 10^{-2} as accuracy on the function values, in order to limit the number of function evaluations. In any case, the results of the implemented algorithm are reliable, showing a consistent applicability on all the functions tested. Moreover it requires little housekeeping and small storage space.

Finally it is of interest to note that in some experiments, a sequence of orders to compute the value of the determinant of the search direction matrix, at each iteration, was introduced into the program. This was included in order to control the linear independence of the search vectors and to introduce a reset, if required. But from the results no reset was necessary, though theoretically required in paper [3] in the proof of convergence.

ACKNOWLEDGMENT

The author wishes to thank J.J.Mckeown for helping in the experiments performed at N.O.C., Hatfield Polytechnic, England.

REFERENCES

[1] C. SUTTI: "Remarks on conjugate direction methods for minimisation without derivatives"; Towards Global Optimisation (L.C.W.Dixon, G.P.Szegö eds), North-Holland, 1975.

[2] C. SUTTI: "A new method for unconstrained minimisation without derivatives"; Towards Global Optimisation (L.C.W.Dixon, G.P.Szegö eds), North-Holland, 1975.

[3] C. SUTTI: "Convergence proof of minimisation algorithms for non-convex functions"; J.O.T.A., vol. 23, No. 2, 1977.

[4] H.H. ROSENBROCK: "An automatic method for finding the greatest or least value of a function"; The Comp. Jour. vol.3, pp. 175-184, 1960.

[5] M.J.D. POWELL: "An iterative method for finding stationary values of a function of several variables"; The Comp. Jour. vol. 5, pp. 147-151, 1962.

[6] A.R. COLVILLE: "A comparative study of non linear programming codes"; I.B.M. New York Scientific Center, Rep. 320-2949, 1968.

[7] J.D. SYLVESTROWIZ: Private communication; London School of Economics and Political Science, Houghton Street, Aldwich, London.

[8] M.J.D. POWELL: "An efficient method for finding the minimum of a function of several variables without calculating derivatives"; The Comp. Jour., vol. 7, pp. 145-157, 1964.

[9] D.G. RHEAD: "Further experiments on Zangwill's method"; Inst. of Computer Science, Working Paper No. 347, University of London 1971.

[10] K.W. BRODLIE: "A new method for unconstrained minimisation without evaluating derivatives"; I.B.M. UKSC 19, 1972.

[11] R. FLETCHER: FORTRAN Subroutines for minimisation by quasi-Newton methods; A.E.R.E.,Report No. 7125, 1972.

```
      SUBROUTINE CNS(XA,N,F,DIR,EPS,MAXFU ,XMU)
      DIMENSION X(50),X0(50),D(50)
      DIMENSION XA(50),XB(50),XA1(50),DIR(50,50),EPS(3),DIRN(50)
      COMMON/ERR/NERR
      COMMON/STEP/SC
      COMMON/COUNT/NF
      SC=XMU
      NERR=0
      NF=0
      AL=0.
      CALL CALFUN(XA,N,FA,INF)
C
      I=0
    1 I=I+1
C
      K=0
    2 K=K+1
C
      DO 3 II=1,N
    3 DIRN(II)=DIR(II,N)
      MAXFUN1=MAXFU-NF
      CALL SEARCH(DIRN,MAXFUN1,EPS,XA,N,FA,NF1,XMU)
      IF(NF.GT.MAXFU) GO TO 1001
      L=0
    4 L=L+1
      IF(L.GT.(N-K)) GO TO 45
      DO 41 II=1,N
   41 DIRN(II)=DIR(II,L)
      ALPHA=AMAX1(ABS(SC)*0.5,0.0001)
      DO 42 II=1,N
   42 XA1(II)=XA(II)+ALPHA*DIRN(II)
      CALL CALFUN(XA1,N,FA1,INF)
      IF(NF.GT.MAXFU) GO TO 1001
      AMAX=AMAX1(EPS(2),EPS(2)*ABS(FA))
      IF((FA-FA1).GT.AMAX*0.001) GO TO 43
      DO 44 II=1,N
      DIRN(II)=-DIRN(II)
   44 XA1(II)=XA(II)+ALPHA*DIRN(II)
      CALL CALFUN(XA1,N,FA1,INF)
      IF(NF.GE.MAXFU) GO TO 1001
      AMAX=AMAX1(EPS(2),EPS(2)*ABS(FA))
      IF((FA-FA1).GT.AMAX*0.001) GO TO 43
      GO TO 4
   45 CONTINUE
C
      L=0
    5 L=L+1
      IF(L.GT.(N-K)) GO TO 1000
      DO 51 II=1,N
   51 DIRN(II)=DIR(II,L)
      DO 551 II=1,N
  551 X0(II)=XA(II)
      F0=FA
      MAXFUN1=MAXFU-NF
      CALL SEARCH(DIRN,MAXFUN1,EPS,X0,N,F0,NF1,XMU)
  433 CONTINUE
      IF(NF.GE.MAXFU) GO TO 1001
      IF(ABS(SC).GT.EPS(1)) GO TO 113
      AMAX=AMAX1(EPS(2),EPS(2)*ABS(FA))
      IF(   (FA-F0).GT.AMAX)  GO TO 113
      GO TO 5
  113 AL=SC
      DO 553 II=1,N
```

```
  553 XA1(II)=X0(II)
      FA1=F0
      GO TO 66
   43 AL=ALPHA
   66 CONTINUE
      J=N-K
    6 J=J+1
      DO 61 II=1,N
   61 DIRN(II)=DIR(II,J)
      MAXFUN1=MAXFU-NF
      CALL SEARCH(DIRN,MAXFUN1,EPS,XA1,N,FA1,NF1,XMU)
      IF(NF.GE.MAXFU) GO TO 1001
      IF(J.EQ.N) GO TO 62
      GO TO 6
   62 CONTINUE
      XMOD=0.0
      DO 8 II=1,N
    8 XMOD=XMOD+(XA1(II)-XA(II))**2
      XMOD=SQRT(XMOD)
      DO 7 JJ=1,N
      IF(JJ.LT.L)  GO TO 7
      IF(JJ.EQ.N) GO TO 71
      DO 72 II=1,N
   72 DIP(II,JJ)=DIR(II,JJ+1)
      GO TO 7
   71 DO 73 II=1,N
   73 DIR(II,JJ)=XA1(II)-XA(II)
    7 CONTINUE
      DO 81 II=1,N
      XA(II)=XA1(II)
   81 DIR(II,N)=DIR(II,N)/XMOD
      FA=FA1
  147 IF(K.EQ.(N-1)) GO TO 1
      GO TO 2
 1001 WRITE(3,104)
  104 FORMAT(* MAXFUN EXCEEDED* )
      GO TO 1002
 1000 WRITE(3,100)NF,I,K
  100 FORMAT(* OPTIMUN FOUND AFTER*,I5,*FUNCTION CALLS,*,
     1I5,*CYCLES,*,I5,*SIMPLE IER,*/)
 1002 CONTINUE
      WRITE(3,101)
  101 FORMAT(* X=*)
      WRITE(3,102) (XA(I),I=1,N)
  102 FORMAT(4(1X,E14.7))
      WRITE(3,103) FA
  103 FORMAT(* MINIMUM FUNCTION VALUE=*,E14.7)
      RETURN
      END
```

```
      SUBROUTINE SEARCH(D,MAXFUN1,EPS,X0,N,F0,NF1,MU)
      DIMENSION X0(50),D(50),EPS(3),X1(50)
      COMMON/ERR/NERR
      COMMON/STEP/X
      REAL MU
      NERR=0
      ITEST=3
      ABSACC=EPS(3)
      RELACC=EPS(3)
      MU=AMAX1(ABS(X),1.5*ABSACC)
      XSTEP=SIGN(MU,X)
      X=0.0
      F=F0
   51 GO TO 2000
 2100 GO TO(1,2,335,336),ITEST
    1 DO 70 I=1,N
   70 X1(I)=X0(I)+X*D(I)
      CALL CALFUN(X1,N,F,INF)
      GO TO 51
    2 CONTINUE
      DO 377 I=1,N
  377 X1(I)=X0(I)+DV*D(I)
      CALL CALFUN(X1,N,FV,INF)
      IF(F-FV)50,50,59
   59 F=FV
      X=DV
   50 IF(F.LT.F0) GO TO 60
      F=F0
      X=0.0
      NERR=4
  123 FORMAT (1X,I4)
      GO TO 2200
   60 DO 30 I=1,N
   30 X0(I)=X0(I)+X*D(I)
      F0=F
      GO TO 2200
  335 WRITE(3,111)
  111 FORMAT(*ACCURACY UNOBTAINABLE*)
      GO TO 50
  336 WRITE(3,122)
  122 FORMAT(* MAXFUN EXCEEDED *)
      GO TO 50
 2000 CONTINUE
      GO TO(91,222,222),ITEST
  222 IS=6-ITEST
      ITEST=1
      IINC=1
      XINC=XSTEP+XSTEP
      MC=IS-3
      IF(MC) 4,4,15
    3 MC=MC+1
      IF(MAXFUN1-MC) 12,15,15
   12 ITEST=4
   43 X=DB
      F=FB
      IF(FB-FC) 15,15,44
   44 X=DC
      F=FC
   15 GO TO 2100
   91 GO TO (5,6,7,8),IS
    8 IS=3
    4 DC=X
      FC=F
```

```
      X=X+XSTEP
      GO TO 3
    7 IF(FC-F) 9,10,11
   10 X=X+XINC
      XINC=XINC+XINC
      GO TO 3
    9 DB=X
      FB=F
      XINC=-XINC
      GO TO 13
   11 DB=DC
      FB=FC
      DC=X
      FC=F
   13 X=DC+DC-DB
      IS=2
      GO TO 3
    6 DA=DB
      DB=DC
      FA=FB
      FB=FC
   32 DC=X
      FC=F
      GO TO 14
    5 IF(FB-FC) 16,17,17
   17 IF(F-FB) 18,32,32
   18 FA=FB
      DA=DB
   19 FB=F
      DB=X
      GO TO 14
   16 IF(FA-FC) 21,21,20
   20 XINC=FA
      FA=FC
      FC=XINC
      XINC=DA
      DA=DC
      DC=XINC
   21 XINC=DC
      IF((DV-DB)*(DV-DC)) 32,22,22
   22 IF(F-FA) 23,24,24
   23 FC=FB
      DC=DB
      GO TO 19
   24 FA=F
      DA=X
   14 IF(FB-FC) 25,25,29
   25 IINC=2
      XINC=DC
      IF(FB-FC) 29,45,29
   29 DV=(FA-FB)/(DA-DB)-(FA-FC)/(DA-DC)
      IF(DV*(DB-DC))33,33,37
   37 DV=0.5*(DB+DC-(FB-FC)/DV)
   56 IF(ABS(DV-X)-ABS(ABSACC)) 34,34,35
   35 IF(ABS(DV-X)-ABS(DV*RELACC)) 34,34,36
   34 ITEST=2
      GO TO 43
   36 IS=1
      X=DV
      IF((DA-DC)*(DC-DV))3,26,38
   38 IS=2
      GO TO (3 ,40),IINC
   33 IS=2
      GO TO (41,42),IINC
   41 X=DC
      GO TO 10
```

```
   40 IF(ABS(XINC-X)-ABS(X-DC)) 42,42,3
   42 X=0.5*(XINC+DC)
      IF((XINC-X)*(X-DC)) 26,26,3
   45 X=0.5*(DB+DC)
      IF((DB-X)*(X-DC)) 26,26,3
   26 ITEST=3
      GO TO 43
 2200 CONTINUE
      RETURN
      END
```

TOWARDS GLOBAL OPTIMISATION 2
L.C.W. Dixon and G.P. Szegö (eds.)

A NEWTON-LIKE METHOD FOR STOCHASTIC OPTIMIZATION(*)

B. BETRO', L. DE BIASE
ISTITUTO DI MATEMATICA
UNIVERSITA' DEGLI STUDI DI MILANO
MILANO, ITALY

In this paper a Newton-like method is proposed for the minimization of a function $f:R^N \to R$ of which only noise-corrupted values are known. Its main feature is that the same function evaluations are used in the construction of estimates of the gradient and of the hessian matrix of f and thus the total number of function evaluations per step is kept as low as 4N. Convergence of the algorithm(with probability one) is proved and its asymptotic properties are discussed.

INTRODUCTION

In many technological, economical, medical and chemical applications, the problem arises of minimizing a function f of the control parameters of the system, which is not available in analytical form; only noise-corrupted realizations of the function value can be obtained for every value of the parameter set.

In mathematical form, this problem can be phrased as follows: let $\tilde{f}(x)$, $x \varepsilon S \subset R^N$ be a random function; find x^* such that

$$f(x^*) = \min_{x \varepsilon S} f(x)$$

where $f(x) = E(\tilde{f}(x))$ and $E(.)$ denotes the expected value of $\cdot$.
We shall refer to this problem as a stochastic optimization problem. The study of this kind of problems began in relatively recent years (say about 1950), but, at the moment, increasing attention is being paid to it.

(*) This work was supported in part by G.N.I.M. (Gruppo Nazionale Informatica Matematica) of CNR (Italian National Research Council).

Some methods have been developed in this field, beginning with the classical Robbins-Monro procedure, whose interest is mainly theoretical; they resemble in spirit the classical gradient method of Numerical Analysis, whose performance,as known, is relatively poor.
Starting from these first ideas, other theoretical results were obtained and a number of general theorems were proposed, which state conditions under which the iterative techniques converge to the solution of the problem. Among these, we recall Dvoretzki's theorem (1956), extended to the multidimensional case by Gray (1962), Kushner's theorem (1972) and Poliak-Tsypkin's theorem (1973).

Only in recent years has there been a major interest in methods that can lead to an efficient implementation. It seems natural to extend to the stochastic case the ideas of deterministic optimization of modifying the direction of the gradient by means of a matrix which is an approximation of the inverse hessian matrix at the minimum point.
The algorithms that have been proposed (Elliot-Sworder (1969)and Fabian (1971)) following this approach, are similar to Newton-Raphson's, as at every step a matrix of second differences is explicitly inverted. In the two quoted works an essential condition to ensure convergence of the method to x^* is the independence of such second differences and the first differences used to evaluate the gradient.
Thus a certain number of function evaluations is needed to preserve such independence.

A way of bypassing these difficulties could be to extend to stochastic optimization the algorithm proposed by Nevel'son-Khas'minskii (1973) to solve a nonlinear system $F(x)=0$, $F:R^N \to R^N$, when the realizations of $F(x)$ are noise-corrupted. But in this extension, at the n-th step,only estimates of the hessian matrix up to the (n-1)-th step would be used, and this, during the first steps, may be ineffective.

In our work, starting from the general framework of Archetti-Betrò's paper (1975), in which it is easy to construct statistical estimates of the gradient and hessian matrix, we propose an algorithm which uses the same function evaluations (4N per step) for the construction of the estimates both of the gradient and of the hessian matrix, thus saving function evaluations with respect to the above algorithms.

A general theorem is proved which gives some wide conditions for the convergence of a Newton-like algorithm to the minimum point x^* in the

case when the hessian matrix, at every step, is not necessarily independent of the approximation of the gradient.
This theorem may be applied to our algorithm without peculiar difficulties; we point out that our approach allows us to impose rather mild conditions on f(x) and on the noise variance.

Finally conditions are exhibited under which the asymptotic normality of the approximation to the minimum point gained by our algorithm is ensured. This allows interesting considerations to be made on the asymptotic speed of convergence and on the effectiveness of introducing a sequence of matrices converging to the inverse hessian matrix at the minimum point.

1. REGULARIZATION BY MEANS OF CONVOLUTION OPERATORS

In Archetti-Betrò (1975) the relevance was pointed out of convolution operators in optimization problems, as they have interesting regularizing properties and allow the statistical estimation, by means of few function evaluations, of those quantities usually needed in the numerical solution of this kind of problem.

The convolution operators approach may be outlined as follows: let $h(\beta,x)$, $x \in R^N$, $\beta>0$ be a real function satisfying conditions C1-C5 in Archetti-Betrò (1975). Such conditions are satisfied, for instance, by the function

$$(1.1) \qquad h(\beta,x) = \frac{1}{(2\pi)^{N/2}\beta^N} \exp\left(-\frac{\|x\|^2}{2\beta^2}\right) \qquad (*)$$

and to it we shall always refer in what follows.

Let's now consider a function $f: R^N \to R$ such that, for any positive β, the convolution product with $h(\beta,x)$

$$(1.2) \qquad \hat{f}(\beta,x) = \int_{R^N} h(\beta,x-u) f(u)\,du$$

exists for every $x \in R^N$.
Then, for f belonging to a wide class of functions (for example, it is easily seen that this class contains all the continuous functions such that $|f(x)| < k\|x\|^{\alpha}$ for real k and α and large x), the function $\hat{f} \in C^{\infty}(R^N)$ for any positive β and, moreover,

---

(*) Here and in what follows, $\|x\|$, $x \in R^N$, is the usual euclidean norm in R^N and $\|A\|$, A NxN matrix, is the norm induced by it.

(1.3) $\hat{f}_x(\beta,x)=\int_{R^N} h_x(\beta,x-u)f(u)du$

(1.4) $\hat{f}_{xx}(\beta,x)=\int_{R^N} h_{xx}(\beta,x-u)f(u)du.$

If f is convex in R^N, then some results are proved in the quoted paper which show the feasibility of replacing the problem

$$\text{find } \min_{x\varepsilon S\subset R^N} f(x)$$

with the problem

$$\text{find } \min_{x\varepsilon S\subset R^N} \hat{f}(\beta,x)$$

for small values of β.

For our purposes, it is sufficient to recall the following results:

i) if f is convex in R^N, then $\hat{f}(\beta,x)$ is, for any $\beta>0$ a convex function of x in R^N;

ii) if f is convex in R^N and x^*, $x^*(\beta)\varepsilon S\subset R^N$ are such that $f(x^*)<f(x)$, $\hat{f}(\beta,x^*(\beta))<\hat{f}(\beta,x)$ for any $x\varepsilon S$, then $x^*(\beta)\to x^*$ and $\hat{f}(\beta,x^*(\beta))\to f(x^*)$ as $\beta\to 0$.

Since in optimization it is important to handle nonsingular hessian matrices, we add here a lemma showing that, under mild conditions on f, $\hat{f}_{xx}(\beta,x)$ is positive definite at any point $x\varepsilon R^N$.

Lemma 1.1 Let $\beta>0$ and let f be a convex function such that $\hat{f}_{xx}(\beta,x)$ exists for every $x\varepsilon R^N$; suppose that for every $v\varepsilon R^N, v\neq 0$, there exists a set U_v such that:

(1.5) the Lebesgue measure of U_v is positive

(1.6) $\liminf_{t\to 0} \dfrac{f(u+tv)-2f(u)+f(u-tv)}{t^2} > 0$

for any $u\varepsilon U_v$.

Then, for every $x\varepsilon R^N$, $\hat{f}_{xx}(\beta,x)$ is positive definite.

The proof will be given in Appendix 1.

We may now define the class Ω of the functions for which the passage to the convolution product for optimization purposes is meaningful, as the class of all the convex functions f such that the convolution (1.2) exists for any $\beta>0$ and $x\varepsilon R^N$; $f\varepsilon C^3(R^N)$, and (1.3)-(1.6) hold. The functions $\hat{f}(\beta,x)$, constructed through (1.2) for f belonging to Ω have all the usual regularity properties which are theoretically relevant in optimization; their actual minimization is connected to the numerical evaluation of their gradients and hessian matrices, defined as multidimensional integrals in (1.3) and (1.4).

2. STATISTICAL ESTIMATION OF THE DERIVATIVES OF $\hat{f}(\beta,x)$

Let $f\in\Omega$. In Archetti-Betrò (1975) a way for the statistical evaluation of the integral (1.3), using only values of f, is indicated, which relies upon the probabilistic interpretation of $h(\beta,x)$ defined in (1.1). Indeed by simple derivation, it is:

$$\hat{f}_x(\beta,x)=-\frac{1}{\beta^2}\int_{R^N}(x-u)h(\beta,x-u)f(u)du$$

and also, by a simple change of variables and adding equal integrals,

$$\hat{f}_x(\beta,x)=\int_{R^N}uh(1,u)\frac{f(x+\beta u)-f(x-\beta u)}{2\beta}du.$$

This integral may be viewed as the mean of the random vector

$$\xi(\beta,u,x)=u\frac{f(x+\beta u)-f(x-\beta u)}{2\beta} \tag{2.1}$$

assuming u as a random vector with independent components N(0,1). Therefore, (2.1) is an unbiased estimator of $\hat{f}_x(\beta,x)$. However, we do not use it directly, but we handle it in such a way that the estimators of $\hat{f}_x(\beta,x)$ and $\hat{f}_{xx}(\beta,x)$ use the same evaluations of the function f. This seems to be particularly convenient in the construction of a Newton -like algorithm, as the one we outline in next section, in order to keep the number of function evaluations per step low.

We start constructing a statistical estimate of $\hat{f}_{xx}(\beta,x)$. Let u^i, i=1,2,...,N be independent random vectors with probability density function given by (1.1), with $\beta=1$, c be a positive number; consider the random matrix

$$\hat{H}=[\hat{H}^1|\hat{H}^2|\ldots|\hat{H}^N], \text{ where} \tag{2.2}$$

$$\hat{H}^i=\hat{H}^i(\beta,u^i,x,c)=\frac{\xi(\beta,u^i,x+ce^i)-\xi(\beta,u^i,x-ce^i)}{2c}$$

with ξ given by (2.1), c>0 and e^i, i=1,2,...,N, the N-dimensional vector with zero components, except for the i-th, which is equal to 1. The mean of $\hat{H}^i$ is easily computed as

$$E(\hat{H}^i)=\frac{\hat{f}_x(\beta,x+ce^i)-\hat{f}_x(\beta,x-ce^i)}{2c}=\hat{f}_{xx}(\beta,x)e^i+c\,\eta_i(\beta,x,c), \tag{2.3}$$

where $\eta_i(\beta,x,c)$ is bounded for bounded x and c, for any $\beta>0$ and i=1,...,N. $\hat{H}$ is not an unbiased estimator of $\hat{f}_{xx}(\beta,x)$, but its bias vanishes for $c\to0$.

It is natural to use the same 2N gradient estimates at the points $x\pm ce^i$ as in (2.2) in order to estimate $\hat{f}_x(\beta,x)$, averaging them in the random vector

$$\hat{g}=\hat{g}(\beta,u^1,u^2,\ldots,u^N,x,c)=\sum_{i=1}^{N}\frac{\xi(\beta,u^i,x+ce^i)+\xi(\beta,u^i,x-ce^i)}{2N} \tag{2.4}$$

whose mean is easily computed as

$$E(\hat{g})=\hat{f}_x(\beta,x)+c^2\tau(\beta,x,c), \tag{2.5}$$

where $\tau(\beta,x,c)$ is bounded for bounded x and c and $\beta>0$.
In a noisy situation, when we can compute only independent noise-corrupted values of $f(x)$, $\tilde{f}(x)=f(x)+\tilde{\psi}(x)$, where $\tilde{\psi}(x)$ is, for any x, a random variable with $E(\tilde{\psi}(x))=0$, then the estimates (2.1), (2.2), (2.4) may be replaced by the estimates

$$\tilde{\xi}(\beta,u,x)=u\frac{\tilde{f}(x+\beta u)-\tilde{f}(x-\beta u)}{2\beta};$$

$$\tilde{H}^i=\tilde{H}^i(\beta,u^i,x,c)=\frac{\tilde{\xi}(\beta,u^i,x+ce^i)-\tilde{\xi}(\beta,u^i,x-ce^i)}{2c} \tag{2.6}$$

$$\tilde{g}=\tilde{g}(\beta,u^1,u^2,\ldots,u^N,x,c)=\sum_{i=1}^{N}\frac{\tilde{\xi}(\beta,u^i,x+ce^i)+\tilde{\xi}(\beta,u^i,x-ce^i)}{2N}. \tag{2.7}$$

The mean of (2.6) and (2.7) are still given, respectively, by (2.3) and (2.5). We remark that only 4N function evaluations per step are used for the construction of $\tilde{H}$ and $\tilde{g}$.

It is of interest to find out conditions, both on f and on the variance of $\tilde{\psi}$, ensuring the boundedness of variance (*) of (2.6) and (2.7), at least for bounded x.
Conditions of this kind are pointed out in the following theorem:
<u>Theorem 2.1</u> Let $\Phi(x)=\{\int_{R^N} h(1,u)\|u\frac{f(x+\beta u)-f(x-\beta u)}{2\beta}\|^2 du\}^{1/2}$

and $\Psi(x)=\{\int_{R^N} h(1,u)\|u\|^2\frac{\sigma^2(\tilde{\psi}(x-\beta u))+\sigma^2(\tilde{\psi}(x+\beta u))}{4\beta^2} du\}^{1/2}$.

Let S be some set in R^N and set
$K=\max_{i=1,2,\ldots,N}\max\{\operatorname{Sup}_{x\in S}\Phi(x\pm ce^i),\ \operatorname{Sup}_{x\in S}\Psi(x\pm ce^i)\}$.
Then $E(\|\tilde{g}\|^2)^{1/2}\leqslant 2K$ and $E(\|\tilde{H}\|^2)^{1/2}\leqslant\frac{2NK}{c}$ for any $x\in S$.

The proof will be given in Appendix 1.

(*) Here and in what follows, by variance $\sigma^2(.)$ of a random vector v or of a matrix A we mean $E(\|v-E(v)\|^2)$ and $E(\|A-E(A)\|^2)$ respectively.

Theorem 2.1 says that, in order to have bounded variance over S,K must be finite. If S is bounded, it may be easily checked that this condition holds if, for instance, $|f(x)| \leq L \| x \|^{\alpha}$, for large x, and $\sigma^2(\tilde{\psi}(x)) \leq \sigma^2$ for any x.

The class of all the random functions of the form

(2.8) $\tilde{f}(x)=f(x)+\tilde{\psi}(x)$

such that $f \in \Omega$, $E(\tilde{\psi}(x))=0$, $\tilde{\psi}(x)$ independent of $\tilde{\psi}(y)$ for $x \neq y$, and the number K of theorem 2.1 is finite for bounded S, will be in the sequel called $\tilde{\Omega}$. The algorithm described in next section will be shown to work within this class.

We conclude this section with a remark.
It may seem more natural to compute an estimate for $\hat{f}_{xx}(\beta,x)$ applying the same argument which leads to (2.1). The fact is that in such a way we may obtain an unbiased estimate of $\hat{f}_{xx}(\beta,x)$, but of larger variance than (2.2).
We stress this fact in a simple case.
Let $x \in R^1$ and $f(x)=x^2$. Then $\hat{f}(\beta,x)$ turns out to be, by simple computations, $x^2+\beta^2$, and thus $\hat{f}_x(\beta,x)=2x$, $\hat{f}_{xx}(\beta,x)=2$.
By (2.1) we have $\xi(\beta,u,x)=2xu^2$, and the same result follows from (2.4).
Using (1.4) we have, replacing u with $x-\beta u$,

$$\hat{f}_{xx}(\beta,x)=\int_{R^N} (u^2-1)h(1,u)\frac{(x-\beta u)^2}{\beta^2}\,du,$$

or equivalently, simulating the usual finite differences formula,

$$\hat{f}_{xx}(\beta,x)=\int_{R^N} (u^2-1)h(1,u)\frac{(x-\beta u)^2-2x^2+(x+\beta u)^2}{2\beta^2}\,du.$$

Thus, assuming u as a random number N(0,1), the random number

$$r(\beta,u,x)=(u^2-1)u^2$$

is an unbiased estimator of $\hat{f}_{xx}(\beta,x)$.
Using the recursion formula for the moments of the normal distribution

$$E(u^n)=\begin{cases} 0 & \text{if } n \text{ is odd} \\ (n-1)E(u^{n-2}) & \text{otherwise,} \end{cases}$$

we have $E(r)=2$, $\sigma^2(r)=74$. Using our estimate (2.6) or equivalently (2.2), it turns out

$$\tilde{H}=\frac{2(x+c)u^2-2(x-c)u^2}{2c}=2u^2,$$

with $E(\tilde{H})=2$ and $\sigma^2(\tilde{H})=8$.
Thus $\tilde{H}$ has a considerably lower variance than the unbiased estimate r.

3. A STOCHASTIC OPTIMIZATION ALGORITHM

Let's now consider $\tilde{f}\varepsilon\tilde{\Omega}$ and the problem of minimizing its mean f, defined by (2.8) over a bounded hyperinterval

(3.1) $$S=\{x\varepsilon R^N: \quad \alpha_i\leq[x]_i\leq\beta_i \quad , \quad i=1,2,\dots,N\}$$

($[.]_i$ means the i-th component of .). We assume that the realizations of the random variable $\tilde{f}(x)$ are available for every $x\varepsilon R^N$.

By the definition of $\tilde{\Omega}$, $f\varepsilon\Omega$ and thus the convolution approach can be applied to it. Then, for the function f, the problem

find $\min_{x\varepsilon S}$ f(x)

can be replaced, for small $\beta>0$, by

find $\min_{x\varepsilon S}$ $\hat{f}(\beta,x)$.

Therefore we consider β fixed in the sequel, and set up an algorithm which converges to $x^*(\beta)$, such that $\hat{f}(\beta,x^*(\beta))=\min_{x\varepsilon S}\hat{f}(\beta,x)$.

The algorithm is as follows:

(3.2) $$x_{n+1}=P_S(x_n-a_n\tilde{B}_n^{-1}\tilde{g}_n),$$

where x_1 is a random point in S, $\{a_n\}$ is a sequence of positive numbers and:

$$[P_S(x)]_i=\begin{cases}\alpha_i & \text{if } [x]_i<\alpha_i\\ \beta_i & \text{if } [x]_i>\beta_i\\ [x]_i & \text{otherwise.}\end{cases}$$

Moreover:

(3.3) $$\tilde{g}_n=\tilde{g}(\beta,u_n^1,\dots,u_n^N,x_n,c_n)$$

(see formula (2.7)), and

(3.4) $$\tilde{B}_n=\begin{cases}(\tilde{A}_n+\tilde{A}_n^T)/2 & \text{if } \tilde{A}_n \text{ is positive definite}\\ \text{I or any positive definite matrix} & \text{otherwise}\end{cases}$$

where it is, for two positive sequences $\{c_n\}$, $\{p_n\}$

(3.5) $$\tilde{A}_n = \frac{\sum_{i=1}^{n} p_i \tilde{H}_i}{\sum_{i=1}^{n} p_i}$$ with

(3.6) $$\tilde{H}_i = [\tilde{H}^1(\beta, u_i^1, x_i, c_i) \mid \tilde{H}^2(\beta, u_i^2, x_i, c_i) \mid \dots \mid \tilde{H}^N(\beta, u_i^N, x_i, c\ )]$$

(see formula (2.6)); u_i^j, $j=1,2,\dots,N$; $i=1,2,\dots$ are independent random vectors with independent components $N(0,1)$. We notice that only realizations of the random variable $\tilde{f}(x)$ are needed and no kind of knowledge of its distribution is required.

Since $\tilde{A}_n$ needs not be symmetric or positive definite, we introduce the symmetric positive definite matrix $\tilde{B}_n$. As will be clear in the proof of convergence, positive definiteness of $\tilde{B}_n$ is a crucial condition. The symmetry, moreover, has the effect of simplifying the inversion required in (3.2).

The algorithm (3.2) is similar to the classical Newton-Raphson's one for deterministic optimization, because of the use of a sequence of matrices which modify the gradient direction at every step; thus it is of interest to ensure that this sequence converges (with probability 1) to $\hat{f}_{xx}^{-1}(\beta, x^*(\beta))$, when $x_n \to x^*(\beta)$.
The effectiveness of this convergence for our algorithm will be investigated in Section 4.

To prove convergence with probability 1 of x_n to $x^*(\beta)$ and of $\tilde{B}_n$ to $\hat{f}_{xx}(\beta, x^*(\beta))$ a lemma and a theorem are stated here in a general form, independent of our approach, but immediately appliable to algorithm (3.2). The proof will be given in Appendix 2.

<u>Lemma 3.1</u> Let $f: R^N \to R$ be a twice continuously differentiable function and $\{x_n\}$ a sequence of points in R^N. Let $G(x)$ denote the hessian matrix of $f(x)$; let $\tilde{A}_n = (1/\sum_{i=1}^{n} p_i) \sum_{i=1}^{n} p_i \tilde{H}_i$, with $p_i > 0$, $\sum_1^{\infty} p_i = \infty$ and $\tilde{H}_i$ independent NxN random matrices with mean H_i, such that:

a) $H_i = E(\tilde{H}_i) = G(x_i) + \varepsilon(x_i, c_i)$, with $c_i \to 0$ and $\varepsilon(x_i, c_i) \to 0$ as $i \to \infty$;

b) $E(\|H_i\|^2) \leq \frac{s^2}{c_i^2}$ for some $s < \infty$ with $\sum_1^{\infty} \frac{p_n^2}{c_n^2 (\sum_{i=1}^{n} p_i)^2} < \infty$.

Then, setting $A_n = E(\tilde{A}_n)$,

i) $\sigma^2(A_n) \leq \frac{s^2}{(\sum_{i=1}^{n} p_i)^2} \sum_{i=1}^{n} \left(\frac{p_i}{c_i}\right)^2$;

ii) $\lim_{n\to\infty} \tilde{A}_n - A_n = 0$ with probability 1 ;

iii) $\lim_{n\to\infty} (\sum_{i=1}^{n} p_i G(x_i))/(\sum_{i=1}^{n} p_i) - \tilde{A}_n = 0$ with probability 1;

iv) $\tilde{A}_n \to G(x_o)$ probability 1, provided that $x_n \to x_o$.

<u>Theorem 3.1</u> Let $f: R^N \to R$ be a twice continuously differentiable function with positive definite hessian matrix G over R^N. Let S be as in (3.1) and such that:

a) at every point of the boundary of S the anti-gradient $-g$ of f is directed inwards S.

Consider a random process of the form:

(3.8) $$x_{n+1} = P_S(x_n - a_n \tilde{\Gamma}_n^{-1} \tilde{g}_n(x_n))$$ where

(3.9) $$\tilde{\Gamma}_n = \begin{cases} \tilde{A}_n & \text{if } \tilde{A}_n \text{ is positive definite} \\ I \text{ or any positive definite matrix} & \text{otherwise,} \end{cases}$$

with $\tilde{A}_n = (\sum_{i=1}^{n} p_i \tilde{H}_i(x_i))/(\sum_{i=1}^{n} p_i)$ with $p_i > 0$, $\sum_1^\infty p_i = \infty$

and $\tilde{H}_i(x)$, $x \in S$, independent symmetric random matrices such that

b) $H_i(x) = E(\tilde{H}_i(x)) = G(x) + \varepsilon(x, c_i)$ with $c_i \to 0$ and $\varepsilon(x, c_i) \to 0$ as $i \to \infty$, uniformly for $x \in S$.

c) $E(\|\tilde{H}_i(x)\|^2) \leq s^2/c_i^2$ for some s and every $x \in S$; $\tilde{g}_n(x)$, $x \in S$ is a random N-dimensional vector, not necessarily independent of $\tilde{A}_n$, such that:

d) $E(\tilde{g}_n(x)) = g(x) + \tau_n(x)$, with $\|\tau_n(x)\| \leq \tau d_n$, $x \in S$, $d_n \to 0$;

e) $E(\|g_n(x)\|^2) < h^2$, for some finite h and $x \in S$;

and if the coefficients a_n, c_n, d_n, p_n satisfy the conditions

f) $a_n > 0$, $\sum_1^\infty a_n = \infty$, $\sum_1^\infty a_n d_n < \infty$, $\sum_1^\infty a_n^2/c_n^2 < \infty$,

$$\sum_1^\infty \left(\frac{a_n^2}{\sum_{i=1}^{n} p_i} \left\{ \sum_{i=1}^{n} \frac{p_i^2}{c_i^2} \right\}^{1/2} \right) < \infty , \quad \sum_1^\infty \frac{p_n^2}{c_n^2 (\sum_{i=1}^{n} p_i)^2} < \infty ,$$

then x_n converges to the minimum point x^* of f over S and $\tilde{\Gamma}_n$ converges to $G(x^*)$ with probability 1.

We remark that our theorem differs from the convergence theorems known in the literature for algorithms of the form (3.8), by the fact that independence of $\tilde{\Gamma}_n$ and $\tilde{g}_n$ is not required.
This allows its application to algorithm (3.2), where, for the construction itself, $\tilde{B}_n$ and $\tilde{g}_n$ are dependent.

In our approach the conditions of theorem 3.1 are substantially equivalent to the condition $\tilde{f}\varepsilon\tilde{\Omega}$. This is easily seen considering the definition of $\tilde{\Omega}$, (2.3) and (2.5). The only additive requirements are a) and f). In particular, because of (2.5), we can take $d_n=c_n^2$; thus conditions f) are satisfied if, for example, $a_n=a/n$, $c_n=c/(n^{1/3})$, $p_n=1$, even if this choice may be non optimal.

4. ASYMPTOTIC NORMALITY

In this section the asymptotic distribution of the process (3.2) will be investigated. Under the conditions stated in theorem 4.1 such distribution turns out to be normal.
The study of asymptotic distribution allows us to make some interesting statements on the effectiveness of modifying the gradient direction at every step by means of a sequence of matrices which converges to the inverse hessian matrix at the minimum point.
The following theorem can be proved:

Theorem 4.1 Suppose that all the assumptions of theorem 3.1 are satisfied and moreover that:

i) $\lim_{n\to\infty} \sqrt{n}\tau_n=0$

ii) $E\{(\tilde{g}_n-g_n-\tau_n)(\tilde{g}_n-g_n-\tau_n)^T|x_n\}\to\Sigma$ as $n\to\infty$ for some matrix Σ (*);

iii) $\lim_{n\to\infty} \frac{1}{n}\sum_{j=1}^{n} \sigma_{jr}^2=0$ for every $r>0$, where

$\sigma_{jr}^2=E(\chi\{\tilde{g}_j-g_j-\tau_j\,|\,\|\tilde{g}_j-g_j-\tau_j\|>rj\}\|\tilde{g}_j-g_j-\tau_j\|^2)$ and for any set C

we define its characteristic function as $\chi(x|C)=\begin{cases}1 \text{ if } x\varepsilon C\\ 0 \text{ if } x\notin C\end{cases}$

iv) The minimum point x^* of f over S is in the interior of S.

Then $\sqrt{n}(x_n-x^*)$ has asymptotic distribution

(*) We write $\tilde{g}_n$ and g_n in place of $\tilde{g}_n(x_n)$ and $g_n(x_n)$.

$N(0,G^{-1}(x^*)\Sigma G^{-1}(x^*))$ if x_n is given by (3.8) with $a_n=1/n$.

This can be proved by simply showing that the hypotheses of Fabian's theorem (see for instance Wasan (1969), page 197) are satisfied and the proof will be given in Appendix 2.

It is of interest to examine an algorithm of the form:

$$x_{n+1}=x_n-\frac{1}{n}\tilde{\Gamma}_n^{-1}\tilde{g}_n ,$$

when $\tilde{\Gamma}_n \to \tilde{\Gamma} \neq G(x^*)$ and $x_n \to x^*$. We limit ourselves to the unidimensional algorithm:

$$x_{n+1}=x_n-\frac{1}{n\tilde{\Gamma}_n}\tilde{g}_n .$$

Provided the requirements of Fabian's theorem are satisfied, the application of the same theorem yields an asymptotic variance proportional to

$$\frac{1}{\tilde{\Gamma}(2f''(x^*)-\tilde{\Gamma})} .$$

The value $\tilde{\Gamma}$ which minimizes this variance is $\tilde{\Gamma}=f''(x^*)$. Thus, in one dimension, the choice of a sequence converging to $1/f''(x^*)$ has the effect of minimizing the asymptotic variance of x_n-x^*.

The extension to the multidimensional case is not immediate, but it is still possible to show the asymptotic effectiveness of choosing $\tilde{\Gamma}_n \to G(x^*)$ (see for example the discussion in Nevel'son-Khas'minskii (1973)).

Not all of the conditions for asymptotic normality are easy to verify; in our approach i) holds if c_n is of the form $c/(n^{\alpha})$ with $\alpha>1/4$.

An exact dependence of the matrix Σ on the distribution of the noise is difficult to phrase; its study, however, is important because it enables us to illustrate how the 'artificial noise', due to the introduction of the random vectors u_n^i, affects the natural noise. At present our interest is just to show a very simple example: we shall examin a quadratic positive definite form

$$f(x)=\frac{1}{2}\langle Ax,x\rangle+\tilde{\psi}(x) ,$$

where $\tilde{\psi}(x)$ is defined as in (2.8). We shall apply to this case theorem 4.1, to the extent of finding an estimate for the matrix Σ and of comparing it with the covariance matrix obtained by means of the finite differences formula for evaluating the gradient.

Let's define some variables $\tilde{\psi}_n^i$, as the sum of natural noises relative to the function evaluations our algorithm performs at every point x_n:

$$\tilde{\psi}_n^i=\tilde{\psi}(x_n+u_n^i+c_n e^i)+\tilde{\psi}(x_n-u_n^i+c_n e^i)+\tilde{\psi}(x_n+u_n^i-c_n e^i)+\tilde{\psi}(x_n-u_n^i-c_n e^i).$$

By simple calculations we have:

$$\tilde{g}_n=\frac{1}{N}\sum_{i=1}^{N} u_n^i\langle Ax_n,u_n^i\rangle+\frac{1}{4N}\sum_{i=1}^{N} u_n^i\tilde{\psi}_n^i$$

with $E(\tilde{g}_n)=g_n=Ax_n$, and consequently τ_n (see its definition in theorem 3.1) is equal to 0.

If we set

$$R_n=(\tilde{g}_n-g_n-\tau_n)(\tilde{g}_n-g_n-\tau_n)^T \quad ,$$

we have:

$$R_n=\frac{1}{N^2}\left(\sum_{i,j=1}^{N} u_n^i u_n^{iT}Ax_n x_n^T A u_n^j u_n^{jT}\right)+\frac{1}{2N^2\beta}\left(\sum_{i=1}^{N} u_n^i u_n^{iT}Ax_n\right)\left(\sum_{i=1}^{N} u_n^i\tilde{\psi}_n^i\right)+$$

$$+\frac{1}{16N^2\beta^2}\left(\sum_{i=1}^{N} u_n^i\tilde{\psi}_n^i\right)\left(\sum_{i=1}^{N} u_n^i\tilde{\psi}_n^i\right)^T.$$

Now, for $x_n\to x^*=0$, $E(R_n)$ approaches the matrix

$$\Sigma=\lim_{n\to\infty}\frac{1}{16N^2\beta^2}E\left(\sum_{i=1}^{N} u_n^i\tilde{\psi}_n^i\right)\left(\sum_{i=1}^{N} u_n^i\tilde{\psi}_n^i\right)^T \quad ,$$

where $\Sigma^{ij}=0$ for $i\neq j$.

If the variance of $\tilde{\psi}(x)$ equals a constant σ^2, then $\Sigma=\frac{\sigma^2}{4\beta^2}I$, which is exactly the same covariance matrix we obtain when using the classical finite differences formula with step β, to evaluate the gradient. Thus, in this case, the introduction of artificial noise asymptotically produces no increase in the variance of the estimator.

The advantage of our approach is that it saves function evaluations at every step, using the same 4N function evaluations in the construction of the gradient and of the hessian matrix and that it is appliable to a large class of problems.

5. CONCLUDING REMARKS

In this work we propose a Newton-like method to solve stochastic optimization problems. The introduction of convolution operators provides a general framework in which the construction of estimates of the gradient and hessian matrix of the function to be minimized is easy and not expensive from the point of view of the total number of function evaluations per step (i.e. 4N, where N is the dimension of the problem).

The loss of independence between the estimates of the gradient and of the hessian matrix is overcome by a general convergence theorem. Asymptotic normality is discussed.

At present no reliable computational results are available of this algorithm; in our intentions this will be the subject for new work, as well as the investigation of a more effective step lenghth and the study of an updating formula for the estimate of the inverse hessian matrix at every step.
These last problems are an almost unexplored field of investigation; for considerations on the subject, see Dixon (1976).

6. ACKNOWLEDGEMENTS

We are very grateful to L.C.W. Dixon for fruitful discussions and to Professor M. Cugiani for valuable help in the final draft of this paper.

APPENDIX 1

Proof of lemma 1.1 For any $x, v \in R^N$ we have

$$v^T \hat{f}_{xx}(\beta,x)v = \lim_{t\to 0} \frac{\hat{f}(\beta,x+tv)-2\hat{f}(\beta,x)+\hat{f}(\beta,x-tv)}{t^2} =$$

$$= \lim_{t\to 0} \int_{R^N} h(\beta,x-u)\frac{f(u+tv)-2f(u)+f(u-tv)}{t^2}\,du \geq$$

$$\geq \int_{U_v} h(\beta,x-u) \liminf_{t\to 0} \frac{f(u+tv)-2f(u)+f(u-tv)}{t^2}\,du.$$

Then, by (1.5), (1.6) and the positivity of h, the positive definiteness of $\hat{f}_{xx}(\beta,x)$ holds.

Proof of theorem 2.1.

We have

$$\{E(\|g\|^2)\}^{1/2} \leq \frac{1}{2N} \sum_{i=1}^{N} \{[E(\|\tilde{\xi}(\beta,u^i,x+ce^i)\|^2)]^{1/2} +$$

$$+[E(\|\tilde{\xi}(\beta,u^i,x-ce^i)\|^2)]^{1/2}\}.$$

Now it is

$$\{E(\|\tilde{\xi}(\beta,u^i,x\pm ce^i)\|^2)\}^{1/2} =$$

$$=\{E(\|u^i \frac{f(x+\beta u^i \pm ce^i)+\tilde{\psi}(x+\beta u^i\pm ce^i)-f(x-\beta u^i \pm ce^i)-\tilde{\psi}(x-\beta u^i\pm ce^i)}{2\beta}\|^2)\}^{1/2} \leq$$

$$\leq\{E(\|u^i \frac{f(x+\beta u^i\pm ce^i)-f(x-\beta u^i\pm ce^i)}{2\beta}\|^2)\}^{1/2}+$$

$$+\{E(\|u^i \frac{\tilde{\psi}(x+\beta u^i\pm ce^i)-\tilde{\psi}(x-\beta u^i\pm ce^i)}{2\beta}\|^2)\}^{1/2}=$$

$$=\{\int_{R^N} h(1,u)\|u \frac{f(x+\beta u\pm ce^i)-f(x-\beta u\pm ce^i)}{2\beta}\|^2 du\}^{1/2} +$$

$$+\{\int_{R^N} h(1,u)\|uE(\frac{\tilde{\psi}(x+\beta u\pm ce^i)-\tilde{\psi}(x-\beta u\pm ce^i)}{2\beta})\|^2 du\}^{1/2} \leq$$

$$\leq\{\int_{R^N} h(1,u)\|u \frac{f(x+\beta u\pm ce^i)-f(x-\beta u\pm ce^i)}{2\beta}\|^2 du\}^{1/2} +$$

$$+\{\int_{R^N} h(1,u) \frac{\|u\|^2}{4\beta^2}[\sigma^2(\tilde{\psi}(x+\beta u\pm ce^i))+\sigma^2(\tilde{\psi}(x-\beta u\pm ce^i)]du\}^{1/2}=$$

$$=\Phi(x\pm ce^i)+\Psi(x\pm ce^i).$$

The argument for $\{E(\|\tilde{H}\|^2)\}^{1/2}$ is quite analogous.

APPENDIX 2

<u>Proof of lemma 3.1.</u>

i) is trivial.

ii) is implied by b) and the law of large numbers for independent random variables. Indeed, let

$$\tilde{R}_n= \frac{p_n}{\sum_{i=1}^{n} p_i} \tilde{H}_n. \quad \text{Then}$$

$$\sigma^2(\tilde{R}_n)= \frac{p_n^2}{(\sum_{i=1}^{n} p_i)^2} \sigma^2(\tilde{H}_n) \leq \frac{s^2}{c_n^2} \frac{p_n^2}{(\sum_{i=1}^{n} p_i)^2} .$$

Thus b) implies $\sum_1^{\infty} \sigma^2(\tilde{R}_n)<\infty$ and from Loeve (1963), it follows that $\sum_1^{\infty} (\tilde{R}_n-E(\tilde{R}_n))$ is convergent with probability one, that is

$$\sum_{1}^{\infty} \frac{p_n}{\sum_{i=1}^{n} p_i} (\tilde{H}_n - H_n)$$ is convergent with probability one.

iii) is implied by ii) and by the relation

$$A_n = \frac{1}{\sum_{i=1}^{n} p_i} \sum_{i=1}^{n} p_i G(x_i) + \frac{1}{\sum_{i=1}^{n} p_i} \sum_{i=1}^{n} p_i \varepsilon(x_i, c_i),$$

since $\frac{1}{\sum_{i=1}^{n} p_i} \sum_{i=1}^{n} p_i \varepsilon(x_i, c_i) \to 0$ for $n \to \infty$.

iv) is implied by iii) and by the continuity of $G(x)$.
Indeed

$$A_n = \frac{1}{\sum_{i=1}^{n} p_i} \sum_{i=1}^{n} p_i G(x_i) + \frac{1}{\sum_{i=1}^{n} p_i} \sum_{i=1}^{n} p_i \varepsilon(x_i, c_i) =$$

$$= \frac{1}{\sum_{i=1}^{n} p_i} \sum_{i=1}^{n} p_i G(x^o) + \frac{1}{\sum_{i=1}^{n} p_i} \sum_{i=1}^{n} p_i \{\varepsilon(x_i, c_i) + G(x_i) - G(x^o)\} =$$

$$= G(x^o) + \frac{1}{\sum_{i=1}^{n} p_i} \sum_{i=1}^{n} p_i \{\varepsilon(x_i, c_i) + G(x_i) - G(x^o)\}.$$

As $\varepsilon(x_i, c_i) \to 0$ and $G(x_i) \to G(x^o)$, iv) follows.

Proof of Theorem 3.1.

For a fixed $n_0 \geq 1$ we consider the set **A** of the matrices of the form

$$\text{(A1)} \qquad A = \frac{1}{\sum_{i=1}^{n} p_i} \sum_{i=n_o}^{n} p_i G(x_i) + G_o$$

for some n-tuple of points $x_i \in S$, $i=n_o, \ldots, n$, and some positive definite matrix G_o.
As every matrix $G(x)$ is continuous and positive definite, matrices $A \in \mathbf{A}$ are 'uniformly positive definite' (i.e. a constant $\mu > 0$ exists such that $\mu < \inf_{A \in \mathbf{A}} \min_{i=1,\ldots,N} \lambda_i(A)$ where $\lambda_i(A)$ are the eigenvalues of A)
Now, let $\delta > 0$; if δ is small enough, every matrix with distance less than δ from **A** will be positive definite and bounded. Let $\mathbf{A}(\delta)$ be the set of such matrices; then there exist two positive constants μ_1, μ_2 such that $\mu_1 I \leq A \leq \mu_2 I$ (we say that a matrix A is less than or equal to

a matrix B if B-A is positive definite) for every $A\varepsilon\mathbf{A}(\delta)$ and consequently there exists a k_1 such that

$$\mathrm{Sup}_{A\varepsilon\mathbf{A}(\delta)} \| A^{-1}\| \leq k_1 .$$

Let's now consider the auxiliary process, defined for $n \geq n_o$

$$\text{(A2)} \qquad z_{n+1}=P_S(z_n-a_n\tilde{T}_n^{-1}g_n(z_n))$$

$$z_{n_o}=x_{n_o}, \qquad\qquad \text{with}$$

$$\text{(A3)} \qquad \tilde{T}_n=\begin{cases}\tilde{B}_n & \text{if } \tilde{B}_n\varepsilon\mathbf{A}(\delta)\\ B_n=E(\tilde{B}_n|\, z_{n_o},\dots,z_n) & \text{if } \tilde{B}_n\not\varepsilon\mathbf{A}(\delta),\det B_n\neq 0\\ \text{I or any nonsingular matrix} & \text{otherwise}\end{cases}$$

where

$$\text{(A4)} \qquad \tilde{B}_n=\frac{1}{\sum_{i=1}^{n} p_i}\Big(\sum_{i=n_o}^{n} p_i\tilde{H}_i(z_i)+\sum_{i=1}^{n_o-1} p_i\tilde{H}_i(x_i)\Big).$$

From (A4), b) and c) we have

$$E(\tilde{B}_n|\, z_{n_o},\dots,z_n)=\frac{1}{\sum_{i=1}^{n} p_i}\Big(\sum_{i=n_o}^{n} p_iH_i(z_i)+\sum_{i=1}^{n_o-1} p_iE(H_i(x_i))\Big)=$$

$$=\frac{1}{\sum_{i=1}^{n} p_i}\Big(\sum_{i=n_o}^{n} p_iG(x_i)+\sum_{i=1}^{n_o-1} p_iE(G(x_i))+\sum_{i=n_o}^{n} p_i\varepsilon(z_i,c_i)+$$

$$+\sum_{i=1}^{n_o-1} p_iE(\varepsilon(x_i,c_i)).$$

Because of the positive definiteness of G(x) for any x, setting $G_o=\sum_{i=1}^{n_o-1} p_iE(G(x_i))$, G_o is positive definite; thus, considering the boundedness and the uniform convergence of $\varepsilon(x_i,c_i)$, we obtain

$$B_n-\frac{1}{\sum_{i=1}^{n} p_i}\sum_{i=n_o}^{n} p_iH_i(z_i)-G_o\to 0 \text{ uniformly with respect to } z_{n_o},z_{n_o+1},\dots$$

Thus, if a sufficiently large n_o is selected, $B_n\varepsilon\mathbf{A}(\delta)$ and hence $\tilde{T}_n\varepsilon\mathbf{A}(\delta)$. Considering such an n_o, we are now going to prove that $z_n\to x^*$; the proof is carried out showing that the hypotheses of theorem 4 in Kushner (1972) are verified for the process $\{z_n\}$.

For shortness, we write g_n for $g_n(z_n)$, $\tilde{g}_n$ for $\tilde{g}_n(z_n)$ and τ_n for $\tau_n(z_n)$.

Setting $z'_{n+1}=z_n-a_n\tilde{T}_n^{-1}g_n$, for some $z\varepsilon S$ and matrix $\bar{G}$ such that $\|\bar{G}\|\leq 2k$, $2k=\mathrm{Sup}_{x\varepsilon S}\|G(x)\|$, we have

$$f(z_{n+1})-f(z_n)=g_n^T(z_{n+1}-z_n)+\frac{1}{2}(z_{n+1}-z_n)^T G(z)(z_{n+1}-z_n)=$$

$$=g_n^T(z'_{n+1}-z_n)+g_{n+1}^T(z_{n+1}-z'_{n+1})+(z_{n+1}-z_n)^T\,\bar{G}(z_{n+1}-z'_{n+1})+$$

$$+\frac{1}{2}(z_{n+1}-z_n)^T G(z)(z_{n+1}-z_n);$$

as easily seen, $\|z_{n+1}-z'_{n+1}\|\leq\|z'_{n+1}-z_n\|$, $\|z_{n+1}-z_n\|\leq\|z'_{n+1}-z_n\|$

and, by assumption a), $g_{n+1}^T(z_{n+1}-z'_{n+1})\leq 0$; thus

$$f(z_{n+1})-f(z)\leq g_n^T(z'_{n+1}-z_n)+3k\|z'_{n+1}-z_n\|^2\leq$$

$$\leq -a_n g_n^T\tilde{T}_n^{-1}\tilde{g}_n+3ka_n^2k_1^2\ \tilde{g}_n\ ^2=$$

$$=-a_n g_n^T B_n^{-1}\tilde{g}_n-a_n g_n^T(\tilde{T}_n^{-1}-B_n^{-1})\tilde{g}_n+3ka_n^2k_1^2\|\tilde{g}_n\|^2.$$

Indicating with $E_n(.)$ the conditional expectation $E(\cdot|z_{n_o},\dots,z_n)$, we have:

$$E_n(f(z_{n+1})-f(z_n))\leq -a_n g_n^T B_n^{-1}g_n-a_n g_n^T B_n^{-1}\tau_n+$$

$$+a_n\|g_n\|E_n(\|B_n^{-1}(\tilde{T}_n-B_n)\tilde{T}_n^{-1}\tilde{g}_n\|)+3ka_n^2k_1^2E_n(\|\tilde{g}_n\|^2).$$

Now, by e) and c), setting $k_2=\operatorname{Sup}_{x\in S}\|g(x)\|$, we have:

$$E_n(\|B_n^{-1}(\tilde{T}_n-B_n)\tilde{T}_n^{-1}\tilde{g}_n\|)\leq$$

$$\leq k_1E_n(\|(\tilde{T}_n-B_n)\tilde{T}_n^{-1}\tilde{g}_n\|)\leq k_1^2\{E_n(\|\tilde{B}_n-B_n\|^2)E_n(\|\tilde{g}_n\|^2)\}^{1/2}\leq$$

$$\leq k_1^2h\frac{s}{\sum_{i=1}^n p_i}\{\sum_{i=1}^n\left(\frac{p_i}{c_i}\right)^2\}^{1/2}.$$

Hence:

$$E_n(f(z_{n+1})-f(z_n))\leq -a_n g_n^T B_n^{-1}g_n-a_n g_n^T B_n^{-1}\tau_n+k_1^2k_2hs\frac{1}{\sum_{i=1}^n p_i}\{\sum_{i=1}^n\left(\frac{p_i}{c_i}\right)^2\}^{1/2}+$$

$$+3ka_n^2k_1^2h^2.$$

The first term in the sum is not greater than $-a_n\|g_n\|^2/\mu_2$, the second than $\tau k_1k_2a_nd_n$.

Hence, by f) the first hypothesis of the quoted theorem is verified. To verify also the second hypothesis, we have to show that

$$z_{n+1}-z_n=a_n\delta_{1n}(z_n)+\gamma_n,\quad \sum_{n_o}^{\infty}\gamma_n<\infty \text{ with probability one}$$

and that there exists a positive function $\delta_1(x)$, bounded on bounded sets such that for every $x \in S$, $\|\delta_{1n}(x)\| \leq \delta_1(x)$.

Indeed, for $0 \leq \theta_n^i \leq 1$, $i=1,2,\ldots,N$, we may write, considering the diagonal NxN matrix Θ_n with diagonal elements θ_n^i,

$$z_{n+1} - z_n = -a_n \Theta_n \tilde{T}_n^{-1} \tilde{g}_n \quad ,$$

with $\theta_n^i = 1$, $i=1,2,\ldots,N$ if $z_n - a_n \tilde{T}_n^{-1} \tilde{g}_n \in S$, $0 \leq \theta_n^i \leq 1$ for some i, otherwise.
Let:

$$t_{n_o - 1} = z_{n_o} \quad , \quad t_n = z_{n+1} - z_n \quad ; \quad \text{then} \quad \sum_{n_o}^{\infty} E(\|t_n\|^2) \leq k_1^2 h^2 \sum_{n_o}^{\infty} a_n^2 \quad .$$

Thus (see for example lemma 1 of Appendix 2 in Wasan (1969)), it follows

$$\sum_{n_o}^{\infty} [t_i - E(t_i \mid t_{n_o - 1}, \ldots, t_{i-1})] < \infty \quad \text{with probablity 1.}$$

This means that, setting $\gamma_n = t_n - E(t_n \mid t_{n_o - 1}, \ldots, t_{n-1})$, $\sum_{n_o}^{\infty} \gamma_n < \infty$ with probability one and

$$z_{n+1} - z_n = a_n E(\Theta_n \tilde{T}_n^{-1} \tilde{g}_n \mid t_{n_o - 1}, \ldots, t_{n-1}) \quad \text{with}$$

$$\|E(\Theta_n \tilde{T}_n^{-1} \tilde{g}_n \mid t_{n_o - 1}, \ldots, t_{n-1})\| \leq k_1 h.$$

Thus also the second hypothesis is satisfied.
As a consequence, we have $z_n \to x^*$, $\tilde{g}(z_n) \to 0$, $f(z_n) \to f^*$ with probability one.

In order to show that also the process (3.8) converges to x^* with probability one, we may proceed as follows.
From Lemma (3.1) we have:

$$\lim_{n\to\infty} (\tilde{A}_n - \frac{1}{\sum_{i=1}^{n} p_i} \sum_{i=1}^{n} p_i G(x_i)) = 0 \quad \text{with probability one, i.e.}$$

$$\tilde{A}_n = \frac{1}{\sum_{i=1}^{n} p_i} \sum_{i=1}^{n} p_i G(x_i) + \Gamma_n \quad ,$$

where $\lim_{n\to\infty} \Gamma_n = 0$ with probability one.

Then, for any $\varepsilon > 0$, there exists $\bar{n}$ such that

$$P \{\sup_{n \geq \bar{n}} \|\Gamma_n\| < \delta \} > 1 - \varepsilon \quad ,$$

and then, writing $\tilde{A}_n$, $n \geq \bar{n}$ as

$$\tilde{A}_n = \frac{1}{\sum_{i=1}^{n} p_i} \sum_{i=\bar{n}}^{n} p_i G(x_i) + \sum_{i=1}^{\bar{n}-1} p_i G(x_i) + \Gamma_n$$

it is immediately seen, taking in (A1) $n_o=\bar{n}$, that

$$P\ \{\tilde{A}_n \varepsilon A(\delta),\ n \geqslant \bar{n}\} > 1-\varepsilon\ .$$

Thus, with probability greater than $1-\varepsilon$, (A2), (A3), (3.9) give $\tilde{T}_n=\tilde{B}_n=\tilde{A}_n=\tilde{\Gamma}_n$, $x_n=z_n$ for $n \geqslant \bar{n}$; hence

$$x_n \to x^*$$

with probability greater than $1-\varepsilon$. As ε is arbitrary, $x_n \to x^*$ with probability one.
Convergence of $\tilde{\Gamma}_n$ to $G(x^*)$ follows from lemma 3.1 applied to $\tilde{A}_n$ and to $x_i \to x^*$, and the positive definiteness of $G(x^*)$.
The proof is completed.

Proof of theorem 4.1

For a proper diagonal NxN matrix Θ_n , with diagonal elements θ_n^i , $i=1,2,\ldots,N$, $0 \leqslant \theta_n^i \leqslant 1$, we may write

$$x_{n+1}-x^*=\Theta_n(x_n-a_n\tilde{\Gamma}_n^{-1}\tilde{g}_n)-x^*=$$
$$=x_n-x^*-a_n\Theta_n\tilde{\Gamma}_n^{-1}\tilde{g}_n+(\Theta_n-I)x_n=$$
$$=x_n-x^*-a_n\Theta_n\tilde{\Gamma}_n^{-1}(\tilde{g}_n-g_n-\tau_n)-a_n\Theta_n\tilde{\Gamma}_n^{-1}\tilde{g}_n-a_n\Theta_n\tilde{B}_n^{-1}\tau_n+(\Theta_n-I)x_n=$$
$$=x_n-x^*-a_n\Theta_n\tilde{\Gamma}_n^{-1}(\tilde{g}_n-g_n-\tau_n)-a_n\Theta_n\tilde{\Gamma}_n^{-1}g(x^*)-a_n\Theta_n\tilde{\Gamma}_n^{-1}\bar{G}_n(x_n-x^*)+$$
$$-a_n\Theta_n\tilde{\Gamma}_n^{-1}\tau_n+(\Theta_n-I)x_n= \qquad (*)$$
$$=(I-a_n\Theta_n\tilde{\Gamma}_n^{-1}\bar{G}_n)(x_n-x^*)-a_n\Theta_n\tilde{\Gamma}_n^{-1}(\tilde{g}_n-g_n-\tau_n)+$$
$$-a_n(\Theta_n\tilde{\Gamma}_n^{-1}\tau_n-(\Theta_n-I)x_n).$$

Setting $\Gamma_n=\Theta_n\tilde{\Gamma}_n^{-1}\bar{G}_n$; $\quad V_n=\tilde{g}_n-g_n-\tau_n$;

$$\Phi_n=\Theta_n\tilde{\Gamma}_n^{-1} \quad ; \qquad T_n=\sqrt{n}\ \Theta_n\tilde{\Gamma}_n^{-1}\tau_n\ ,$$

we have, considering that, by iv) $\Theta_n \to I$ and that $a_n=\frac{1}{n}$,

$$\Gamma_n \to I\ , \qquad \Phi_n \to G^{-1}(x^*)\ , \qquad T_n \to 0$$

(*) $\bar{G}_n$ is the hessian matrix in a suitable point on the segment $[x_n,x^*]$.

and hence all the assumptions of Fabian's theorem (see for example Wasan (1969), page 107) are verified with $\lambda=1$, $\alpha=\beta=1$, $P=I$, $\Phi=G^{-1}(x^*)$ and $\Lambda=I$.

REFERENCES

Archetti, F. and Betrò, B. (1975). Convex programming via stochastic regularization; Quaderni del Dipartimento di Ricerca Operativa e Scienze Statistiche, N. 17, Università di Pisa.

Dixon, L.C.W. (1976). On on-line variable metric methods; N.O.C. Tech. Rept., N. 79.

Dvoretzky, A. (1956). On stochastic approximation; Proceedings of the Third Berkeley Symposium on Mathematical Statistics and Probability, I; 39, 55.

Elliott, D.E. and Sworder, D.D. (1969). A variable metric technique for parameter optimization; Automatica IFAC, 5, N. 5.

Fabian, V. (1971). Stochastic approximation; in: Optimizing Methods in Statistics, Rustagi ed. (Academic Press, New York); 439, 470.

Gray, K.B. (1962). The application of stochastic approximation to the optimization of random circuits; Amer. Math. Soc., Proceedings of Symposium on Applied Mathematics, 16; 178, 192.

Kushner, H.J. (1972). Stochastic approximation algorithms for the local optimization of functions with non unique stationary points; IEEE Trans. on A.C., AC 17, N. 5; 646, 654.

Loeve, M. (1963). Probability theory (Van Nostrand); 236.

Nevel'son, M.B. and Khas'minskii, R.Z. (1973). An adaptive Robbins - Monro procedure; Autom. Rem. Control 34, N. 10.

Poliak, B.T. and Tsypkin, Ya.Z. (1973). Pseudogradient adaptation and training algorithms; Autom. Rem. Control 34, N. 3; 377, 397.

Wasan, M.T. (1969). Stochastic approximation (Cambridge at the University Press).

PART THREE

LOCAL CONSTRAINED OPTIMISATION

TOWARDS GLOBAL OPTIMISATION 2
L.C.W. Dixon and G.P. Szegö (eds.)

A NUMERICAL COMPARISON BETWEEN TWO APPROACHES TO THE NONLINEAR PROGRAMMING PROBLEM

Dr. M.C. Bartholomew-Biggs
The Numerical Optimisation Centre
The Hatfield Polytechnic
Hertfordshire, England

This report deals with two recent proposals for solving constrained minimisation problems. One approach, suggested independently by Powell, Hestenes and Rockafellar, involves the successive unconstrained minimisation of augmented Lagrangian functions. The other approach, first devised by Murray and modified by Biggs, entails the solution of a sequence of quadratic programming subproblems. These two ideas are briefly discussed and compared. The main purpose of the paper, however, is to present some numerical evidence about the relative performance of these two methods when applied to a selection of realistic problems.

1. INTRODUCTION

Consider the nonlinear programming problem

$$\begin{aligned} &\text{Min } F(\underline{x}) \qquad \underline{x} \in E^n \\ &\text{s.t. } b_i(\underline{x}) = 0 \qquad i=1, \ldots, m \\ &\qquad b_j(\underline{x}) \geq 0 \qquad j=m+1, \ldots, q \, . \end{aligned} \tag{1}$$

A method which, since 1964, has been widely and successfully used for solving this problem is the penalty function technique (see [10]). This involves the function

$$P(\underline{x}, r) = F(\underline{x}) + \frac{1}{r} \sum_{i=1}^{m} b_i(\underline{x})^2 + \frac{1}{r} \sum_{j=m+1}^{q} (\min(0, b_j(\underline{x})))^2 \tag{2}$$

If $\underline{x}(r)$ denotes the unconstrained minimum of (2), for some value of the penalty parameter r, then it can be shown under fairly mild conditions that

$$\lim_{r \to 0} \underline{x}(r) = \underline{x}^*, \quad \text{a solution of (1).}$$

Hence penalty function algorithms require $P(\underline{x},r)$ to be minimised for some sequence of positive values of r decreasing towards zero. A trajectory of penalty function minima is thereby obtained which (possibly with the aid of some extrapolation technique) can be used to predict the solution of (1).

Experience with the penalty function method led in 1968 to some suggestions for obtaining even more efficient constrained minimisation algorithms. Both Murray [19] and Powell [21] noted that as $r \rightarrow 0$ the second derivative matrix of $P(\underline{x},r)$ tends to become ill-conditioned, thus making the unconstrained minimisations increasingly difficult near a solution of (1). Powell suggested a new form of penalty function for the equality constrained problem, namely

$$R(\underline{x}, \underline{\sigma}, \underline{\theta}) = F(\underline{x}) + \sum_{i=1}^{m} \frac{1}{\sigma_i} (b_i(\underline{x}) - \theta_i)^2 \qquad (3)$$

A similar idea was put forward by Hestenes [16]. Powell shows that there exist values of $\underline{\theta}$ such that the minimum of (3) coincides with a solution of (1) and moreover $\underline{\sigma}$ need not approach zero in order for this to happen. Hence the unconstrained minimisations of (3) are likely to be easier, in practice, than minimisations of (2).

Murray suggested that explicit minimisation of (2) should be avoided and devised a quadratic programming subproblem, the solution of which provides an approximation to a point on the trajectory of penalty function minima. In [19] he gives an algorithm which approaches a solution of (1) via a sequence of such QP subproblems.

There has been further development of both these ideas. A generalisation of (3) to include inequality constraints has been considered by Rockafellar [23] and a detailed computational algorithm is given by Fletcher [11]. This algorithm is described in the next section. Murray's method has been modified by Biggs [3], [4]. The modified algorithm - described in section 3 - includes a QP subroutine which (unlike Murray's) does not rest on the assumption that the search is close to the trajectory of penalty function minima. Furthermore the modified subproblem is set up as a QP with equality constraints whereas Murray's method is based on a QP with inequality constraints.

Methods based on (3) - or some generalisation of (3) - will be called augmented Lagrangian methods. This is fairly easy to explain. Suppose that, for the equality constrained problem, λ^*_i is the Lagrange multiplier associated with the constraint $b_i(\underline{x}) = 0$. The Lagrangian function is then

$$\mathcal{L}(\underline{x}) = F(\underline{x}) - \sum_{i=1}^{m} \lambda^*_i \, b_i(\underline{x}) \qquad (4).$$

Now it can be shown that the values of $\underline{\theta}$ and $\underline{\sigma}$ which cause $R(\underline{x}, \underline{\sigma}, \underline{\theta})$ to have a minimum at the solution of (1) are such that

$$2 \frac{\theta_i}{\sigma_i} = \lambda^*_i \qquad i=1, \ldots, m \qquad (5)$$

Hence the function (3) which gives the solution of (1) can be rewritten as

$$R(\underline{x}, \underline{\sigma}, \underline{\theta}) = F(\underline{x}) - \sum_{i=1}^{m} \lambda^*_i \, b_i(\underline{x}) + \sum_{i=1}^{m} \left\{ \frac{1}{\sigma_i} (b_i(x))^2 + \frac{1}{\sigma_i} \theta_i^2 \right\}$$

which is precisely the Lagrangian function augmented by a penalty term.

Methods based on a sequence of QP subproblems will simply be referred to as Recursive quadratic programming methods.

A brief theoretical discussion of the two approaches will be given in section 4. This is followed by a numerical comparison in which an implementation of each is tested on a range of examples. The computational results provide a fairly clear indication of the factors which govern the relative efficiency of the two approaches.

2. THE AUGMENTED LAGRANGIAN APPROACH

A generalisation of the function (3) for dealing with problem (1) when inequality constraints are present is

$$R(\underline{x}, \underline{\sigma}, \underline{\theta}) = F(\underline{x}) + \sum_{i=1}^{m} \frac{1}{\sigma_i}(b_i(\underline{x}) - \theta_i)^2 + \sum_{j=m+1}^{q} \frac{1}{\sigma_j} \text{Min}(0, b_j(\underline{x}) - \theta_j))^2 \quad (6).$$

If $\hat{\underline{x}}(\underline{\sigma}, \underline{\theta})$ denotes the unconstrained minimum of (6) for some choice of $\underline{\sigma}$ and $\underline{\theta}$ then it can be shown that $\hat{\underline{x}}(\underline{\sigma}, \underline{\theta}) = \underline{x}^*$, a solution of (1), if

$$\begin{aligned} & b_i(\hat{\underline{x}}(\underline{\sigma}, \underline{\theta})) = 0 && i=1, \ldots, m \\ \text{and either} \quad & b_j(\hat{\underline{x}}(\underline{\sigma}, \underline{\theta})) > \theta_j && \\ \text{or} \quad & b_j(\hat{\underline{x}}(\underline{\sigma}, \underline{\theta})) = 0 && j=m+1, \ldots, q \end{aligned} \quad (7)$$

The vector $\underline{\theta}$ is therefore chosen before each minimisation so that some prediction of the constraint values at the next minimum satisfies (7). An important feature is that it is not necessary for the elements of $\underline{\sigma}$ to approach zero in order for (7) to hold. It is however necessary for each σ_i to be sufficiently small for $R(\underline{x}, \underline{\sigma}, \underline{\theta})$ to have a minimum: and moreover the rate of convergence to a solution of (1) can often be improved by reducing some σ_i during the algorithm. Fletcher [11] gives some rules for obtaining suitable values of $\underline{\sigma}$ and the algorithm that he recommends is stated briefly below. Fuller details of the motivation behind the various steps will be found in Fletcher's paper.

An augmented Lagrangian algorithm

Initially $\underline{\theta} = \underline{\theta}^{(1)}$ which are externally specified parameter values. The superscript k denotes iteration number. A further parameter $V^{(k)}$ is used to monitor the rate of convergence and initially $V^{(o)} = \infty$. In the algorithm we will write μ_i for $\frac{1}{\sigma_i}$.

Step 1 Calculate $\underline{\xi} = \hat{\underline{x}}(\underline{\sigma}, \underline{\theta})$ from (6).

Step 2 Calculate $\underline{\beta}$ so that

$$\beta_i = |b_i(\underline{\xi})| \qquad i=1, \ldots, m$$
$$\beta_j = |\min(\theta_j, b_i(\underline{\xi}))| \qquad j=m+1, \ldots, q .$$

If $\|\underline{\beta}\|_\infty \geq v^{(k)}$ set $\mu_i = 10\mu_i$ for all i s.t. $\beta_i \geq v^{(k)}$ and return to step 1.

Step 3 Set $k = k+1$, $\underline{\theta}^{(k)} = \underline{\theta}$, $\underline{\mu}^{(k)} = \underline{\mu}$, $v^{(k)} = \|\underline{\beta}\|_\infty$.

If $v^{(k)} \leq \epsilon$, some preset tolerance, stop.

Step 4 Obtain a new vector $\underline{\theta}$.

Let $M(\underline{\theta}^{(k)}) = \{i : i \leq m \text{ or } b_i(\underline{\xi}) < \theta_i^{(k)}, \quad i=m+1, \ldots, q\}$ define a set of, say t, active constraints. Let the t-vector of constraint values at $\underline{\xi}$ be denoted by $\underline{g}$ and let the normals to the active constraints be the rows of the t x n matrix A. Let B be an estimate, obtained during the last minimisation of (6), of

$\nabla^2 R(\underline{\xi}, \underline{\sigma}^{(k)}, \underline{\theta}^{(k)})$.

$$\text{Calculate } \underline{\phi} = -(AB^{-1}A^T)^{-1}\underline{g} \tag{8}$$

set $j = 0$.

For $i=1, \ldots, q$

set $\left.\begin{array}{l} j = j-1 \\ \Delta\theta_i = -\theta_i^{(k)} \end{array}\right\}$ if $i \notin M(\underline{\theta}^{(k)})$

$\Delta\theta_i = \phi_{j+i}/\mu_i$ if $i \in M(\underline{\theta}^{(k)})$.

Set $\underline{\theta} = \underline{\theta}^{(k)} + \Delta\underline{\theta}$.

Step 5 Obtain a new vector $\underline{\mu}$.

This is based on the difference between the correction $\Delta\underline{\theta}$ used in step 4 and the alternative correction $\delta\underline{\theta}$ (Powell [21]) given by

$$\delta\theta_i = -b_i(\underline{\xi}) \qquad i=1, \ldots, m$$

$$\delta\theta_j = -\min(\theta_j^{(k)}, b_j(\underline{\xi})) \qquad j=m+1, \ldots, q.$$

The new $\underline{\mu}$ vector is given by

$$\mu_i = \mu_i^{(k)} \max\left(1, 4\left|\frac{\delta\theta_i - \Delta\theta_i}{\delta\theta_i}\right|\right) \qquad i=1, \ldots, q.$$

Return to step 1.

Further comments on this algorithm will be made in section 4.

3. The Recursive Quadratic Programming approach

We have already mentioned that this method was derived in order to give approximations to points in the trajectory of penalty function minima. It has the effect of separating the contributions made by the function and the constraints in determining the position of the minimum of $P(\underline{x}, r)$.

In describing the method briefly we shall refer to $M(\underline{x})$ as the set of active constraints at the point x. These constraints can be thought of, for example, as being the equality constraints together with the violated inequalities. In the algorithm below however it will be seen that this definition is slightly extended since this has been found to be advantageous in practice. Let $\underline{g}$ be the vector of active constraint values and let A denote the matrix whose rows are the normals of the active constraints. Also let $\underline{f} = \nabla F(\underline{x})$.

Consider the Taylor series prediction

$$\nabla P(\underline{x} + \underline{p}, r) = f + (\nabla^2 F)p + \frac{2}{r} A^T g + \frac{2}{r} A^T A\, p + \frac{2}{r} \sum_{i \in M} g_i \nabla^2 g_i .$$

Let $L = \nabla^2 F + \frac{2}{r} \sum_{i \in M} g_i \nabla^2 g_i$.

Suppose that $\underline{x} + \underline{p}$ is the position of the minimum of $P(\underline{x}, r)$. Then it can be shown that to determine $\underline{p}$ as the solution of the QP

$$\text{Min } \tfrac{1}{2} \underline{p}^T L \underline{p} + f^T \underline{p}$$

$$\text{s.t. } A\underline{p} = -\frac{r}{2} \hat{\underline{\mu}} - \underline{g} \qquad (9)$$

$$\text{where } (\tfrac{r}{2} I + A L^{-1} A^T) \hat{\underline{\mu}} = AL^{-1} \underline{f} - \underline{g} \qquad (10)$$

is equivalent to using the Newton-like prediction

$$(L + \frac{2}{r} A^T A)\, \underline{p} = -\underline{f} - \frac{2}{r} A^T g \qquad (11).$$

There are several advantages in using (9) rather than (11). One is that the matrix $(L + \frac{2}{r} A^T A)$ will tend to become ill-conditioned as $r \to 0$. Problem (9) does not involve the use of this matrix and instead we may work with L and $(\frac{r}{2} I + AL^{-1}A^T)$ neither of which need be badly-conditioned (unless, for instance, the active constraint normals in problem (1) are not linearly independent). When using (9) it is also easy to make use of updated approximations to the matrix L. It can in fact be shown that as the solution of (1) is approached <u>along the trajectory of penalty function minima</u> the matrix L tends to the Hessian matrix of the Lagrangian function. Similarly the vector $\hat{\underline{\mu}}$ given by (10) will tend to the vector of Lagrange multipliers associated with the binding constraints of problem (1). Finally, problem (9) brings out the role of the parameter r in forcing the solution estimates towards the feasible region. This makes it easy to vary r from iteration to iteration so as to maintain good progress.

An algorithm implementing the recursive quadratic programming approach will now be described. This has already been reported in references [4] and [7].

<u>A recursive quadratic programming algorithm</u>

Initially a penalty parameter $r = r^{(1)}$ is specified. A scalar $\alpha < 1$ must also be given for use in step 1.

An initial point $x^{(1)}$ is given together with a <u>positive definite</u> estimate $H^{(1)}$ of the inverse Hessian matrix of the Lagrangian function.

The active constraint set at $\underline{x}^{(1)}$ consists of the equality constraints together with any violated inequality constraints.

The superscript k denotes iteration number.

Step 1 If there are no active constraints set $\hat{\underline{\mu}}^{(k)} = 0$ and go to step 2. Otherwise calculate $\hat{\underline{\mu}}^{(k)}$ by solving

$$(\frac{r}{2} I + A^{(k)} H^{(k)} A^{(k)T}) \hat{\underline{\mu}}^{(k)} = A^{(k)} H^{(k)} \underline{f}^{(k)} - \underline{g}^{(k)}.$$

check that $\frac{r}{2} (\underline{g}^{(k)T} \hat{\underline{\mu}}^{(k)} + \underline{g}^{(k)T} \underline{g}^{(k)}) \geq 0$.

If not, set $r = - 2\alpha\ \underline{g}^{(k)T} \underline{g}^{(k)} / \underline{g}^{(k)T} \hat{\underline{\mu}}^{(k)}$ and recalculate $\hat{\underline{\mu}}^{(k)}$.

Step 2 Calculate

$$\underline{p} = H^{(k)} (A^{(k)T} \hat{\underline{\mu}}^{(k)} - \underline{f}^{(k)})$$

(the solution to (9) with $H^{(k)}$ replacing L^{-1}).

Perform a line search along $\underline{p}$ to obtain a new point

$$\underline{x}^{(k+1)} = \underline{x}^{(k)} + s\ \underline{p}$$

where

$$|P (\underline{x}^{(k+1)},r) - P (\underline{x}^{(k)},r) - s\ p^T \nabla P (\underline{x}^{(k)},r)\ | \geq \epsilon_1 |s\underline{p}^T \nabla P(\underline{x}^{(k)},r)|$$

and

$$|P (\underline{x}^{(k+1)},r) - P (x^{(k)},r)| \leq \epsilon_2\ sp^T \nabla P (\underline{x}^{(k)},r)$$

for some prescribed $\epsilon_1, \epsilon_2 > 0$.

Step 3 Update the matrix $H^{(k)}$ so that

$$H^{(k+1)} (x^{(k+1)} - x^{(k)}) = f^{(k+1)} - A^{(k+1)T} \hat{\underline{\mu}}^{(k)} - f^{(k)} + A^{(k)T} \hat{\underline{\mu}}^{(k)} .$$

$H^{(k+1)}$ is obtained from $H^{(k)}$ using the Broyden-Fletcher-Shanno formula. The update is only applied when it will yield a positive definite matrix $H^{(k+1)}$.

Step 4 Determine the active constraint set at $\underline{x}^{(k+1)}$. As before, the equality constraints and any violated inequality constraints are regarded as active. The active set must also include all inequality constraints which were active at $\underline{x}^{(k)}$ and for which the corresponding elements of $\hat{\underline{\mu}}^{(k)}$ was positive.

Step 5 Obtain a new value of r.

set $r^{(k)} = r$

if $\underline{g}^{(k+1)T} \hat{\underline{\mu}}^{(k)} \geq 0, \quad r = r^{(k)}$

if $\underline{g}^{(k+1)T} \hat{\underline{\mu}}^{(k)} < 0, \quad r = -2\alpha\ \underline{g}^{(k+1)T} \underline{g}^{(k+1)} / \underline{g}^{(k+1)T} \hat{\underline{\mu}}^{(k)}$.

Replace k by k+1 and return to step 1.

Briefly to comment on this algorithm, it may be said that the use of a positive definite matrix to estimate the (possibly indefinite) matrix L has been justified in [5]. The test on r in step 1 is intended to force the search directions to reduce the active constraint penalty, (i.e. $\underline{g}^{(k)T} A^{(k)} \underline{p} \leq 0$). Since $\hat{\underline{\mu}}^{(k)}$ can be regarded as a vector of Lagrange multiplier approximations the updating of H in step 3 is consistent with this matrix being an estimate of the inverse Hessian of the Lagrangian function. Again, near the solution at least, it should be helpful to take account of the sign of these Lagrange multiplier estimates in choosing the active constraint set in step 5.

4. Discussion and numerical comparisons

Two particular algorithms have been presented in some detail and these will be used in the numerical tests later in this section. We now wish to consider these algorithms as representatives of two different approaches. Besides the references already cited, methods like that of section 2 have been discussed by Hartmann [15], Bertsekas [2], Kort [17] and Sayama et al [24]. Other algorithms based on QP subproblems (not necessarily related to the penalty function approach) have been proposed by Fletcher [12], [13], Han [14] and Powell [22].

In both the augmented Lagrangian methods and the recursive quadratic programming algorithms an important feature is the calculation and use of a vector of Lagrange multiplier estimates. The difference between the approaches is that the QP methods obtain new Lagrange multiplier estimates at every new point while the augmented Lagrangian techniques retain the same estimates throughout a complete minimisation of (6). Now it can be seen from (8) and (10) that the calculation of Lagrange multiplier estimates involves the formation of a matrix product and a factorization of the result. If there are t active constraints, therefore, the cost of obtaining Lagrange multiplier estimates is $O(t^2 n + tn^2 + t^3)$ multiplications. The cost is incurred during the calculation of every point reached by the QP methods. On the other hand the augmented Lagrangian methods obtain most of their new points at a cost of $O(n^2)$ multiplications (if a variable metric minimisation method is being used) and the more expensive Lagrange multiplier calculation is only performed a few times. Of course it is to be expected that there will be some advantage from continually revising the Lagrange multiplier estimates: and it can be seen that it may be inefficient to devote a lot of effort to the minimisation of $R(\underline{x}, \underline{\sigma}, \underline{\theta})$ when the $\underline{\theta}$ vector is far from its desired value. When $t << n$ then the expense of calculating multiplier estimates will not be great: but as t and n become large we shall be concerned to see whether the greater cost per point of the QP methods is balanced by a significant advantage in the number of points tried.

This is the background for the numerical tests. The problems chosen are specified in some detail in the appendix. Here they are merely summarised in terms of dimensions.

Problem name	n	m_e	m_i	m_b
Test 1	4	1	9	3
Transformer design [1]	6	-	8	2
Static Power Scheduling [9]	9	6	8	8
Colville Problem 3 [8]	5	-	16	5
Shell Dual [8]	15	-	20	11
Shell Dual (equalities)	15	5	15	11
Aircraft design (linear objective) [20]	14	5	27	10
Aircraft design (nonlinear objective) [20]	14	5	27	10
Dynamic Power Scheduling [6]	15	-	65	15

n = number of variables

m_e = number of equality constraints

m_i = number of inequality constraints

m_b = number of binding constraints

Table 1 - Summary of problems

	Augmented Lagrangian Method		Recursive quadratic programming method	
	function evals.	gradient evals.	function evals.	gradient evals.
Test 1	100	68	21	18
Transformer design*	92	65	41	33
Static Power Scheduling	94	59	11	10
Colville problem 3	63	43	12	12
Shell Dual	141	107	44	40
Shell Dual (equalities)	147	96	25	25
Aircraft design* (linear objective)	167	125	123	79
Aircraft design* (nonlinear objective)	124	92	47	42
Dynamic Power Scheduling (version 1)	136	92	33	29

* Numerical derivatives were used

Table 2

Comparison between the methods in terms of function and gradient evaluation

These problems were attempted using computer programs written at the N.O.C. and implementing the algorithms given in sections 2 and 3. Table 2 above shows the performance of these algorithms measured in terms of number of function and gradient evaluations required for convergence. (The line search routine used by both the tested methods only required function values at trial points and gradient values were only computed when a new solution estimate had been accepted.) The calculations were performed using single precision FORTRAN on the Hatfield Polytechnic PDP-10 computer.

It should first be pointed out that the result for our implementation of the augmented Lagrangian method applied to the Colville problems compares well with the figures quoted by Fletcher [11]. Hence there is evidence that an efficient code is being used to represent this class of methods. Secondly, to put these results in context, we may mention that a conventional penalty function method, based on (2), requires 113 gradient evaluations to solve problem Test 1 and 188 gradient evaluations to solve the transformer design problem. This provides a measure of the improvement offered by the techniques discussed here compared with earlier optimisation methods. Finally - and of chief interest in the present report - we see that the recursive quadratic programming method almost invariably requires significantly fewer function and gradient evaluations that the augmented Lagrangian method. This is in line with the remarks made above which suggested that the QP methods would benefit from revising the Lagrange multiplier estimates at every point. It becomes more important therefore that the methods should also be compared in terms of the time taken to solve the problems. This will reveal the extent to which the greater number of function evaluations needed by the augmented Lagrangian methods is compensated for by the cheapness of each iteration. The results will also give an indication of how far the relative efficiencies of the two approaches is governed by problem size. The comparison of computer times is given in Table 3.

Table 3 shows that, for these test problems, the advantage in function evaluations of the recursive QP methods is not always accompanied by a corresponding advantage in computer time. In fact the two algorithms may be much more closely matched than table 2 would suggest. The QP methods seem to enjoy a clear superiority for the smaller problems: but as numbers of variables and constraints increase then the augmented Lagrangian approach may be faster because it requires less "housekeeping" on each iteration. Clearly, however, the cost of evaluating the function will also influence the relative efficiencies of the two approaches. A very expensive function calculation might completely outweigh the effect of the more costly iterations performed by the recursive quadratic programming techniques.

Problem	Augmented Lagrangian method	Recursive quadratic programming method
	Time (sec)	Time (sec)
Test 1	3.5	1.7
Transformer design*	3.8	3.0
Static Power Scheduling	12.6	3.8
Colville problem 3	3.3	1.7
Shell Dual	31.5	47.0
Shell Dual (equalities	26.4	27.0
Aircraft design* (linear objective)	119.0	114.0
Aircraft design* (nonlinear objective)	121.5	64.0
Dynamic Power Scheduling	40.7	43.7

* Numerical derivatives were used

Table 3

To be more precise about the relative speeds of the two methods, let n_r denote the number of function evaluations needed by the recursive quadratic programming algorithm to solve a problem and let n_a denote the number of function evaluations performed by the augmented Lagrangian method. Let τ denote the time taken for each complete function and gradient evaluation. Then a good estimate of the time taken by the augmented Lagrangian method is given by

$$t_a = n_a \, (k_a \, n^2 + \tau)$$

where k_a is a machine dependent constant and n is the number of variables in the problem. Similarly the time taken by the recursive quadratic programming method can be estimated by

$$t_r = n_r \, (k_r \, mn^2 + \tau)$$

where k_r is a machine dependent constant and m represents the number of binding constraints in the problem. Even though k_a and k_r are machine dependent we should expect the ratio k_a/k_r to be approximately constant for all types of computer. From the experiments performed in this report it appears that suitable values for the PDP-10 computer are $k_a = .001$ and $k_r = .0004$.

In considering the practical implications of the differences in execution time for the two algorithms it will be helpful to distinguish four cases.

i) If $\tau << k_a n^2$ (function evaluation very cheap)

$$\frac{t_a}{t_r} = \frac{k_a}{k_r m} \quad \frac{n_a}{n_r} \quad .$$

ii) If $\tau = k_a \, n^2$ (function evaluation comparable with an iteration of the augmented Lagrangian method)

$$\frac{t_a}{t_r} = \left(\frac{2\,k_a}{k_r m + k_a}\right) \frac{n_a}{n_r}$$

iii) If $\tau = \sqrt{k_a\, k_r m}\; n^2$ (function evaluation more expensive than an iteration of the augmented Lagrangian method)

$$\frac{t_a}{t_r} = \frac{k_a + \sqrt{k_a\, k_r m}}{k_r m + \sqrt{k_a\, k_r m}} \quad \frac{n_a}{n_r} \ .$$

iv) If $\tau = k_r m\, n^2$ (function evaluation comparable with an iteration of the recursive QP method)

$$\frac{t_a}{t_r} = \frac{k_a + k_r m}{2 k_r m} \cdot \frac{n_a}{n_r} \ .$$

Using the values of k_a and k_r calculated above, it can be seen that in case (1) the recursive quadratic programming approach can be expected to be faster if $n_a > .4m\, n_r$. The results in Table 2 suggest that the ratio n_a/n_r typically lies between 2 and 4. Hence for problems where the function value is very inexpensive we may anticipate that when m is greater than about 10 the augmented Lagrangian approach is likely to be faster. Treating the other cases in the same way we can present the following Table 4 which gives guidelines for the selection of algorithm to solve particular problems.

Type of function evaluation	(i)	(ii)	(iii)	(iv)
Range of m for which QP methods are likely to be faster.	m < 10	m < 18	m < 40	all m

Table 4

The choice of methods based on the cost per function evaluation and the number of active constraints in the problem.

In order to check the estimates in Table 4 some further test examples were devised, which would allow the effect of dimensionality to be observed. The first problem (SPTEST) was intended to represent case (i) and consisted of a simple function and some sparse equality constraints. (The problems are given in more detail in the appendix.) In the second problem (DNTEST) the Jacobian matrix of constraint normals was dense: and thus the function and gradient evaluations were rather more costly. For the third example the problem DNTEST was run with numerically estimated derivatives. It frequently happens in practical problems that derivatives are approximated by differences and hence this is a quite realistic way of providing a case where the function and gradient evaluation is rather expensive.

m	augmented Lagrangian method	Recursive quadratic programming method
8	13.2	10.6
10	25.2	18.3
15	42.2	56.0
20	120.0	146.0

Table 5: Computer times (sec) for problem SPTST

m	augmented Lagrangian method	Recursive quadratic programming method
5	7.2	4.3
8	21.4	16.5
10	44.5	32.8
15	121.0	118.0
20	218.5	406.3

Table 6: Computer times (sec) for problem DNTST

m	augmented Lagrangian method	Recursive quadratic programming method
5	16.1	10.0
10	143.9	50.5
11	187.6	78.4
12	249.6	93.5
15	> 600	191.5

Table 7: Computer times (sec) for problem DNTST with numerically estimated derivatives.

All these problems could be run with different number of constraints, m, and the dimension was set at n = 2m. The tables given above show how solution times for the two methods varied with problem size. The results can be seen to be consistent with the estimates made in Table 4. For the cheapest function (table 5) the augmented Lagrangian approach becomes faster for $10 < m < 15$. Table 6 shows that for slightly more expensive function evaluations the cross-over is between $m = 15$ and $m = 20$. For the problem with numerical derivatives table 7 demonstrates the consistent superiority of the QP method.

5. Conclusion

We have shown in this paper that the augmented Lagrangian methods and the recursive quadratic programming methods are both very effective approaches to the nonlinear programming problem. A numerical comparison has revealed that the QP methods frequently require fewer function evaluations than augmented Lagrangian methods but perform more overhead calculations per iteration. It is therefore possible to give some general guidelines on the choice of algorithm for solving a particular optimisation problem.

The augmented Lagrangian methods should be favoured if the dimensions are large and the function is fairly inexpensive to calculate. On the other hand, the QP methods are to be prefered for small problems or those where the function evaluation is very costly.

6. References

1. Ballard, D.H., Jelinek, C.O. & Schinzinger, R. An algorithm for the solution of constrained generalised polynomial programming problems. Comp. J. Vol 17, No. 3 p 261 (1974).
2. Bertsekas, D.P. On penalty and multiplier methods for constrained minimisation. In Nonlinear Programming 2 edited by O.L. Masgasarian, R.R. Meyer, & S.M. Robinson (Academic Press) (1975).
3. Biggs, M.C. Constrained minimisation using recursive equality quadratic programming. In Numerical Methods for Nonlinear Optimization edited by F.A. Lootsma (Academic Press) (1972).
4. Biggs, M.C. Constrained minimisation using recursive quadratic programming: some alternative subproblem formulations. In Towards Global Optimisation edited by L.C.W. Dixon & G.P. Szegö (North Holland) (1975).
5. Biggs, M.C. Constrained minimisation using recursive quadratic programming: some convergence properties. Numerical Optimisation Centre Technical Report No. 52, The Hatfield Polytechnic (1974).
6. Biggs, M.C. An approach to optimal scheduling of an electric power system. Numerical Optimisation Centre Technical Report No. 63, Hatfield Polytechnic (1975).

7. Biggs, M.C. Some improvements to the 'OPTIMA' subroutines. Numerical Optimisation Centre Technical Report No. 69, The Hatfield Polytechnic (1975).

8. Colville, A.R. A comparative study on nonlinear programming codes. I.B.M. Report 320-2949 (1968).

9. Dillon, T.S. & Morstzyn, K. Active and reactive load scheduling in a thermal power system in the presence of tap changing transformers using nonlinear programming. Trans. IEAEE Vol. EE8 No. 2 (1972).

10. Fiacco, A.V. & McCormick, G.P. Nonlinear programming - sequential unconstrained minimisation techniques. (John Wiley) (1968).

11. Fletcher, R. An ideal penalty function for constrained optimisation. In Nonlinear Programming 2 edited by O.L. Mangasarian, R.R. Meyer & S.M. Robinson (Academic Press) (1975).

12. Fletcher, R. An algorithm for solving linearly constrained optimisation problems. Math. Prog. Vol. 2 No. 2 p 133 (1972).

13. Fletcher, R. An exact penalty function for nonlinear programming with inequalities. Math. Prog. Vol. 5 No. 2 p 129 (1973).

14 Han, S.P. Superlinearly convergent variable metric algorithms for general nonlinear programming problems. Dept. of Computer Science Technical Report 75-233, Cornell University (1975).

15. Hartman, J.K. Iterative determination of parameters for an exact penalty function. Journ. Opt. Theory & Applics. Vol. 16 p 49 (1974).

16. Hestenes, M.R. Multiplier and gradient methods. Journ Opt. Theory & Applics Vol. 4 p 303 (1969).

17. Kort, B. Rate of convergence of the method of multipliers with inexact minimisation. In Nonlinear Programming 2 edited by O.L. Mangasarian, R.R. Meyer & S.M. Robinson (Academic Press) (1975).

18. Murray, W. Ill conditioning in penalty and barrier functions arising in constrained nonlinear programming. In proceedings of the 6th International Symposium on Mathematical Programming (1967).

19. Murray, W. An algorithm for constrained minimisation. In Optimisation edited by R. Fletcher (Academic Press) (1969).

20. Piggott, B.A.M. & Taylor, B.E. Application of numerical optimisation techniques to the preliminary design of a transport aircraft. Royal Aircraft Establishment Technical Report 71074 (1971).

21. Powell, M.J.D. A method for nonlinear constraints in minimisation problems. In Optimisation edited by R. Fletcher (Academic Press) (1969).

22. Powell, M.J.D. Algorithms for nonlinear constraints that use Lagrangian functions. Presented at IX International Symposium on Math. Prog. (1976).

23. Rockafellar, R.T. A dual approach to solving nonlinear programming problems by unconstrained optimisation. Math. Prog. Vol. 5 No. 3 p 354 (1973).

24. Sayaria, H., Kameyama, Y., Nakayana, H. & Sawaragi, Y. The generalised Lagrangian functions for mathematical programming problems. Institute for systems design and optimisation Report 55, Kansas State University (1975).

Appendix Details of test problems

Test 1 - a small example

$$\text{Minimise} \quad x_1 x_4 (x_1 + x_2 + x_3) + x_3$$

$$\text{s.t.} \quad x_1^2 + x_2^2 + x_3^2 + x_4^2 = 40$$

$$x_1 x_2 x_3 x_4 \geq 25$$

$$5 \geq x_i \geq 1 \qquad i=1, \ldots, 4$$

starting point (1, 5, 5, 1)
solution (1, 4.743, 3.82115, 1.37941)

Transformer design problem [1]

Minimise

$$.0204\, x_1 x_4 (x_1 + x_2 + x_3) + .0187\, x_2 x_3 (x_1 + 1.57\, x_2 + x_4)$$

$$+ .0607\, x_1 x_4 x_5^2 (x_1 + x_2 + x_3) + .0437\, x_2 x_3 x_6^2 (x_1 + 1.57\, x_2 + x_4)$$

$$\text{s.t.} \quad x_i \geq 0 \qquad i=1, \ldots, 6$$

$$.001\, (x_1 x_2 x_3 x_4 x_5 x_6) \geq 2.07$$

$$1 - .00062\, x_1 x_4 x_5^2 (x_1 + x_2 + x_3) - .00058\, x_2 x_3 x_6^2 (x_1 + 1.57\, x_2 + x_6) \geq 0$$

where x_1, x_2, x_3, x_4 are physical dimensions of winding and core; x_5 is magnetic flux density and x_6 is current density. The objective function represents the worth of the transformer, including operating costs; and the constraints, other than the simple bounds, refer to the rating of the transformer and the allowable transmission loss.

Feasible starting point	Solution
$x_1 = 5.54$, $x_2 = 4.4$	$x_1 = 5.33$ $x_2 = 4.66$
$x_3 = 12.02$, $x_4 = 11.82$	$x_3 = 10.43$ $x_4 = 12.08$
$x_5 = .702$, $x_6 = .852$	$x_5 = .752$ $x_6 = .878$.

Static power scheduling problem [10]

Minimise

$$3000\, x_1 + 1000\, x_1^3 + 2000\, x_2 + 666.667\, x_2^3$$

s.t.

$$.4 - x_1 + 2C\,x_5^2 + x_5\,x_6\,(D \sin(-x_8) - C \cos(-x_8)) + x_5\,x_7\,(D \sin(-x_9) - C \cos(-x_9)) = 0$$

$$.4 - x_2 + 2C\,x_6^2 + x_6\,x_5\,(D \sin(x_8) - C \cos(x_8)) + x_6\,x_7\,(S \sin(x_8 - x_9) - C \cos(x_8 - x_9)) = 0$$

$$.8 + 2C\,x_7^2 + x_7\,x_5\,(D \sin(x_9) - C \cos(x_9)) + x_7\,x_6\,(D \sin(x_9 - x_8) - C \cos(x_9 - x_8)) = 0$$

$$.2 - x_3 + 2D\,x_5^2 - x_5\,x_6\,(C \sin(-x_8) + D \cos(-x_8)) - x_5\,x_7\,(C \sin(-x_9) + D \cos(-x_9)) = 0$$

$$.2 - x_4 + 2D\,x_6^2 - x_6\,x_5\,(C \sin(x_8) + D \cos(x_8)) - x_6\,x_7\,(C \sin(x_8 - x_9) + D \cos(x_8 - x_9)) = 0$$

$$-.337 + 2D\,x_7^2 - x_7\,x_5\,(C \sin(x_9) + D \cos(x_9)) - x_7\,x_6\,(C \sin(x_9 - x_8) + D \cos(x_9 - x_8)) = 0$$

$$x_i \geq 0 \qquad i=1,\ 2$$

$$1.0909 \geq x_i \geq 0.90909 \qquad i=5,\ 6,\ 7$$

where $C = \sin(0.25)\ 48.4 / 50.176$

$D = \cos(0.25)\ 48.4 / 50.176$.

In this problem x_1 and x_2 are the real power outputs from two generators; x_3 and x_4 are the reactive power outputs; x_5, x_6 and x_7 are voltage magnitudes at three nodes of an electrical network and x_8 and x_9 are voltage phase angles at two of these nodes. The equality constraints are the real and reactive power balance equations, stating that the power flowing into a node must balance the power flowing out. The remaining constraints are simple limits on the real power and the voltage magnitudes.

The starting point used is

$x_1 = x_2 = .8,\ \ x_3 = x_4 = .2,\ \ x_5 = x_6 = x_7 = 1.0454,\ \ x_8 = x_9 = 0.$

The solution is

$x_1 = .667, \quad x_2 = 1.0224, \quad x_3 = .2283, \quad x_4 = .1848, \quad x_5 = x_6 = 1.0909$

$x_7 = 1.0691, \quad x_8 = .1066, \quad x_9 = -.3388.$

The <u>Shell Dual</u> problem is given in full by Colville [8]. The problem entitled <u>Shell Dual (equalities)</u> is identical except that the nonlinear constraints are treated as equalities. The starting point used for both problems is

$x_i = 10^{-4}$ i=1. 2, ..., 6, 8, ..., 15

$x_7 = 60.$

Aircraft design problem [20]

This problem can be expressed as:

$$
\begin{array}{lrl}
\text{Min} & x_{14} & \\
\text{s.t.} & x_{14} & = c_o\,(\underline{x}) \\
& x_2 & \geq c_1\,(\underline{x}) \\
& c_2(\underline{x}) & \geq x_4 \\
& x_1 & \geq c_3\,(\underline{x}) \\
& x_1 & \geq c_4\,(\underline{x}) \\
& x_1 & \geq c_5\,(\underline{x}) \\
& 1 & \geq x_1 \geq 0 \\
& 2 & \geq x_2 \geq 0 \\
& .35 & \geq x_3 \geq .15 \\
& .15 & \geq x_4 \geq .06 \\
& .35 & \geq x_5 \geq .25 \\
& .2 & \geq x_6 \geq .05 \\
& .8 & \geq x_7 \geq .3 \\
& .35 & \geq x_8 \geq .05 \\
& 1 & \geq x_9 \geq .5 \\
& .45 & \geq x_{10} \geq 0 \\
& .45 & \geq x_{11} \geq 0 \\
& .3 & \geq x_{12} \geq 0 \\
& .3 & \geq x_{13} \geq 0.
\end{array}
$$

The variables have the following meanings.

$x_1 = \dfrac{\text{mass of engines}}{10000}$ (kg) ; $x_2 = \dfrac{\text{wing area}}{100}$ (m^2)

$x_3 = \dfrac{\text{sweepback angle}}{100}$ (deg) ; x_4 = wing thickness-chord ratio

x_5 = wing taper tatio ; $x_6 = \dfrac{\text{chord of leading edge high-lift device}}{\text{wing chord}}$

$x_7 = \dfrac{\text{span of flaps}}{\text{wingspan}}$; $x_8 = \dfrac{\text{chord of trailing edge high-lift device}}{\text{wing chord}}$

x_9 = aspect ratio ; $x_{10} = \dfrac{\text{flap angle at landing threshold}}{100}$ (deg)

$x_{11} = \dfrac{\text{flap angle at take off}}{100}$ (deg) ;

$x_{12} = \dfrac{\text{deflection of leading edge device at landing}}{100}$ (deg) ;

$x_{13} = \dfrac{\text{deflection of leading edge device at take off}}{100}$ (deg) ;

$x_{14} = \dfrac{\text{total take off mass of aircraft}}{100000}$ (kg) .

The expressions for the constraint function $c_o(\underline{x})$, ..., $c_5(\underline{x})$ are given by Perry in RAE Technical Memorandum Aero 1237 (1970).

The scaling applied to the physical quantities to yield the optimisation variables listed above are not precisely the same as those used by Piggott & Taylor. Their scaling sought to bring all the optimisation variables to order unity. We found that the problem could be solved more rapidly if the scaling was related to the (practically significant) tolerances associated with each physical variable. For instance the tolerance on the mass of the engines is 5 kg and that on the wing sweepback is .05 degrees. With the scaling used here both these tolerances correspond to changes of the same order (.0005) in the variables x_1 and x_3. This kind of "balancing" among the variables appears to improve the ultimate convergence of the optimisation.

The problem as posed above has a linear objective function. It is interesting to compare this formulation with the equivalent problem of minimising $c_o(\underline{x})$ (subject to the same set of constraints). The aircraft design problem has in fact been attempted in this paper using both the linear and the nonlinear objective function. The feasible starting point is

$x_1 = .5461$ $x_2 = .9856$ $x_3 = .3248$ $x_4 = .101$ $x_5 = .3$ $x_6 = .119$

$x_7 = .63$ $x_8 = .299$ $x_9 = .75$ $x_{10} = .3$ $x_{11} = .299$ $x_{12} = .15$

$x_{13} = .15$ $x_{14} = .41118$.

The solution is

$x_1 = .3505$ $x_2 = .9209$ $x_3 = .3169$ $x_4 = .127$ $x_5 = .25$ $x_6 = .05$

$x_7 = .8$ $x_8 = .058$ $x_9 = .581$ $x_{10} = .45$ $x_{11} = .314$ $x_{12} = .278$

$x_{13} = .236$ $x_{14} = .34096$.

5. Dynamic power scheduling problem [6]

Minimise

$$\sum_{k=0}^{4} 2.3\, x_{3k+1} + .0001\, x_{3k+1}^2 + 1.7\, x_{3k+2} + .0001\, x_{3k+2}^2 + 2.2\, x_{3k+3} + .00015\, x_{3k+3}^2$$

s.t.

$$90 \geq x_{3k+1} \geq 0 \qquad k=0, \ldots, 4$$
$$120 \geq x_{3k+2} \geq 0 \qquad k=0, \ldots, 4$$
$$60 \geq x_{3k+3} \geq 0 \qquad k=0, \ldots, 4$$
$$6 \geq x_1 - 15 \geq -7$$
$$6 \geq x_{3k+1} - x_{3k-2} \geq -7 \qquad k=1, \ldots, 4$$
$$7 \geq x_2 - 50 \geq -7$$
$$7 \geq x_{3k+2} - x_{3k-1} \geq -7 \qquad k=1, \ldots, 4$$
$$6 \geq x_3 - 10 \geq -7$$
$$6 \geq x_{3k+3} - x_{3k} \geq -7 \qquad k=1, \ldots, 4$$
$$x_1 + x_2 + x_3 \geq 60$$
$$x_4 + x_5 + x_6 \geq 50$$
$$x_7 + x_8 + x_9 \geq 70$$
$$x_{10} + x_{11} + x_{12} \geq 85$$
$$x_{13} + x_{14} + x_{15} \geq 100$$

Starting point (feasible)

$x_1 = 20 \quad x_2 = 55 \quad x_3 = 15 \quad ; \quad x_4 = x_7 = x_{10} = x_{13} = 20$

$x_5 = x_8 = x_{11} = x_{14} = 60 \quad ; \quad x_6 = x_9 = x_{12} = x_{15} = 20$

Solution

$x_1 = 8 \quad x_2 = 49 \quad x_3 = 3 \quad x_4 = 1 \quad x_5 = 56 \quad x_6 = 0 \quad x_7 = 1$

$x_8 = 63 \quad x_9 = 6 \quad x_{10} = 3 \quad x_{11} = 70 \quad x_{12} = 12 \quad x_{13} = 5 \quad x_{14} = 77$

$x_{15} = 18.$

This is a representation of the problem of scheduling three generators to meet the demand for power over a period of time. The variable x_{3k+i} denotes the output from the i-th generator at time $t^{(k)}$. The constraints in the problem are upper and lower limits on the power available from each generator, bounds on the amount by which the output from a generator can change from time $t^{(k)}$ to $t^{(k+1)}$, and the condition that at each time $t^{(k)}$ the power generated must at least satisfy the demand.

Problem SPTST

$$\min_x \sum_{i=1}^{2m} x_i^2 + (x_1 - x_n)^2$$

$$\text{s.t. } a_i\, x_{2i-1}^2 + b_i\, x_{2i} = c_i \qquad i=1, \ldots, m$$

for fixed a_i, b_i, c_i.

Problem DNTST

$$\min_x (x_1 - x_n)^2 + \sum_{i=1}^{n-1} (x_{i+1} - x_i)^2 \qquad n = 2m$$

$$\text{s.t. } \sum_{j=1}^{n} a_{ij} x_j^2 = b_i \qquad i=1, \ldots, m$$

TOWARDS GLOBAL OPTIMISATION 2
L.C.W. Dixon and G.P. Szegö (eds.)

AN INTRODUCTION TO FUNCTIONAL LINEAR PROGRAMMING

Dr. J.J. McKeown
The Numerical Optimisation Centre
The Hatfield Polytechnic
Hertfordshire, England

Functional Linear Programming is a technique originally devised [2, 3] to solve a particular class of practical problems, but which is certainly capable of much wider application. The aim of this paper is to provide a straightforward description of the method, and to relate it to other algorithms of a similar character.

1. INTRODUCTION

Consider the following classical linear programming problem:

$$\min_{\underline{x}} \quad W = \underline{c}^t \underline{x} \qquad x \in R^n$$

$$\text{s.t. } \underline{Ax} = \underline{b} \tag{1}$$

$$\underline{x} \geq 0$$

Where $\underline{A}$ is an mxn matrix and $m < n$.

If there is a solution to (1), then it will be of the following form:

$$\underline{x}^* = \{ \underline{x}_1 \ \underline{x}_2 \} \qquad \text{(where } \{ \} \text{ denotes a column vector)}$$

$$\underline{x}_1 \in R^m, \ \underline{x}_2 \in R^{(n-m)}$$

with $\underline{A} = [A_1 \ A_2]$, $\underline{c} = [c_1 \ c_2]$
partitioned accordingly
and $\underline{x} = \underline{A}_1^{-1}\underline{b}$, $\underline{x}_2 = \underline{0}$.
The Lagrange multipliers, also known as the Dual Variables and the Shadow Prices, are given by:

$$\underline{\pi} = (\underline{A}_1^t)^{-1}\underline{c}_1$$

At any point $\underline{x}^k$ in the space R^n we may make an arbitrary division of the x_i^k into m 'basic' $\underline{x}_1^k$ and (n-m) 'non-basic' variables $\underline{x}_2^k$, and $\underline{A}$ and $\underline{c}$ may be correspondingly partitioned. $\underline{x}^k$ will then satisfy the equality constraints in (1) if the following values are assigned to $\underline{x}_1^k$ and $\underline{x}_1^k$:

$$\underline{x}_1^k = \underline{A}_1^{k^{-1}} \ \underline{b}, \ \underline{x}_2^k = 0.$$

If x_1^k also satisfies the positivity constraints, it is a feasible solution to the problem and is also 'basic'. The effect of increasing one of the variables in the $\underline{x}_2^k$ set from zero is given by:

$$\frac{\partial W}{\partial x_{2,i}^{k}} = (\underline{c}_{2,i} - \underline{\pi}^t \underline{a}_{2,i})^k \equiv c'^{\,k}_{2,i} \qquad (2)$$

Where $\underline{a}_{2,i}^k$ is the i'th column of $\underline{A}_2^k$.

This partial derivative is understood to be confined to the non-basic space spanned by $\underline{x}_2^k$; it is the rate of change of W with respect to $x_{2,i}^k$ when all other variables $x_{2,j}^k$ are held constant, but the $x_{1,j}^k$ are allowed to vary in such a way as to maintain feasibility. If $c'^{\,k}_{2,i}$ is negative, it will pay to increase $x_{2,i}^k$ to the point where one of the currently basic variables, say $x_{1,j}^k$, becomes zero. $x_{2,i}^k$ can then be seen as replacing $x_{1,j}^k$ in the basic set to produce a new basic feasible solution. The matrices $\underline{A}_2^k$ and $\underline{A}_1^k$ interchange a column, so that $\underline{A}_1^{k+1^{-1}}$ is easily computed by updating $\underline{A}_1^{k^{-1}}$. This process is the strategy behind the well-known Simplex and Revised Simplex algorithms.

Consider now a problem differing from (1) in that one of the column vectors, say $\underline{a}_n$, and the corresponding coefficient c_n are not constant but are functions of some independent variable θ which is free to take on any value in the discrete set H. It is clear that the objective function must now be minimised over $\underline{x}$ and θ, while a value of $\underline{x}$ must be chosen such that the constraints are satisfied for all possible values of θ in H. The variable coefficients therefore imply the introduction of a new linear variable for every value of θ in H; the complete set of variables for the new problem is therefore:

$$x \equiv \{x_1, x_2 \dots x_{n-1}, x_1^n \dots x_l^n\}$$

where l is the number of discrete values of θ in H.

The new problem is therefore:

$$\min_{\underline{x}} \; W \quad \sum_{i=1}^{n-1} c_i x_i + \sum_{i=1}^{l} c_n(\theta_i)\, x_i^n$$

$$\text{s.t.} \quad \sum_{i=1}^{n-1} \underline{a}_i x_i + \sum_{i=1}^{l} \underline{a}_n(\theta_i)\, x_i^n = b$$

$$x_i \geq 0, \quad i=1 \dots n$$

$$x_i^n \geq 0, \quad i=1 \dots l$$

Although an additional l variables have been introduced into the problem, the maximum number of nonzero variables at the solution remains equal to m, the number of equality constraints. The difference is that in the final answer some, or indeed all, of the nonbasic variables may be drawn from the set x_i^n. If all the possible values of $c_n(\theta i)$ and $\underline{a}_n(\theta i)$ were computed the problem could be solved by any of the standard linear programming methods without modification. However, the solution would provide some additional information over and above that usually obtained. Specifically, if L* of the a_i^n variables appear in the basis at the solution, then this number of values of θ_i will have been chosen from H ; the solution can be viewed as optimal with respect to $\underline{x}$, x_i^n, θ_i and L. Indeed, we can introduce the quantity $X(\theta)$ which can be seen as a distribution of x_n over θ as shown in figure 1.

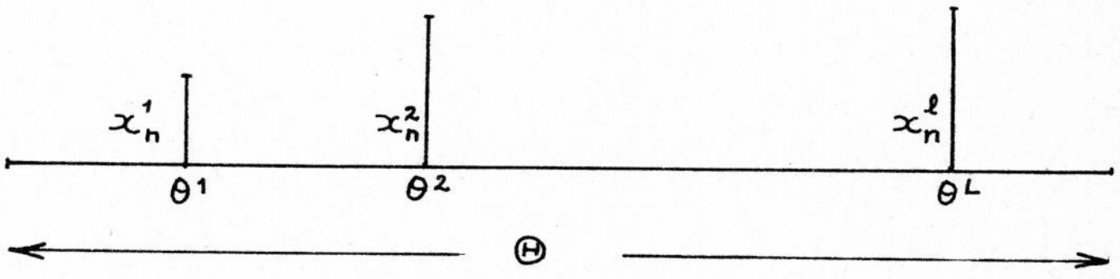

Figure 1

(For a discussion of distributions, see [5]).

The problem can thus be written as follows:

$$\begin{aligned} &\underset{\underline{x} \in R^{n-1},\ X(\theta)}{\text{Min}} \quad W = \sum_{i=1}^{n-1} c_i \, x_i + \int_{\theta \in H} c_n(\theta) \, d\,X(\theta) \\ &\text{s.t.} \quad \sum_{i=1}^{n-1} \underline{a}_i \, x_i + \int_{\theta \in H} \underline{a}_n(\theta) \, d\,X(\theta) = \underline{b} \qquad (3) \\ &\underline{x},\ X \geqslant 0 \end{aligned}$$

So long as H is a discrete set, this is still an ordinary linear programming problem and the integrals are simply summations. Only the notation has changed, and we shall now consider how the operations carried out during the application of the Simplex algorithm can be expressed in the new notation.

At the end of the k'th iteration the solution will consist of basic variables $\underline{x}_1^k$ and nonbasic variables $\underline{x}_2^k$. In general, the basic set will include some components of X, say L^k in number. The operations to be carried out at this point are to determine whether $\underline{x}^k$ is optimal and, if not, to find a new variable to enter the basis and an old one to remove. There are various criteria which might be used to determine the new basic variable, but once this is done the outgoing variable is (in the absence of degeneracy) uniquely defined. Let us consider, for example, the Simplex strategy for chosing a new basic variable. This involves selecting the nonbasic variable having the minimum reduced gradient; if this minimum is non-negative then the solution is already optimal. The reduced gradient for the i'th non-basic variable is given by equation 2; that corresponding to X is easily seen to be:

$$c'_n(\theta)^k = c_n(\theta) - \underline{\pi}^{t^k} \underline{a}_n(\theta) \tag{4}$$

Clearly, the minimum value of this over the set of all θ_i in H is required, excluding those values corresponding to components already in the current basis. The resulting value, say $c'_n(\theta_{min})$, is then compared with the reduced gradients of the 'ordinary' non-basic variables and the overall minimum chosen. There is thus a double minimisation involved. If the overall minimum is in fact $c'_n(\theta_{min})$, then X gains a component at θ_{min} whose value, x^n_{min}, is the value of the new basic variable. This is obtained in the usual way by forming $\underline{A}_1^{-1}\underline{a}_n(\theta_{min})$ and $\underline{A}_1^{-1}\,\underline{b}$ ($\underline{\alpha}$ and $\underline{\beta}$ respectively) and eliminating from $\underline{A}_1$ the column corresponding to I, where:

$$r_I = \min_i \left\{ \beta_i/\alpha_i \,\middle|\, \alpha_i > 0 \right\}$$

The basis vector $\underline{x}_1$ and the inverse basis $\underline{A}_1$ may now be updated in the usual way, and L_1 is increased by one if the new component in X does not displace one already present.

This algorithm is simply the Simplex algorithm with some new notation, required by the inclusion of the variable X. Clearly, there is no need to mix 'ordinary' variables with distributions in the problem statement, since any column $\underline{a}_i$ in $\underline{A}$ may be thought of as a column vector $\underline{a}_i(\theta^i)$ whose variable θ^i is confined to a set H^i having only one element. The new formulation is therefore:

$$\min_{X_1 \ldots X_n} \quad W = \sum_{i=1}^{n} \int_{\theta \in H} c_i(\theta^i)\, dX_i(\theta^i)$$

$$\text{s.t.} \qquad \sum_{i=1}^{n} \int_{\theta \in H} \underline{a}_i(\theta^i)\, d\, X_i(\theta^i) = \underline{b} \tag{5}$$

$$X_i \geq 0 \ .$$

The only advantage to this formulation in the case where the H^i are discrete sets is that the columns $\underline{a}_i(\theta^i)$ need not be computed a priori for all values of θ^i in the set H^i. Instead, each iteration requires the following minimization to be carried out:

$$c'^k_{imin}(\theta^{imin}_{min})^k = \min_i \left\{ \min_{\theta^i \in H^i} \left(c_i(\theta^i) - \underline{\pi}^{t^k} \underline{a}_i(\theta^i)\right)\right\} \tag{6}$$

In many cases the evaluation of $\underline{\pi}^t \underline{a}_i(\theta^i)$ for a set of values of θ^i will involve less computation than would be needed for the evaluation of $\underline{a}_i$ itself for the same set of values. However, the algorithm is still the Simplex algorithm, a finite process. It will, nevertheless, determine as a byproduct the optimal sets of values of the parameters θ^i and L_i, the positions and numbers of the components in the optimal distribution $X_1^* \ldots X_n^*$. It is clear that the following relationship is satisfied by all basic solutions to (5), including the optimum if one exists:

$$\sum_{i=1}^{n} L_i \leqslant m \; . \tag{7}$$

This follows from the basic properties of the Linear Porgramming Problem and is completely independent of the number of elements in the sets H^i.

2. Functional Linear Programming

Thus far, a form of the Simplex algorithm has been derived which allows some advantage to be taken of the case in which the coefficients are not completely independent of one another, but can be grouped together and indexed by the parameters θ^i which are drawn from discrete sets H^i. The simplex algorithm itself has not been changed by this process, of course, but a notation has been derived which is convenient for this kind of problem.

The real value of the new notation becomes apparent when a new class of problem is considered, which is not solvable by the classical Simplex algorithm. This is the case in which the sets H^i are continuous rather than discrete, either wholly or partly. The coefficients are therefore functions of the θ^i, which may or may not themselves by continuous. The notation developed above allows the Simplex technique to be generalised into a new algorithm which allows this problem to be handled, although we have now moved away from the classical linear programming problem, from a finite to an infinite process. Indeed the new case may be seen as a Linear Program with an infinite number of variables or as a linear program with functional variables (Functional Linear Programming). The only practical difference in a typical iteration, however, is that the inner minimisation expressed by (6) is now the minimisation over a set of functions rather than of a discrete set. In general, n such minimisation will be required on each iteration. The solution will consist of a set of distributions $X_1^*, X_2^* \ldots X_n^*$, the individual features of which will be discussed below. First, however, consider the following illustrative problem (ref. 2).

A force, applied at point A on a plane, is to be supported by a set of straight pin-jointed bars connecting it to a rigid foundation B-B, considered to be infinite in extent. The bars are to be made from a linear material of given stiffness E, and the point A is to displace by a given amount whose components are δ_x, δ_y under the force whose components are F_x, F_y in the coordinate system x , y. Find the number of bars, their disposition and their cross-sectional areas such that the volume of the structure meeting these constraints is minimal.

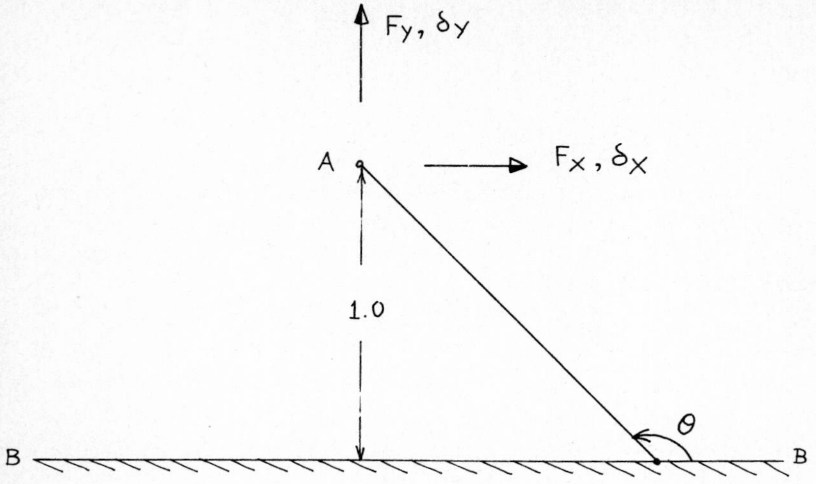

Figure 2

The strain in any bar from A meeting B-B at angle θ will be determined completely by the given displacement δ_x and δ_y, and the force exerted by it will be proportioned to its cross-sectional area x_i. The total force exerted by all such bars must balance the applied forces, so that the problem to be solved is as follows (noting that the length of a bar meeting B-B at angle θ is equal to $1/\sin\theta$):

$$\underset{X}{\text{Min}}\ W = \int \frac{1}{\text{Sin}\,\theta}\ dX(\theta)$$

$$\text{s.t.}\quad E\int_H \left\{(\text{Cos}^2\theta\ \text{Sin}\,\theta)\,\delta_x + (\text{Cos}\,\theta\,\text{Sin}^2\theta\,)\,\delta_y\right\} dX(\theta) = F_x$$

$$E\int_H \left\{(\text{Cos}\,\theta\ \text{Sin}^2\theta)\,\delta_x + (\text{Sin}^3\theta\,)\,\delta_y\right\} dX(\theta) = F_y$$

$$X(\theta) \geq 0, \qquad H|0 \leq \theta \leq \pi$$

(δ_x, δ_y, F_x, F_y are given).

For any particular choice of δ_x and δ_y this problem is in the form (5), with n = 1 and m = 2. Thus, inequality (7) is immediately applicable and it follows that the maximum number of bars is two, regardless of the assigned displacement. The solution will therefore be a distribution like that shown in figure 3, where the cross-sectional areas of the bars are represented by the lengths x_1 and x_2, and the corresponding angles are θ_1 and θ_2.

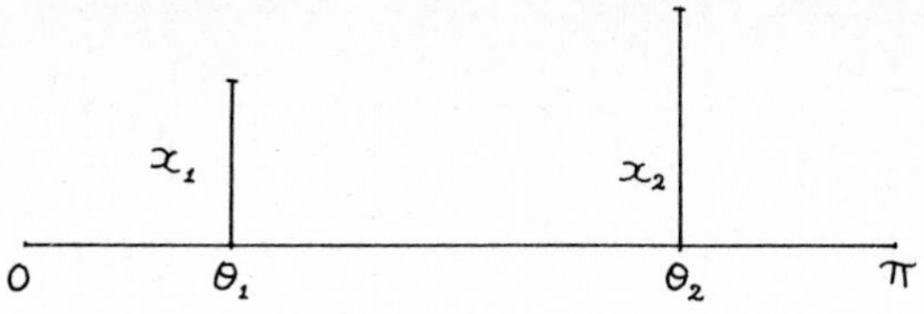

Figure 3

The problem will be solved for the case $\delta_y = F_y = 0$, $F_x = 1/E$, $\delta_x = 1$. Since no basic feasible solution is known a priori, the usual phase 1 solution of Linear Programming must be performed. The variables s_1, s_2 and $\mathcal{S} = s_1 + s_2$ are introduced into the problem, which becomes, after a little manipulation:

$$\min_{s_1, s_2, X} \mathcal{S}$$

$$\begin{aligned} \text{s.t.} \quad & \int \delta_x \cos^2\theta \sin\theta \, dX(\theta) + s_1 &&= 1 \\ & \int \delta_x \sin^2\theta \cos\theta \, dX(\theta) + s_2 &&= 0 \\ & -\int \frac{1}{\sin\theta} \, dX(\theta) + W &&= 0 \\ & \int \delta_x \left\{ \cos^2\theta \sin\theta + \sin^2\theta \cos\theta \right\} dX(\theta) + \mathcal{S} &&= 1 \end{aligned} \tag{8}$$

$$s_1, s_2, X \geq 0, \quad \{H \mid 0 \leq \theta \leq \pi\}$$

The sequence of operations involved in solving the problem are summarised in the following tables.

	$X(\theta)$	s_1	s_2	W	$\mathcal{S}$	F
	$\cos^2\theta \sin\theta$	1	0	0	0	1
	$\sin^2\theta \cos\theta$	0	1	0	0	0
$-\frac{\partial W}{\partial X}$	$-{}^1/\sin^2\theta$	0	0	1	0	0
$-\frac{\partial S}{\partial X}$	$\cos^2\theta \sin\theta + \sin^2\theta \cos\theta$	0	0	0	1	1

STEP 1

	$X(\theta)$	s_1	$x_1, \pi/4$	W	$\mathcal{S}$	F
	$\cos^2\theta \sin\theta$	1	0	0	0	1
	$2\sqrt{2}\sin^2\theta\cos\theta$	0	1	0	0	0
$-\frac{\partial W}{\partial X}$	$-{}^1/\sin\theta + 4\sin\theta\cos\theta$	0	0	1	0	0
$-\frac{\partial S}{\partial X}$	$-\sin^2\theta\cos\theta + \cos^2\theta\sin\theta$	0	0	0	1	1

STEP 2

	$X(\theta)$	$x_2, 3\pi/4$	$x_1, \pi/4$	W	$\mathcal{S}$	F
	$\sqrt{2}(\cos^2\theta\sin\theta - \sin\theta\cos\theta)$	1	0	0	0	2
	$\sqrt{2}(\sin^2\theta\cos\theta + \cos^2\theta\sin\theta)$	0	1	0	0	2
	$-{}^1/\sin\theta + 4\cos^2\theta\sin\theta$	0	0	1	0	4
	0	0	0	0	1	0

STEP 3

Table 1

STEP 1 Find the value of θ minimising $\partial S/\partial X = \cos^2\theta\sin\theta + \sin^2\theta\cos\theta$. This turns out to be $\theta = \pi/4$, $\partial S/\partial X = -1/\sqrt{2}$. We therefore introduce a column corresponding to $\theta = \pi/4$ into the basis. Substituting this value of θ into the first column, we have: $\{1/2\sqrt{2},\ 1/2\sqrt{2},\ -\sqrt{2},\ 1/\sqrt{2}\}$. To determine which of the basic variables to eliminate, we compare the ratio of each element in the RHS column with the corrsponding element of the new one; the minimum ratio corresponds to the special variable s_2 which is therefore eliminated by pivoting. When this operation is performed the second table is produced.

STEP 2 Repeating the operation, we find $\partial S/\partial X = -1/\sqrt{2}$ at $\theta = 3\pi/4$. The new column is $\{1/\sqrt{2}, -1, 2\sqrt{2}, 1/\sqrt{2}\}$ and s_1 is eliminated to give the last table.

STEP 3 $\partial S/\partial X$ is now identically zero, as is $\mathcal{S}$; we have therefore found a basic feasible solution with bars at angles $\pi/4$ and $3\pi/4$, each with a cross-sectional area of $\sqrt{2}$, and a total structure volume of 4 units. The next step is to drop the last row and the column corresponding to $\mathcal{S}$ and to minimise $\partial W/\partial X$. This produces a minimum value of zero, at $\theta = \pi/4$ and $3\pi/4$; the first basic feasible solution is therefore also optimal. This is of course an exceptional circumstance, brought about by the particular values chosen for $\underline{\delta}$ and $\underline{F}$. The resulting structure is shown in figure 4.

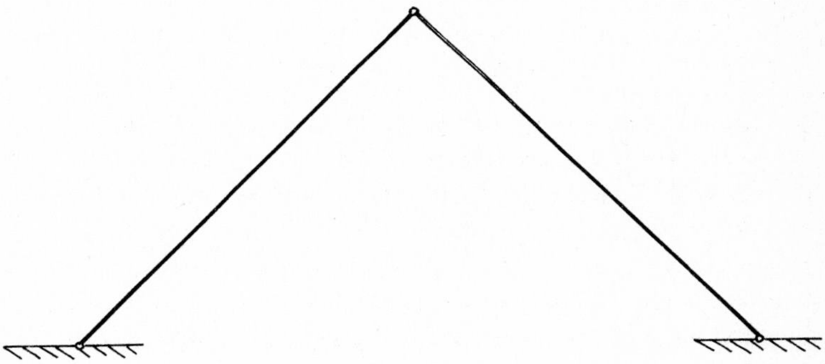

Figure 4

In the simple example above, the number of components in X* was restricted, by inequality (7), to either one or two. In the more general case in which $n > 1$ the values of $L_1, \ldots, L_n$ are not so easily predictable, although the algorithm will determine them. In fact, it is the ability to determine optimal values for these variables which seems to be the main feature distinguishing Functional Linear Programming (FLP) from the Dantzig-Wolfe Generalised Programming algorithm (GLP). The difference is perhaps best illustrated by considering two distinct ways in which a variable coefficient vector, say $\underline{a}_n(\theta)$, may be interpreted when it appears in the statement of a Linear Programming problem. In the FLP algorithm described above, it is interpreted as an infinite number of columns any number of which may appear independently in the final basis. The number of occurences of $\underline{a}_n(\theta)$ in the basis is of course the optimal value of L, and is a variable of the problem. However, $\underline{a}_n(\theta)$ might be viewed as a single column, associated with a value of θ which is not known a priori; this appears to be the interpretation underlying Generalised Linear Programming. The difference between these two interpretations is that whereas in the first case L is a variable to be determined, in the second its value is assumed to be 1. An example of the GLP approach can be found in [1], pp 440-443. Here the author makes no attempt to use the functional form of $\underline{a}_n(\theta)$, replacing it instead by a piecewise-linear approximation interpolating between a discrete set of values $\underline{a}_n(\theta_i)$. Any solution is therefore valid only if $\underline{a}_n(\theta)$ occurs in the basis either at one of these pre-set values of θ_i, or as a linear combination of two adjacent such columns (if it appears at all). The original Generalised Linear Programming algorithm had no means for ensuring that this does in fact happen for general $\underline{a}_n(\theta)$, although more recent methods involving such concepts as special ordered sets [4] can be used to overcome this difficulty. Interestingly, Dantzig [1] does suggest that if $\underline{a}_n$ does appear in a final basis with nonadjacent values of θ, then some physical justification might sometimes be found; however, he does not take this any further and in particular does not make the point that when such a physical situation exists there is no need to use a linear approximation to $\underline{a}_n(\theta)$ in the first place.

In order to fix these differences between Functional and Generalised Linear Programming, consider the simple problem already used to illustrate FLP. An approach to this in the spirit of Generalised Linear Programming would involve dividing the range $[0,\pi]$ into a number of intervals by a discrete set of angles θ_i. Applying the technique exactly as described in ref. 1, the true answer could not be obtained because it would involve non-adjacent columns in the basis and such a solution would violate the assumptions underlying the piecewise-linear approximations. Later and more sophisticated methods using branch and bound techniques to ensure adjacency would only make the failure more certain. This inability on the part of Generalised Programming to solve problem (5) has nothing to do with the fact that the vector function $\underline{a}_n(\theta)$ is replaced by an approximation. Even if some method of interpolation were to be used to determine solutions to any required degree of accuracy, the situation would in no way be altered. The point being made here is not that GLP is unable to solve such a problem, since a slight modification would probably enable it to do so, but rather that Generalised Linear Programming is geared to the solution of a different kind of variable-coefficient Linear Programming problem than FLP .

The dual of problem (5) is the following:

$$\max_{\underline{\lambda} \in R^m} \quad \underline{b}^t \underline{\lambda}$$

$$\text{s.t.} \quad \underline{a}_i^t(\theta^i)\,\underline{\lambda} \le c^i(\theta^i) \tag{9}$$

$$\text{for all} \quad \theta^i \in H^i, \quad i=1,2, \dots, n$$

(9) can be derived by regarding (5) as the limiting case of a Linear Program with an increasing number of variables, and considering the usual LP dual form. The

mutual duality of (5) and (9) is more rigorously proved by Gomulka [3].

Problem (9) has a finite number of variables but an infinite number of linear constraints. Regardless of the functional form of the $a_{ij}(\theta)$, the constraint set is clearly convex. Returning for a moment to the comparison between Functional and Generalised Linear Programming, it might appear that the latter, in the form of Dantzig's Convex Programming algorithm, might therefore be brought to bear on (5) via (9). However, once again some modification to the algorithm would appear to be made necessary by the fact that (9) has an infinite number of constraints, albeit linear.

Having established that the main distinguishing feature of Functional Linear Programming is its ability to determine optimal values of the integer variables L_i, it is interesting to consider these variables in more detail. Inequality (7) already sets an upper limit on the sum of the L_i, but tighter bounds can often be established by a little analysis, and even by inspection. The following result is useful in this context.

Property 1 Let $X_1^*, X_2^*, \ldots, X_n^*$ be the solution to (5); define $W^i(X^i) \equiv \int_{H^i} c_i(\theta^i)\, dX^i(\theta^i)$. Then every X_i^* must satisfy, separately, the following conditions:

$$W^i(X_i^*) = \min_{X^i} W(X^i) \tag{10(i)}$$

$$\text{s.t.} \int_{H^i} a_i(\theta^i)\, d X^i(\theta^i) = b_i \tag{10(ii)}$$

$$X_i \geq 0 \tag{10(iii)}$$

where
$$\underline{b}_i^* = \underline{b}_i - \sum_{j \neq i} \int_{H^j} \underline{a}_j(\theta^j)\, d X^j(\theta^j) \tag{10(iv)}$$

Proof The m-vector $\underline{b}^{i*}$ is the RHS of (5) minus the contribution of the distributions X^{i*}, $j \neq i$. Any change in $\underline{X}^{i*}$ which leaves this vector unchanged will therefore continue to be feasible for the overall problem (5) so long as it also leaves it satisfying 10(iii). It follows that if a new value, say X^{i**}, can be found which satisfies 10(ii) and 10(iii) and is such that $W^i(X^{i**}) < W^i(X^{i*})$, then $W(X_1^*, X_2^*, \ldots, X_i^{**}, \ldots, X_n^*) < W(X_1^*, \ldots, X_n^*)$. This would violate the hypothesis of optimality. Hence the statement is true.

This result shows that each distribution X_i separately solves a subsidiary Functional Linear Program, whose RHS vector $\underline{b}_i^*$ is not of course known a priori. This fact often enables us to place upper bounds on L_i. Consider the general vector function $\underline{a}(\theta) \equiv \{a_1(\theta), \ldots, a_m(\theta)\}$. As θ takes on all possible values in its range, $\underline{a}(\theta)$ will generate a set of vectors spanning a space of dimension $m' \leq m$. This value is clearly the maximum dimension of the Functional Linear Program defined by (8), so that

$$L_i^* \leq m_i' \leq m.$$

$\underline{a}(\theta)$ can always be represented in the following form:

$$\underline{a}(\theta) = \underline{A}\,\underline{f}\,(\theta)$$

where $\underline{A}$ is a fixed matrix and $\underline{f}$ is a set of independent functions. For example, if the functions $a_i(\theta)$ are all independent, $\underline{A}$ is simply the unit matrix. Let M_f be the number of such independent functions. Clearly, the space spanned by $\underline{a}(\theta)$ cannot exceed the rank of $\underline{A}$. This is clearly not greater than M_f, on the one hand, and is also dependent upon the number of the $a_i(\theta)$ which are identically zero. In a large problem this would normally be a significant number. Additionally, in particular cases there may be some extra information about $\underline{A}$ available which would allow the value of m_i to be known more accurately.

One particularly interesting class of problems, of course, is that for which $L_i \leq 1$, that is, the class for which Functional Linear Programming and Generalised Linear Programming will produce identical results. Insight into this situation may be gained by considering the dual of (9) of the problem once again.

Problem (9) has a finite number of variables; the number of constraints is strictly infinite, but if the $a_{ij}(\theta^i)$ are continuous functions they will define envelopes around a feasible region. In any case, the constraint set is always convex. The following statement may be made:

<u>Statement 2</u> Let $X_1^*, X_2^*, \ldots, X_n^*$ be the optimal solution to problem (5); then a necessary but not sufficient condition for L_i^* to exceed 1 is that the convex set $\{\lambda \mid \underline{a}_i^t(\theta^i)\,\underline{\lambda} \leqslant c^i(\theta^i),\ \theta^i \in H^i\}$ has at least one vertex.

<u>Proof</u> Assume that $L_i^* > 1$ for some i, and let θ_j^i, j=1, ..., L_i^* be the corresponding values of θ^i associated with the components of X_i^*. For these values of θ^i, the inequality constraints in (10) are satisfied as equalities, so that:

$$\underline{a}_i^t(\theta_j^i)\,\underline{\lambda} - c^i(\theta_j^i) = 0, \qquad j=1, \ldots, L_i^*.$$

It follows that the solution lies at the intersection of L_i^* hyperplanes which are tangent to the feasible boundary. Since the feasible set is convex, it follows that such hyperplanes can only intersect at their points of contact with the boundary. The result follows.

This result can be easily illustrated by reference to the structural problem already discussed. The dual of this problem is (using the case $\delta_y = F_y = 0$)

$$\max_{\lambda_1,\lambda_2} \quad \lambda_1 F_X$$

$$\text{s.t.} \quad \lambda_1 (\cos^2\theta \sin\theta)\delta_x + \lambda_2 \cos\theta \sin^2\theta \leq 1/\sin\theta \qquad (11)$$

$$0 \leq \theta \leq \pi$$

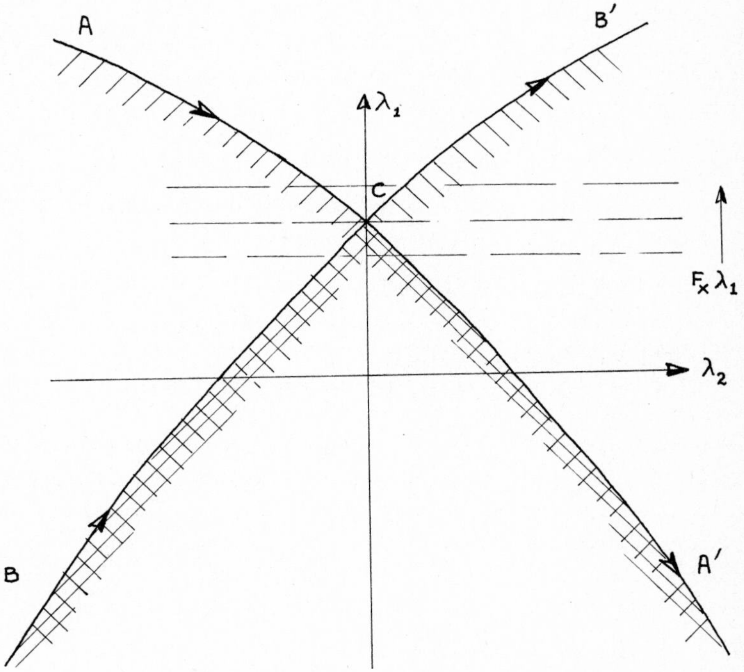

Fig. 5

Figure 5 shows the form of the constraints. Starting with a small positive value of θ corresponding to a point A on the envelope, the linear constraints trace out the boundary A - A' as θ approaches $\pi/2$ radians. At that value it passes through infinity and returns through B to trace out the branch B - B' as θ proceeds towards π. There are thus two fairly distinct 'feasible' regions (shown hatched) and their intersection is the feasible set for problem (11) (shown cross-hatched). The contours of the objective function consist of horizontal lines, and its maximum value corresponds to the contour passing through C, the apex of the feasible region. This point corresponds to the design of figure 4, with $L^* = 2$; it is formed by the intersection of two tangents, to A - A' and B - B' respectively. Notice, however, that a different choice of objective function could lead to an optimal design corresponding to a single tangent at some point on the boundary of the feasible set. This demonstrates that the existence of a vertex does not guarantee that L^* shall exceed 1; it depends upon the objective function of the dual problem, that is, upon the RHS vector of the primal.

Finally, as an illustration of the kind of real problem which has been solved by the FLP algorithm, consider the following (Ref. 3):

$$\min_{X_1, X_2, \ldots, X_n} \sum_{i=1}^{n} A_i \int_{\theta^i \in H^i} dX_i(\theta^i)$$

$$\text{s.t.} \quad \sum_{i=1}^{n} \int \{ \underline{p}_0^i + \underline{p}_1^i \cos 2\theta^i + \underline{p}_2^i \sin 2\theta^i + \underline{p}_3^i \cos 4\theta^i + \underline{p}_4^i \sin 4\theta^i \} \, dX(\theta^i) = \underline{P} \tag{12}$$

$$X_i \geq 0, \quad 0 < \theta^i < \pi$$

Once again, this problem has been drawn from the field of optimal structures. A typical distribution X_i has the following physical interpretation: L_i, the number of components in X_i, is the optimal number of layers of material in the i'th region of the structure; θ_j^i is the orientation of the reinforcing fibres in the j'th such layer; and x_j^i is the thickness of this layer. It can be seen that the integer variables L_i are of the utmost importance in this context. Numerous problems of this type have been solved to this date, although so far they have been of fairly small size - n and m up to 33 and 43 respectively.

Conclusion

Functional Linear Programming is a technique for solving linear programming problems whose variables are distributions rather than scalars. It has been introduced in this paper as a straightforward generalisation of the Simplex method of linear programming from which, apart from being an infinite process, it differs computationally mainly in requiring functional minimisations as well as the usual list search to determine new basic columns. The method has proved successful in solving the problems for which it was originally devised; its application to more general convex programming problems is described in the paper in this book by Resta, Treccani and Sideri.

Acknowledgement

The author is indebted to the Italian Consiglio Nazionale Delle Ricerche (C.N.R.) for the award of a Visiting Professorship at the Mathematics Department of the University of Genoa during April 1976, when some of the work described in the paper was carried out.

References

1. Dantzig, G.B. (1963). Linear Programming and extensions, Princeton University Press.

2. McKeown, J.J. (1974). A Quasi-Linear Programming Algorithm for Optimising Fibre-Reinforced Structures of Fixed Stiffness. Computer Methods in Applied Mechanics and Engineering 6 (1975) pp 123-154.

3. Gomulka, J., McKeown, J.J. & Treccani, G. (1975). Functional Linear Programming. Technical Report No. 70, Numerical Optimisation Centre, Hatfield Polytechnic.

4. Beale, E.M.L. (1975). Optimization Techniques Based on Linear Programming. In Optimization in Action edited by L.C.W. Dixon, Academic Press.

5. Dunford, N. & Schwartz, J.P. (1958). Linear Operators Part I. Interscience Publishers.

TOWARDS GLOBAL OPTIMISATION 2
L.C.W. Dixon and G.P. Szegö (eds.)

A NEW METHOD FOR CONVEX PROGRAMMING

GIOVANNI RESTA - ENRICO SIDERI - GIULIO TRECCANI (University of Genoa)

INTRODUCTION

The aim of this work is to construct a new method for solving some convex programming problem. The method was originally suggested by Mc Keown and Gomulka as an application of Functional Linear Programming, and the basic idea of the method is to use the revised simplex technique for approaching convex programming; this approach is due to Dantzing [1] and Wolfe [2].
In our method, however, we try to solve the dual of a convex programming problem as a Functional Linear Programming problem [5]; this implies a remarkable reduction of the computational work involved in the algorithm but also some difficulty about convergence, which is deeply investigated in this paper.
The method is also related to "Cutting Plane" methods of Kelley [3] and Topkis [4], but as it does not require the solution, at each iteration, of a subproblem which is a programming problem with linear constraints, it is much simpler from a computational point of view.

STATEMENT OF THE PROBLEM

The method we present can be used for solving the following problem of convex programming.
Let $\varphi_i : \mathbb{R}^n \to R$, $i = 1,\dots,s$, continuously differentiable convex functions and $q \in \mathbb{R}^n$.
(*) Maximize $q^T x$ subject to the constraints $\varphi_i(x) \leq 0$, $i=1,\dots,s$.
In the following, we shall use the notations and make the assumptions that follow.

NOTATIONS AND BASIC ASSUMPTIONS

Let $S_i = \{x \in \mathbb{R}^n : \varphi_i(x) < 0$ and $S = \bigcap_{i=1}^{s} S_i$.
For every $x \in \partial S$ let $I(x) = \{n \in N : n \leq s, \varphi_n(x) = 0\}$.
Assume that S is noempty and bounded and, without loss of generality, that it contains the origin.

We denote the gradient of φ_i by $\nabla\varphi_i$ and we observe that from the assumptions stated above it follows that the problem (*) has a solution and that, $\forall x \in \partial S$, $x^T \nabla\varphi_i(x) > 0$, $i \in I(x)$.

Problem (*) can be rewritten as follows

(*) Maximize $q^T x$ subject to the constraint $x \in \bar{S}$.

Lemma 1. $x^* \in \bar{S}$ is a solution of problem (*) if and only if $x^* \in \partial S$ and there exist $\lambda_i \geq 0$, $i \in I(x^*)$, such that $q = \sum_{i \in I(x^*)} \lambda_i \nabla \varphi_i(x^*)$.

Def. 1. Given a set of k pairs (x_j, i_j), $j=1,\dots,k$, $x_j \in \partial S$, $i_j \in I(x_j)$, this set generates T, a q-oriented set on S, if the following properties hold:

1.1. $T = \bigcap_{j=1}^{k} H_j = \bigcap_{j=1}^{k} \left\{ x \in \mathbb{R}^n : \nabla \varphi_{i_j}(x_j)^T (x - x_j) \leq 0 \right\}$

1.2. There exists $\ell \in \mathbb{R}^k$, $\ell = \left\{ \lambda_1, \dots, \lambda_k \right\}$, such that $\lambda_j \geq 0$ and $q = \sum_{j=1}^{k} \lambda_j \nabla \varphi_{i_j}(x_j)$.

Def. 2. A q-oriented set on S, T is said to be basic if:

2.1. $k = n$

2.2. The $n \times n$ matrix B(T), whose columns are the vectors $b_j = \nabla \varphi_{i_j}(x_j)$, where (x_j, i_j) are the pairs that generate T, is non singular.

B(T) is said the basic matrix and the components λ_j of the vector ℓ defined by 1.2 are said the T-components of q.

Def. 3. Let T be a q-oriented set on S generated by the pairs (x_j, i_j), $j=1,\dots,k$. Let $P_j = \left\{ x \in \mathbb{R}^n : \nabla \varphi_{i_j}(x_j)^T (x - x_j) = 0 \right.$. Any point $z \in \bigcap_{j=1}^{k} P_j$ is said a vertex of T.

Remark 1. If T is basic, there exists a unique vertex z(T).

Def. 4. Let T be a q-oriented set on S. T is said nondegenerate if T is basic and the T-components of q are strictly positive.

Property 1. Let T be a q-oriented set on S generated by the pairs (x_j, i_j), $j=1,\dots,k$. Then

1.1. $S \subseteq T$

1.2. $S \cap P_j \neq \emptyset$, $j=1,\dots,k$ (see def.3)

Lemma 2. $x^* \in \partial S$ is a solution of problem (*) if and only if there exists a set T, q-oriented on S, which has a vertex in x^*.

Proof. Sufficiency. Assume that T is generated by the pairs (x_j, i_j), $j=1,\ldots,k$. As x^* is a vertex of T, it belongs to P_j (see def. 3 for $j=1,\ldots,k$. On the other hand P_j is a tangent hyperplane to the set $\bar{S}_{i_j}$ at the point x_j, while $x^* \in \partial S \subseteq \bar{S}_{i_j}$; this implies that P_j is a tangent hyperplane to $\bar{S}_{i_j}$ at the point x^*. It follows that $\nabla\varphi_{i_j}(x_j) = \mu_j \nabla\varphi_{i_j}(x^*)$, μ_j 0 and $q = \sum_{j=1}^{k} \lambda_j \mu_j \nabla\varphi_{i_j}(x^*)$, while $\lambda_j \mu_j \geq 0$; the sufficiency follows then from lemma 1.

Necessity. If $x^* \in \partial S$ is a solution of problem (*), lemma 1 implies that $q = \sum_{i \in I(x^*)} \mu_i \nabla\varphi_i(x^*)$, $\mu_i \geq 0$. Then it is possible to construct a q-oriented set T on S, which is generated by pairs with the first element x^*, and this set T has obviously a vertex at x^*.

Def. 5. A sequence T_k of basic q-oriented sets on S has the property $\mathcal{P}$ if the following relationship s hold, $\forall$ k, between T_k and T_{k+1}.

5.1. Let $x_j(T_k)$, $i_j(T_k)$, $J=1,\ldots,n$, the pairs that generate T_k. Then there exists a unique index $J(k)$ such that $x_j(T_k) = x_j(T_{k+1})$, $i_j(T_k) = i_j(T_{k+1})$ for $J \neq J(k)$.

5.2. There exists $\alpha > 0$ such that

$$\vartheta_K = \nabla\varphi_{i_{j(k)}}(T_{k+1}))^T (z(T_k) - x_{j(k)}(T_{k+1})) \geq$$

$$\alpha \text{ dist } (L_k, x_{j(k)}(T_{k+1}))$$

where

$$L_k = \bigcap_{J \neq J(k)} P_j(T_k) = \bigcap_{J \neq J(k)} P_j(T_{k+1})$$

For simplicity of notations, we denote in the following the gradient $\nabla\varphi_{i_j(T_{k+1})}(x_j(T_{k+1}))$ by b_j^{k+1}, the j-th column of the basic matrix $B(T_{k+1})$

Lemma 5. Let T_k be a sequence of q-oriented basic sets on S, satisfying property $\mathcal{P}$. Then the following properties hold:

5.1. $\forall x \in S \quad q^T z(T_k) > q^T x$

5.2. $q^T[z(T_k) \quad z(T_{k+1})] = \lambda_{j(k)}(T_{k+1})\, \vartheta_k$

where $\lambda_{j(k)}(T_{k+1}) \geq 0$ is the $J(k)$-th T_{k+1}-component of q.

Proof. Since T_k is basic, $z(T_k)$ is the only maximum point of the linear function $q^T y$, $y \in T_k$. As, for property 1, $S \subsetneq T_k$ and S is open, the inequality $q^T z(T_k) > q^T x$ must hold $\forall x \in S$.

If we denote by λ_j the T_{k+1}-components of q, def.5 implies:

$$q = \sum_{j=1}^{n} \lambda_j b_j^{k+1} = \sum_{j \neq j(k)} \lambda_j b_j^k + \lambda_{j(k)} b_{j(k)}^{k+1}$$

$$(z(T_k) - x_j(T_k))^T b_j^k = (z(T_{k+1}) - x_j(T_k))^T b_j^k = 0 \quad \text{for } j \neq j(k);$$

$$(z(T_k) - z(T_{k+1}))^T b_j^k = 0 \quad \text{for } j \neq j(k);$$

$$q^T(z(T_k) - z(T_{k+1})) = \sum_{j \neq j(k)} \lambda_j b_j^k (z(T_k) - z(T_{k+1})) +$$

$$+ \lambda_j(k) b_{j(k)}^{k+1} (z(T_k) - z(T_{k+1})) = \lambda_{j(k)} b_{j(k)}^{k+1} (z(t_k) - z(T_{k+1})) =$$

$$= \lambda_{j(k)} b_{j(k)}^{k+1} (z(T_k) - x_{j(k)}(T_{k+1}) + x_{j(k)}(T_{k+1}) - z(T_{k+1}))$$

$$= \lambda_{j(k)} b_{j(k)}^{k+1} (z(T_k) - x_{j(k)}(T_{k+1})) = \vartheta_k \lambda_{j(k)}.$$

Cor. 1. Let T_k be a sequence of q-oriented basic sets on S, satisfying property $\mathcal{P}$. Then

1.1. The sequence $q^T z(T_k)$ is non increasing and bounded from below.

1.2. If $\alpha^* = \lim_k q^T z(T_k)$, we have

$$0 < \sum_{k=1}^{\infty} q^T(z(T_k) - z(T_{k+1})) = q^T z(T_1) - \alpha^* < +\infty$$

1.3. Any limit point of the sequence $z(T_k) - z(T_{k+1})$ belongs to the orthogonal complement of q.

Theorem 1.

Let T_k be a sequence of q-oriented basic sets on S, satisfying property $\mathcal{P}$ and assume that $\inf\{\vartheta_k\} = 0$. Then either there exists k_1 such that $x_{j(k_1)}(T_{k_1+1})$ is a solution of problem (*), or there is a subsequence $\{T_{k_m}\}$ such that $x_{j(k_m)}(T_{k_m+1}) \rightarrow x^*$, a solution of problem (*).

Proof. If $\inf\{\vartheta_k\} = 0$, then either $\vartheta_{k_1} = 0$ for some k_1 or $\vartheta_{k_m} \rightarrow 0$ for some sequence $\{T_{k_m}\}$.

If $\vartheta_{k_1} = 0$, then 5.2 implies $x_{j(k_i)}(T_{k_1+1}) \in L_{k_1} \cap P_{j(k_1)}(T_{k_1+1}) =$

$= \bigcap_{j \neq j(k_1)} P_j(T_{k_1}) \cap P_{j(k_1)}(T_{k_1+1}) = \bigcap_{j=1}^{n} P_j(T_{k_1+1}) = z(T_{k_1+1})$

Lemma 2 implies then that $x_{j(k_1)}(T_{k_1+1})$ is a solution of problem (*).

Assume now that there exists $\{T_{k_m}\} \subseteq \{T_k\}$ such that $\vartheta_{k_m} \to 0$.

For any k, we have $(z(T_k) - x_j(T_k))^T b_j^k = 0$, $j=1,\dots,n$, and

$q^T z(T_k) = \sum_{j=1}^{n} \lambda_j(T_k)\, x_j(T_k)^T b_j^k$.

Corollary 1 gives $0 < \sum_{j=1}^{n} \lambda_j(T_k)\, x_j(T_k)^T b_j^k < q^T z_1$, since $x_j(T_k)^T b_j^k =$

$= x_j(T_k)^T \nabla\varphi_{i_j(T_k)}(x_j(T_k)) > 0$. We have then that $0 \leq \lambda_j(T_k) \leq \Lambda$ for some $\Lambda > 0$.

There will exist a subsequence, that for simplicity of notations we denote by $\{T_k\}$ such that:

(i) $x_{j(k)}(T_k) \to x^* \in \partial S$

(ii) $j(k) \to J^*$

(iii) $x_j(T_k) \to x_j^*$, $j \neq j^*$

(iv) $b_j^k \to \nabla\varphi_{i_j^*}(x_j^*)$ for $j \neq j^*$, with $i_j^* \in I(x_j^*)$

(v) $b_{j(k)}^k \to \nabla\varphi_{i_{j^*}^*}(x^*)$, with $i_{j^*}^* \in I(x^*)$

(vi) $\vartheta_k \to 0$.

(vii) $\lambda_j(T_k) \to \lambda_j^* \geq 0$, $j=1,\dots,n$.

As dist $(L_k, x_{j(k)}(T_{k+1})) \geq \text{dist}(P_j(T_k), x_{j(k)}(T_{k+1}))$, $J \neq J(k)$, $\vartheta_k \to 0$ implies that

$x^* \in P_j^*$, $j \neq j^*$, where $P_j^* = \left\{ x \in \mathbb{R}^n : (x - x_j^*)^T \nabla\varphi_{i_j^*}(x_j^*) = 0 \right\}$

We have now $x^* \in P_j^* \cap S_{i_j^*}$, $j \neq j^*$, and P_j^* is a tangent hyperplane to $S_{i_j^*}$ at the point x^*, hence

$$\nabla\varphi_{i_j^*}(x_j^*) = \mu_j \nabla\varphi_{i_j^*}(x^*), \quad \mu_j > 0, \; j \neq j^*.$$

We conclude that:

$$q = \sum_{j=1}^{n} \lambda_j(T_k) b_j^k \to \sum_{j \neq j^*} \lambda_j^* \nabla\varphi_{i_j^*}(x_j^*) + \lambda_{j^*}^* \nabla\varphi(x^*) =$$

$$= \sum_{j \neq j^*} \mu_j \lambda_j^* \nabla\varphi_{i_j^*}(x^*) + \lambda_{j^*}^* \nabla\varphi_{i_{j^*}^*}(x^*),$$

and the theorem follows from Lemma 1.

Lemma 4. Let $\{T_k\}$ be a sequence of q-oriented basic sets on S satisfying property $\mathcal{P}$ and assume that $\inf\{\vartheta_k\} = \vartheta > 0$.

Then we have:

$$\lim_{k \to +\infty} \lambda_{j(k)}(T_{k+1}) = 0$$

Proof. It is a consequence of Lemma 3.3.2 and Cor.1.

Lemma 5. Let $\{T_k\}$ be a sequence of q-oriented basic sets on S satisfying property $\mathcal{P}$.

Let η^k be the vector $B^{-1}(T_k)\, b^{k+1}_{j(k)}$, where as usual we denote by $B(T_k)$ the basic matrix of T_k and by $b^{k+1}_{j(k)}$ the $j(k)$-th column of the basic matrix of T_{k+1}.

Then we have:

$$\det B(T_{k+1}) = \eta^k_{j(k)} \det B(T_k)$$

$$\lambda_{j(k)}(T_k) = \eta^k_{j(k)} \lambda_{j(k)}(T_{k+1})$$

Proof. As $b^{k+1}_{j(k)} = B(T_k)\, B^{-1}(T_k)\, b^{k+1}_{j(k)} = B(T_k)\eta^k = \sum_{j \neq j_k} \eta^k_j b^k_j + \eta^k_{j(k)} b^k_{j(k)}$

and $b^k_j = b^{k+1}_j$ for $j \neq j(k)$, it must be $\det B(T_{k+1}) = \eta^k_{j(k)} \det B(T_k)$.

On the other hand $q = B(T_k)\, B^{-1}(T_k) q = \sum_{j \neq j(k)} \lambda_j(T_k) b^k_j + \lambda_{j(k)}(T_k)\, b^k_{j(k)} =$

$= \sum_{j \neq j(k)} \lambda_j(T_{k+1})\, b^k_j + \lambda_{j(k)}(T_{k+1})\, b^{k+1}_{j(k)}$

We have then $\sum_{j \neq j(k)} \left[\lambda_j(T_k) - \lambda_j(T_{k+1}) \right] b^k_j + \lambda_{j(k)}(T_k) b^k_{j(k)} -$

$- \lambda_{j(k)}(T_{k+1})\, b^{k+1}_{j(k)} = 0.$

Using the equality $b^{k+1}_{j(k)} = \sum_{j \neq j(k)} \eta^k_j b^k_j + \eta^k_{j(k)} b^k_{j(k)}$, it must be:

$$\sum_{j \neq j(k)} \left[\lambda_j(T_k) - \lambda_j(T_{k+1}) - \lambda_{j(k)}(T_{k+1}) \eta^k_j \right] b^k_j +$$

$$+ \left[\lambda_{j(k)}(T_k) - \eta^k_{j(k)} \lambda_{j(k)}(T_{k+1}) \right] b^k_{j(k)} = 0$$

As $B(T_k)$ is nonsingular, its columns are linearly independent, which implies:

$$\lambda_{j(k)}(T_k) = \eta^k_{j(k)} \lambda_{j(k)}(T_{k+1})$$

Lemma 6. Let $\{T_k\}$ be a sequence of q-oriented basic sets on S satisfying property $\mathcal{P}$ and assume that $\inf\{\vartheta_k\} = \vartheta > 0$. Then there exists a subsequence $\{T_{k_m}\}$ such that:

$$\lim_m \lambda_{j(k_m)}(T_{k_m}) = 0$$

Proof. Det $B(T_k)$ is bounded and Lemma 5 implies that there exists a subsequence $\{T_{k_m}\}$ such that $0 \leq \eta^{k_m}_{j(k_m)} \leq 1$, $\forall m$, hence $\lambda_{j(k_m)}(T_{k_m}) \leq$ $\leq \lambda_{j(k_m)}(T_{k_m+1})$; the result follows then from Lemma 4.

Def.6. A sequence $\{T_k\}$ of q-oriented basic sets, satisfying property $\mathcal{P}$, satisfies property $\mathcal{Q}$, if $\forall$ k we have $j(k) \neq j(k+1)$, where j(k) is defined by def. 5.5.1.

Theorem 2.
Assume that either $S \subseteq \mathbb{R}^2$ or $S \subseteq \mathbb{R}^3$. Then theorem 1 holds if we substitute the hypothesis $\inf\{\vartheta_k\} = 0$ by the assumption that $\{T_k\}$ satisfies property $\mathcal{Q}$.

Proof. Assume that $S \subseteq \mathbb{R}^2$ and that $\mathcal{Q}$ is satisfied. Assume then, by contradiction, that $\inf\{\vartheta\}_k = \vartheta > 0$. Lemma 4 and 6 imply that there is a subsequence $\{T_{k_m}\}$ such that $\lim_m \lambda_{j(k_m-1)}(T_{k_m}) =$ $= \lim_m \lambda_{j(k_m)}(T_{k_m}) = 0$.

As $j(k_m-1) \neq j(k_m)$, $\forall$ m, it would be:

$$q = \sum_{j=1}^{2} \lambda_j(T_{k_m}) = \lambda_{j(k_m-1)}(T_{k_m}) b_{j(k_m-1)} + $$
$$+ \lambda_{j(k_m)}(T_{k_m}) b_{j(k_m)} \longrightarrow 0.$$

Assume now $S \subseteq \mathbb{R}^3$. By the same argument as before, if we denote by $i(k_m)$ the index different from both $j(k_m)$ and $j(k_m-1)$, we have
$\lambda_{i(k_m)} b_{i(k_m)} (T_{k_m}) \longrightarrow q$ and the proof follows as in theorem 1.

Our aim is now to introduce an algorithm for solving problem (*), which uses the results obtained so far.

The solution x* is computed iteratively by means of a sequence $\{T_k\}$ of q-oriented basic sets on S, which satisfies properties $\mathcal{P}$ and $\mathcal{Q}$, starting from an initial basic set T_o.

For any q-oriented basic set T_k on S, generated by the pairs (x^k_j, i^k_j), j=1,...,n, we can compute a (n 1) x (n+1) nonsingular matrix D_k of the form:

$$D_k = \begin{pmatrix} 1 & -C^T_k \\ 0 & B_k \end{pmatrix}$$

where B_k is the basic matrix of T_k, i.e. the $n \times n$ nonsingular matrix whose j-th column b^k_j is the vector $\nabla_{i^k_j}(x^k_j)$, and C_k is the n-vector whose components are $\nabla_{i^k_j}(x^k_j)^T x^k_j$.

If $H_k = B^{-1}_k$, then $M_k = D^{-1}_k$ has the form:

$$M_k = \begin{pmatrix} 1 & C^T_k H_k \\ 0 & H_k \end{pmatrix}$$

and since the vertex $z(T_k)$ of the set T_k satisfies the equation $z(T_k) = (B^T_k)^{-1} C_k$, we have:

$$M_k = \begin{pmatrix} 1 & z(T_k)^T \\ 0 & H_k \end{pmatrix}$$

On the other hand, setting $\tilde{q} = \begin{pmatrix} 0 \\ q \end{pmatrix} \in \mathbb{R}^{n+1}$, we have that the (n+1)-vector $M_k\tilde{q}$ has $z(T_k)^T q$ as the first component, and the T_k-components of q as the last n components.

Consider now a new pair (u, ℓ) $u \in S$, $\in I(u)$, which correspond to a column vector:

$$\begin{pmatrix} - \nabla\varphi_\ell(u)^T u \\ \nabla\varphi_\ell(u) \end{pmatrix}$$

Using the procedure of the revised simplex, we compute directly the matrix $M_{k+1} = D^{-1}_{k+1}$, where D_{k+1} is the nonsingular matrix whose columns are the same as D_k, except the j(k)-th column, which is now $(-\nabla\varphi_\ell(u)^T u, \nabla\varphi_\ell(u))^T$, in such a way the last n components of the (n+1)-vector $M_{k+1}\tilde{q}$ are non-negative.

It follows that the pairs:

$$\begin{cases} (x_j^{k+1}, i_j^{k+1}) = (x_j^k, i_j^k), & j \neq j(k) \\ (x_{j(k)}^{k+1}, i_{j(k)}^{k+1}) = (u, \ell) \end{cases}$$

generate a q-oriented basic set T_{k+1} on S, which satisfies def. 5.5.1. By a suitable choice of the pair (u, ℓ) it will be possible to satisfy also def. 5.5.2. and property α.

(A) <u>A sketch of the algorithm</u>.

Assume that: (i) a q-oriented basic set T_o on S, (ii) a point $w \in S$, (iii) a real number $0 < \varepsilon < 1$, are given.

Compute B_o, H_o, D_o, M_o and set $\tilde{q} = \begin{pmatrix} 0 \\ q \end{pmatrix} \in \mathbb{R}^{n+1}$

(1) $K = 0$, $m = 0$

(2) $M_k = M_o$

(3) $\xi_k = (\lambda \in \mathbb{R}: \varphi_i(w+\lambda(z_k - w)) \leq 0,\ i=1,\ldots,s,$

where $z_k = z(T_k)$ is the n-vector whose components are the last n components of the first row in M_k.

(4) If $1-\varepsilon \leq \xi_k \leq 1+\varepsilon$, then STOP.

(5) $u_k = w + \xi_k(z_k - w)$

(6) Compute $\ell \in I(u_k)$ such that $\nabla\varphi_{\ell}(u_k)^T(z_k - u_k) \geq \nabla\varphi_j(u_k)^T(z_k - u_k),\ \forall j \in I(u_k)$

(7) Compute M^* by the revised simplex, introducing in the inverse D_k of M_k the column $(-\nabla\varphi_{\ell}(u_k))^T u_k,\ \nabla\varphi_{\ell}(u_k))^T$ and dropping the j-th column in D_k.

(8) $j(k) = \bar{j}$

(9) If $j(k) = m$, go to (12)

(10) $m = j(k)$

(11) $k = k+1$

(12) $M_k = M^*$

(13) go to 3.

We observe that, whenever the algorithm stops (step 4), the vertex of the actual q-oriented set T_k is optimal in the limit of the precision ε, as a consequence of lemma 2.

<u>Theorem 3</u>. The procedure described above either stops after a finite number of iterations, or it produces an infinite sequence of matrices.

5.1. $M_o, M_1, \ldots, M_1, M_2, \ldots, M_k, \ldots, M_{k+1}, \ldots$

such that there does not exist a subsequence of 5.1 with constant index K.

Proof. We have to prove that for every $k \in N$, after a finite number of cycles of the procedure, one of the following must hold: (i) either the algorithm stops, (ii) or it goes from step (9) to step (10), that is, k is increased to k+1

Assume, by contradiction, that the procedure generates an infinite sequence of matrices:

5.2. $M_o, M_1, \ldots, M_k, M_k, \ldots$

such that for some $k \in N$, k is never increased to k+1 (step 11).

We observe that, in this case, the straight line L_k considered in 5.2 of def.1 is the same for all the basic q-oriented sets T_k corresponding to the matrices M_k.

Denote by $\{M^i\}_{i=1}^{\infty}$ the subsequence $M_k, \ldots, M_k, \ldots$ of 5.2, by y^i the vertex of the basic q-oriented set corresponding to M^i, and by $b^i_{j(k)}$ the j(k)-th column of the basic matrix B^i corresponding to M^i (see def.1 and 2 and the introduction to this section). It is important to observe that y^i belongs to the straight line L_k for every $i \in N$.

Since by the rules of the revised simplex $(y^{i+1}-y^i)^T b^{i+1}_{j(k)} =$ $= \nabla\varphi_\ell(u)^T(u-y^i) < 0$ (step 6) and $b^{i+1}_{j(k)} = \sum_j \eta_j b^i_j$ with $\eta_{j(k)} > 0$ (dropping rule of the revised simplex), we get

$$0 > (y^{i+1}-y^i)^T b^{i+1}_{j(k)} = \sum_j \eta_j (y^{i+1}-y^i)^T b^i_j = \eta_{j(k)} (y^{i+1}-y^i)^T b^i_{j(k)}$$

as $(y^{i+1}-y^i)^T b^i_j = 0$, $j \neq j(k)$.

Setting $p = y^2-y^1$, $y^{i+1}-y^i = \tau_i p$, it follows that $\tau_i > 0$, $\forall i \in N$ and $p^T b^i_{j(k)} < 0$, $\forall i \in N$.

Now we see that $\sum_j \tau_j$ can be neither convergent nor divergent, which is a contradiction.

Assume that $\sum_j \tau_j < \infty$ and set $\vartheta^* = \operatorname{Sup}\{\vartheta \in \mathbb{R}^+ : w + \vartheta p \in S\}$, $u^* = w + \vartheta^* p \in \partial S$.

Step (5) of the procedure constructs a point $u_{i+1} \in \partial S$ while from steps (6) and (7) we have that $b^{i+1}_{j(k)} = \nabla\varphi_\ell(u_{i+1})$ for some $\ell \in I(u_{i+1})$.

We observe that $y^{i+1} - y^1 = \sum_{j=1}^{i} \tau_j p$, which implies:

$$u_{i+1} = w + \frac{\sum_{j=1}^{i-1} \tau_j p + y^1 - w}{\sum_{j=1}^{i-1} \tau_j} \left(\xi_i \sum_{j=1}^{i-1} \tau_j\right)$$

Since $\sum_j^{\infty} \tau_j = +\infty$, we get $\dfrac{\sum_{j+1}^{i-1} \tau_j p + y^1 - w}{\sum_{j=1}^{i-1} \tau_j} \longrightarrow p.$

On t e other hand, S is bounded and convex, which implies $\xi_i \sum_{j=}^{i-} \tau_i \to \vartheta_*$, $u_{i+1} \longrightarrow u^*$.

For very $\ell \in I(u^*)$, we have $\nabla\varphi_\ell(u^*)^T p > 0$, as S is open and conve and w u* = w+ϑ*p.

Thi is a contradiction, since any cluster point of the sequence $\{p^T b^{i+1}_{j(\cdot)}\}$ bel gs to the set $\{p^T \nabla\varphi_\ell(u^*) : \ell \in I(u^*)$ and $p^T b^{i+1}_{j(k)} < 0,\ \forall i \in N$.

Ass now that $\sum_j \tau_j = \tau < +\infty$, set $y^* = y^1 + \tau p = \lim_j y^{i+1}$, $\xi^* =$ $= \ \ \{\xi \in \mathbb{R}^+ : w + \xi(y^* - w) \in S\}$, $u^* = w + \xi^*(y^* - w) \in \partial S$.

if u_{i+1}, is, as before, the point constructed in step (5), we ge $u_{i+1} \to u^*$, $\xi_i \to \xi^* > 0$.

From the simplex rules we have:

$$(y^{i+1})^T b^{i+1}_{j(k)} = (u_{i+1})^T b^{i+1}_{j(k)}$$

and en, for every cluster point b* of the sequence $\{b^i_{j(\ )}\}$, $b^* \in \{\nabla\varphi_\ell(u^*) : \ell \in I(u^*)\}$, we have:

$$0 \ \lim_i (y^{i+1} - u_{i+1})^T b^{i+1}_{j(k)} = \lim_i (y^{i+1} - w - \xi_i (y^i - w))^T b^{i+1}_{j(k)} =$$

$$(y^* - w - \xi^*(y^* - w))^T b^* = \frac{1 - \xi^*}{\xi^*} (u^* - w)^T b^*.$$

This clearly implies that $\xi^* = 1$, since $w \in S$, $u^* \in \partial S$, $b^* \in \{\nabla\varphi_\ell(u^*),\ \ell \in I(u^*)\}$ and S is open and convex.

On the other hand, this is impossible as the test contained in step (4) stops the procedure whenever $1-\varepsilon \leq \xi_i \leq 1+\varepsilon$, $\varepsilon > 0$.
Since neither $\sum_{j=1}^{\infty} \tau_j = +\infty$, nor $\sum_{j=1}^{\infty} \tau_j < +\infty$ and $\tau_j > 0$, the assumption that the procedure generates an infinite sequence of matrices M_k with the same index k is contradictory

Theorem 4.

If the procedure (A) does not stop after a finite number of steps, then the subsequence $\{\bar{M}_k\}$ of the sequence of matrices (3.3.1) defined as follows;

4.1. $\bar{M}_k$ is the first matrix in (3.3.1) of index k, $\forall\, k \in N$, corresponds to a sequence of q-oriented basic sets which satisfy properties $\mathcal{P}$ and $\mathcal{Q}$.

Proof. By construction, property $\mathcal{Q}$ is satisfied and 5.1 of def.5 holds.
We have only to prove 5.2 of def.5 for the sequences of vertices and generators corresponding to the sequence M_k defined by 4.1.

Denote by $\underline{M}_k$ the sequence of the last matrices in (3.3.1) with index k and, for simplicity of notations, by z, y and z* respectively the vertices corresponding to $\bar{M}_k$, $\underline{M}_k$ and $\bar{M}_{k+1}$; it is easy to see that the straight line L_k defined in def.5 passes through z, z* and contains y and every vertex corresponding to matrices of index k in (3.3.1).
Denote by u the point constructed in step (5) of the procedure, $u = w + \xi(y-w)$, and by b the vector $\nabla\varphi_\ell(u)$ introduced in steps (6) and (7).

Set

4.2. $$\alpha = \min_{\substack{x \in \partial S \\ \in I(x)}} \left(\nabla\varphi_\ell(x)^T \frac{x-w}{\| x-w \|} \right)$$

Clearly $\alpha > 0$, since $w \in S$, an open, bounded and convex set.
Obviously $y \in L_k$, $u \in \partial S$ imply:

$$\| y-u \| \geq \mathrm{dist}(L_k, u))$$

and since $y-u = (1-\xi)(y-w) = \frac{1-\xi}{\xi}(u-w)$

we get

$$b^T(y-u) = \frac{b^T(y-u)}{\| y-u \|} \| y-u \| = \frac{b^T(u-w)}{\| u-w \|} \| y-u \| \geq \alpha\, \mathrm{dist}\,(L_k, u)$$

On the other hand,

$$b^T(z-u) = b^T(z-y) + b^T(y-u) \geqslant b^T(z-y) + \alpha \, dist(L_k,u)$$

We observe that $b^T(y-z^*) = b^T(y-u) > 0$ and, as we have seen in the proof of theorem 3:

$$z-y = \rho(y-z^*) \quad \text{for some } \rho \geqslant 0$$

which implies $b^T(z-y) \geqslant 0$ and

4.3. $\quad b^T(z-u) \geqslant dist(L_k,u)$.

Using the notations of def.5, (4.4.3) is precisely (5.5.2) and property holds.

6. Conclusions and Remarks

We have proved that the procedure (A) enables us to construct a sequence of q-oriented basic sets satisfying properties $\mathcal{P}$ and $\mathcal{Q}$. Thus if $S \subseteq R^2$ or $S \subseteq R^3$, the vertices of these sets approximate the optimal solution of problem (*). If $S \subseteq R^n$, a sufficient condition for convergence is that $\inf_k \{\vartheta_k\} = 0$, where ϑ_k is given by 5.2 of def.5.
It is an open qestion if properties $\mathcal{P}$ and $\mathcal{Q}$ are sufficient to ensure convergence for $S \subseteq \mathbb{R}^n$, $n \geqslant 3$.
The answer is probably connected to the possibility of cycling in ordinary linear programming problems.

Three problems arise in the implementation of procedure A in a real algorithm and will be the subject of a future investigation:

6.1. It is necessary to produce a point $w \in S$.

6.2. It is necessary to produce an initial basis.

6.3. It may happen that at some iteration K, the point $u \in \partial S$ computed in step (5) of the procedure is the optimal solution, while the method does not stop. This does not prevent convergence, but in the implementation of the method it will be obviously preferable that whenever some optimal point is computed, the procedure stops, even if the vertex of the actual q-oriented basic set is exterior to the closure S of the feasible set.

1 DANTZIG - Linear Programming and extensions
Princeton Univ. Press, Princeton (1963)

2 WOLFE - Accelerating the cutting plane method for nonlinear programming
SIAM J. Appl. Math. 9 (1961)

3 WOLFE - Some simplex-like nonlinear programming procedures.
Op. Res. 10 (1962)

4 KELLEY - The cutting-Plane Method for solving convex programs.
J. Soc. Indus. and Appl. Math. 8 (1960)

5 TOPKIS - Cutting-Plane methods without nested constraint sets.

6 MC KEOWN - An introduction to Functional Linear Programming.
In Towards Global Optimisation II. Dixon-Szego (1977)

7 GOMULKA-MC KEOWN-TRECCANI - Functional Linear Programming
Hatfield Polytechnic N.O.C. Technical Report 70 (1975)

TOWARDS GLOBAL OPTIMISATION 2
L.C.W. Dixon and G.P. Szegö (eds.)

CONVEX PROGRAMMING VIA FUNCTIONAL LINEAR PROGRAMMING

BY G. RESTA

ABSTRACT

Functional linear programming has been investigated in its theoretical implications [7], [8] and [11], as a development of an "ad hoc method" to solve linear problems with an infinite number of linear constraints, due to J. Mc Keown [5] and [9].

One of the possible numerical strategies for convex programming, based on functional linear programming, is developed and described in [12]. In this paper I present some numerical experiencies with this algorithm and compare them with the performances of other algorithms [4] and [10]. It seems that at least for a subclass of problems, this algorithm is very efficient.

1) GENERALITIES OF FUNCTIONAL LINEAR PROGRAMMING

The revised simplex method [1] in its dual formulation is a well-known and frequently used method for linear programming. Suppose that the solution of L.P.1 is unique and I is a finite set of indices.

LP 1
$$\max c^T x \quad , x \in R^n$$
$$\text{s.t } a_i^T x - \beta_i \leq 0 \quad , i \in I$$

We can observe that, since the solution z^* is unique and the set I is finite, at z^* at least n constraints must be active.

Furthermore, because the constraints are linear, by the Kuhn-Tucker theorem, a set I^* of indices must exist such that:

(1.1) - I^* contains n elements, $I^* = \{i_1, i_2 ... i_n\}$

(1.2) - The matrix B^* having as columns $\{a_{i_j}\}_{j=1...n}$ is non singular

(1.3) - $\exists \lambda_j$ such that $\sum_{j=1}^{n} \lambda_j a_{i_j} = c$ and $\lambda_j > 0 \quad j=1....n$

(1.4) - $a_{i_j}^T z^* - \beta_{i_j} = 0 \quad j = 1...n$

(1.5) - $a_i^T z^* - \beta_i \leq 0 \quad \forall i \in I$

Define c_{B^*} to be a vector having as components $\{\beta_{i_j}\}_{j=1....n}$

Using matricial notations, provided that (1.2) holds we can transform (1.3) to (1.3') and (1.4) to (1.4')

(1.3') if $\ell = B^{*-1}c$, every component of ℓ is positive

(1.4') $B^{*T}z^{*} = c_{B^*}$ or $z^{*} = (B^{*T})^{-1} c_{B^*}$

Let E be the family of all sets of indices I_* satisfying (1.1), (1.2) and (1.3).

A point z_* is related by definition (1.4) to I_*, if $I_* \in E$.

A subset $\Omega_* \subset R^n$ is related to I_* by the following definition, if $I_* \in E$.

$$\Omega_* = \{ x \in R^n \ : \ a_i^T x - \beta_i \leq 0 \ , \ i \in I_* \}$$

We can observe that, by the Kuhn-Tucker theorem, z_* is the unique solution of the problem $LP1_*$.

$$LP1_* \qquad \max c^T x \quad x \in R^n$$
$$\text{s.t. } a_i^T x - \beta_i \leq 0 \qquad i \in I_*$$

In fact the set Ω_* is a piecewise-linear cone centered in z_*, and (1.3) ensures that the line $\Gamma = \{ x = z_* + \lambda c \ , \ \lambda \in R^- \}$ is contained in the set Ω_*.

Let Ω be the set $\{ x \in R^n \ : \ a_i^T x - \beta_i \leq 0, \ i \in I \}$

We can observe that the set Ω is contained in Ω_*, so $c^T z^* \lneq c^T z_*$ if z^* is the solution of LP1 and z_* is the solution of $LP1_*$, and that the equality holds if $I_* = I^*$.

By a duality result LP1 is equivalent to LP2, in the sense that they have the same solution z^*, where

$$LP2 \qquad \min c^T z^*$$
$$\text{s.t. } I_* \in E$$

If I_1 and I_2 belong to E, and respectively Ω_1 and Ω_2 are the subsets related to them, and respectively z_1 and z_2 are the points related to them by (1.4), we say that Ω_2 fits the set Ω better than Ω_1 if $c^T z_2 < c^T z_1$.

We can approach the solution of LP2 by generating a sequence of sets of indices $\{I_m\}_{m \in N}$ such that $\forall m \in N \quad I_m \in E$, and Ω_{m+1} fits the set Ω better than Ω_m does.

This strategy can be geometrically interpreted in the following way: given Ω_m , a convex cone centered in z_m , containing the set Ω , containing the line $\Gamma = \{ x=z_{*m} + \lambda c, \lambda \in R^- \}$, and being a piecewise linear hat, we construct a new Ω_{m+1} by substituting one of the indices in the set I_{+m}.

The strategy of the dual simplex method consists in generating a sequence of sets $\{\Omega_m\}_{m \in N}$ in this way.

Let us now provide a brief sketch of the dual revised simplex technique. Given a set of indices I_{*i} , satisfying (1.1),(1.2), (1.3) and defining z_{*i} as the point defined by (1.4) we compute different $a_k^T z_{*i} - \beta_k = \gamma_k$, $k \in I$, $k \notin I_{*i}$. If all the γ_k are non positive (1.5) is also satisfied, and so z_{*i} is the solution of LP1 if not we select $\gamma_{\underline{k}}$ to be the maximum of the $\gamma_k, k \in I, k \notin I_{*i}$.

The set $I_{*(i+1)}$ will contain $\underline{k}$ and (n - 1) of the elements of I_{*i}, and will be such that it satisfies (1.1), (1.2), (1.3). The above mentioned choice of the index $\underline{k}$ between $I - I_{*i}$ determines the column to be inserted in the simplex.

The choice of the column to be dropped out of the simplex must be made in such a way that (1.1),(1.2) and (1.3) are still satisfied and ensures that $c^T z_{i+1} < c^T z_i$.

In the following we will refer to these choices, respectively as the "inserting" and "dropping" rules.

Unfortunately at some iteration (1.3) may hold in a weaker form such that $c^T z_{i+1} = c^T z_i$. This is termed a "degenerate basis" and for adetailed investigation of it we recommend the reader to [1] .

We observe that this method is still applicable when the solution z is not unique.

Since it is not the intention of this paper to give a general treatment of the revised simplex technique, but to give the reader an introduction to it, we suppose that at every iteration (1.3) is satisfied. This implies also that the solution is unique.

Roughly speaking we can say that the dual revised simplex technique is a procedure which builds a sequence of sets of indices for which (1.1),(1.2) are satisfied, (1.3) is satisfied in a weaker form, and the non-satisfaction of (1.5) allows an improved set to be constructed.

Let us now consider the convex programming problem. Suppose now φ_j , $j \in J_1$ are convex non-linear differentiable functions, and $\varphi_j(x) = a_j^T x - \beta_j$ are linear functions.

Define $S_j = \{x \in R^n : \varphi_j(x) \leq 0\}$, $j \in J_1 \cup J_2$; $S = \bigcap_j S_j$.

Consider the following problem:

CP1

$$\max c^T x , \quad x \in R^n$$
$$\text{s.t. } \varphi_j(x) \leq 0, \quad j \in J_1 \cup J_2$$

Let $\Gamma_j = \{u \in R^n : \varphi_j(u) = 0\}$, $j \in J_1$

Using the fact that a convex set (namely S_j , $j \in J_1$) can be interpreted as the envelope of its tangent hyperplanes, by force of the pseudo-convexity relations, CP1 is equivalent to CP2.

CP2

$$\max c^T x , \quad x \in R^n$$
$$\text{s.t. } \nabla \varphi_j(u)^T (x - u) \leq 0, u \in \Gamma_j, j \in J_1$$
$$a_j^T x - \quad _j \quad , j \in J_2$$

Denoting by $i = (j,u)$, where $u \in \Gamma_j$ and $j \in J_1$;

$$a_i = \nabla \varphi_j(u) \quad ;$$
$$\beta_i = \nabla \varphi_j(u)^T u;$$
$$I = \Big(\bigcup_{j \in J_1} (j \times \Gamma_j)\Big) \cup J_2$$; CP2 can be stated as LP3

LP3

$$\max c^T x , \quad x \in R^n$$
$$\text{s.t. } a_i^T x - \beta_i \leq 0, \quad i \in I$$

The only differences between LP1 and LP3 are that the solution of the first is by assumption unique, and the set I is finite, while LP3 is more general.

The reason why LP3 is more difficult than LP1 is that when applying the "inserting" rule of the revised simplex, it is no longer sufficient at any iteration to compare the values of a finite set of numbers, instead it is necessary to maximize over an infinite set.

If we suppose the functions φ_j, $j \in J_1$ to be quadratic, we have proved that such a maximisation is not necessary.

Let us now define s_j to be a solution of $\nabla\varphi(s_j)=0, \varphi_j(y) < 0$, then the point u_j in which the segment joining y with s_j intersects the set Γ_j has the following property:

1) $\nabla \varphi_j(u_j)^T (y - u_j) = \max_{u \in \Gamma_j} \nabla \varphi_j(u)^T (y - u) \quad , j \in J_1.$

For a proof of this fact we refer the reader to the appendix.

Because of this result the choice of $\underline{k}$, the index to be "inserted", $\underline{k} \in I - I_{*i}$ is not very difficult when all the φ_j, $j \in J_1$ are quadratic.

In fact if when choosing the index $\underline{k}$, we have at our disposal $\{s_j\}_{j \in J_1}$, stationary points of φ_j , $j \in J_1$, we can procede as follows:

Intersect the segments joining z_{*m} with s_k, $k \in J_1$ respectively with the sets Γ_k, $k \in J_1$. This is a numerically very easy as we may use the parabolic interpolation formula to find the point in which $\varphi_k(u) = 0$ in the segment joining z_{*m} with s_k.

Furthermore if the function is quadratic there is a formula for u_k given z_{*m} and $\varphi_k(z_{*m})$, since $\varphi_k(s_K)$ is known and the directional derivative in s_k is 0 , as s_K is a stationery point for φ_k.

So if we denote by $\ell_j = (\dot{J}, u_j)$, $j \in J_1$ and correspondingly

$$a_{\ell_j} = \nabla \varphi_j(u_j)$$

$\beta_{\ell_j} = \nabla\varphi_j(u_j)^T u_j$; in order to choose the index to be inserted it is not necessary to maximise over I , but sufficient to calculate $a_{\ell_j}^T z_{*m} - \beta_{\ell_j}$, $j \in J_1$, with $a_j^T z_{*m} - \beta_j, j \in J$ and to compare the values of this finite set of numbers.

This is an extension of the dual revised simplex technique to particular cases in which infinite linear constraints are present.

For a more detailed discussion of the relationships between funct. linear programming, generalized linear programming and linear programming we refer the reader to paper [11] in which functional linear programming is examined in more detail both in the primal (infinite variable subject to linear constraints), and in the dual (finite number of variables subject to an infinite set of linear constraints) modes.

In this paper we will present an algorithm which is one possible application of functional linear programming in its dual formulation. For the relationship between LP3 (primal formulation) and LP2 (dual formulation) we refer the reader to papers [7] and [8]. In fact if an infinite number of linear constraints is present, LP3 is nolonger equivalent to LP2.

If we suppose that the solution of LP3 is unique the following weaker result still holds:

The solution of LP3 can be reached by a sequence of sets of indices $\{I_{*m}\}_{m \in N}$ such that $I_{*m} \in E$ $m \in N$ and $c^T z_{*m} > c^T z_{*(m+1)}$, and that $\lim_m d(z_{*m}, \Omega) = 0$.

Our method will reach the solution of LP3 by constructing a sequence of $\{I_{*m}\}_{m \in N}$ satisfying the above conditions.

The general idea is the use of the sequence of hats (piecewise linear convex cones) centered at z_{*m}, containing $\Gamma_m = \{x = z_{*m} + \lambda c,\ \lambda \leq 0\}$, and tending to fit the set Ω.

The numerical technique is the same as the one of the dual revised simplex method except that the "inserting" rule is changed.

The resulting method is radically different from the revised simplex method, because even on linear problems they do not coincide and in some respects which will be examined later, it seems an improvement even on linear problems.

We have transformed CP1 to CP2 by introducing many linear constraints. We observe that some of them are useless.
In fact to build the intersection of the sets $\Omega_j = \{x \in R^n : \varphi_j(x) \leq 0\}$, where $_j$ is a convex function we have replaced the level se by one defined by the envelope of Ω_j which consists of an infinite number of linear constraints.
CP2 is than the result of combining all the resulting conditions.

Some of the constraints, however, are useless, in the sense that the set Ω is well described by a subset of these constraints and also the solution of CP1 can be approached by sets of indices belonging to this last subset.

Consider in fact $\quad \varphi(x) = \max\limits_{j \in J_1 \cup J_2} \varphi_j(x)$

Let

$$\Gamma = \{x \in R^n : \varphi(x) = 0\};$$
$$L_j = \Gamma_j \cap \Gamma;$$
$$J_3 = \{j \in J_1 : L_j \neq \emptyset\}$$
$$J_4 = \{j \in J_2 : L_j \neq \emptyset\}.$$

Because a convex set Ω is the envelope of its tangent hiperplanes, problem CP1 is equivalent to the following CP3

CP3
$$\max \; c^T x \quad , x \in R^n$$
$$\text{s.t.} \; \nabla\varphi_j(u)^T (x-u) \leq 0 \;, \; u \in L_j \;, j \in J_3$$
$$a_j^T x - \beta_j \leq 0 \;, \; j \in J_4$$

Denoting by $\quad i = (j,u)$, where $u \in L_j$ and $j \in J_1$;

$$a_i = \nabla\varphi_j(u) \;;$$
$$\beta_i = \nabla\varphi_j(u)^T u \;;$$

$I = (\bigcup\limits_{j \in J_3} (j \times L_j)) \cup J_4$; CP3 can be stated as LP3

We can now state the "inserting" rulefor our method.
Suppose we are at iteration m and we have z_{*m}, suppose w is a point internal to the set .We find a point u by locating the intersection of the segment joining z_{*m} and w with Γ (numerically obtained by interpolating the function φ between z_{*m} and w, to reach a point in which $\varphi(u) = 0$). At this point we choose an index such that $u \in L_j$ and evaluate $\nabla \varphi_j(u)$, if $j \in J_3$ or we use a_j if $j \in J_4$. This will be the index to be inserted in $I_{*(m+1)}$.
In paper [12] more details of the algorithm are given and it is also proved that under very restrictive assumptions the strategy is convergent.

This technique has the advantage, in respect to the above stated generalisation of the revised simplex technique, of requiring only one gradient evaluation per iteration and a few function evaluations.

The interpolation subroutine is constructed to be very efficient on problems of the form $\psi(\lambda)=\max_{j \in J} \psi_j(\lambda)$, since it does not compute all the $\psi_j(\lambda)$, $j \in J$, at every λ.
The main theoretical advantage of the proposed technique is that it is completely invariant with respect to any kind of convex scaling of the functions involved.

2) PROBLEMS

The more general problem we can solve with this strategy can be stated as follows. If θ_j, φ_j and γ_j are any of the following:

$\theta_j : R^n \to R$ convex differentiable functions, $j \in J_1$
$\theta_j : R^n \to R$ linear differentiable functions, $j \in J_2$

$\varphi_j : R^n \to R$ convex differentiable functions, $j \in J_3$

$\varphi_j : R^n \to R$ linear functions, $j \; J_4$

$\gamma_j : R^n \to R$ linear functions, $j \; J_5$

and $\theta : R^n \to R$, $\theta(x) = \max_{j \in J_1 \cup J_2} \theta_j(x)$, we can define

CP1

$$\min \theta(x) \qquad x \in R^n$$
$$\text{s.t. } \varphi_j(x) \leq 0 \qquad j \in J_3 \cup J_4$$
$$\text{s.t. } \gamma_j(x) = 0 \qquad j \in J_5$$

This problem is equivalent to the following

CPL1

$$\max \; -y \qquad y \in R, \; x \in R^n$$
$$\text{s.t. } \theta_j(x) - y \leq 0 \qquad j \in J_1 \cup J_2$$
$$\varphi_j(x) \leq 0 \qquad j \in J_3 \cup J_3$$
$$\gamma_j(x) = 0 \qquad j \in J_5$$

Now we define $t \; R^{n+1}$ $t = \begin{pmatrix} x \\ y \end{pmatrix}$, $c = \begin{pmatrix} 0 \\ -1 \end{pmatrix}$,

$\bar{\theta}_j(t) = \theta_j(x) - y \qquad j \in J_1 \cup J_2$

$\bar{\varphi}_j(t) = \varphi_j(x) \qquad j \in J_3 \cup J_4$

$\bar{\gamma}_j(t) = \gamma_j(x) \qquad j \in J_5$

We can observe that given a feasible point $\underline{x}$ for CP1, we can obtain a feasible point for CP2 by adding an additional component

With these definitions

CPL1 becomes CPL2

CPL2

$$\max \; c^T t \qquad t \in R^{n+1}$$
$$\text{s.t. } \bar{\theta}_j(t) \leq 0 \qquad j \in J_1 \cup J_2$$
$$\text{s.t. } \bar{\varphi}_j(t) \leq 0 \qquad j \in J_3 \cup J_4$$
$$\text{s.t. } \bar{\gamma}_j(t) = 0 \qquad j \in J_5$$

Thus the class of applicability of this method is very wide.
As far as unconstrained minimisation is concerned $J_3 = J_4 = J_5 = \emptyset$, problems of minimax can be easyly solved by this technique. The algorithm is provided with an automatic safeguard such that if we have a problem with $J_3 = J_4 = \emptyset$, (minmax problem with equality linear constraint)modifies the internal point w at any iteration in order to accelerate the procedure.

As far as constrained optimisation is concerned we can handle problems with a very high number of constraints because at any iteration of the method the gradient of only one of the costraints is required. Unfortunately the method can only be used for convex programming problems and no extension to non-convex problems has been devised.

3) TEST PROBLEMS

1) The objective function is the maximum between eight non-convex functions, the constraints are nine linear constraints, one of which is an equality constraint.

This non-convex problem is a minmax formulation of an economic equilibrium problem in [2].

$$\text{Minimize} \quad \varphi(x) = \max \varphi_i(x) \;, \quad i = 1,\dots 8$$

$$\text{s.t.} \quad \sum_{J=1}^{8} x_j = 1$$

$$x_j \geq 0 \qquad J = 1\dots 8$$

$$\text{Where} \quad \varphi_j(x) = \sum_{\ell=1}^{5} \quad (x) \qquad j=1\dots 8$$

$$\text{and} \quad \varphi_{j,\ell}(x) = (A_{\ell,i} \sum_{k=1}^{8} w_{\ell,k} x_k)(x_i^{b_\ell} \sum_{k=1}^{8} A_{\ell,k} x_k^{(1-b_\ell)}) - W_{\ell,i} \;, \quad \ell = 1\dots 5$$

The five x eight matrices A and W , together with the five-dimensional vector b , are given in [2].

The standard starting point is $x_j = .125$ $J = 1\dots 8$, while the solution is at (.27,.03,.06,.09,.07,.31,.10,.07).

In practice we must add the constraints $x_j \geq .01$, $J = 1\dots 8$, since the computer can not evaluate 0 elevated to a real exponent.

2) The objective function is non-convex and the constraints are 15 linear inequality constraints.

It is the problem of the Shell Development Company in its primal formulation, given as test problem number 3 in [3]

$$\text{minimize} \qquad \varphi(x) = \ell^T x + x^T C x + \sum_{j=1}^{5} d_j x_j^3$$

$$\text{subject to} \qquad Ax \geq b$$

$$x_j \geq 0 \qquad J = 1 \ldots 5$$

Here the five x five matrix C, the ten x five matrix A, the five dimensional vectors e and d , and the ten dimensional vector b are given in [3]. The standard starting point is (0.,0.,0.,0.,1.) and the solution occurs at (.3000,.3335,.4000,.4285,.224).

3) The objective function is a quadratic strictly convex function, and the constraints are sixtyfive linear inequality constraints.

It is a problem related to electric power scheduling [6]

$$\text{Minimize} \qquad \varphi(x) = \sum_{k=0}^{4} 2.3\, x_{3k+1} + .0001 x_{3k+2}^2 + + 2.2 x_{3k+3} + .00015\, x_{3k+3}^2$$

subject to	$x_{3k+1} \geq 0$	$k = 0,\ldots 4$	[2,6]
	$x_{3k+2} \geq 0$	$k = 0,\ldots 4$	[7,11]
	$x_{3k+3} \geq 0$	$k = 0,\ldots 4$	[12,16]
	$90 \geq x_{3k+1}$	$k = 0,\ldots 4$	[17,21]
	$120 \geq x_{3k+2}$	$k = 0,\ldots 4$	[22,26]
	$60 \geq x_{3k+3}$	$k = 0,\ldots 4$	[27,31]
	$x_{3k+1} - x_{3k-2} \geq -7$	$k = 1,\ldots 4$	[32,35]
	$x_{3k+2} - x_{3k-1} \geq -7$	$k = 1,\ldots 4$	[36,39]
	$x_{3k+3} - x_{3k} \geq -7$	$k = 1,\ldots 4$	[40,43]
	$6 \geq x_{3k+1} - x_{3k-2}$	$k = 1,\ldots 4$	[44,77]
	$7 \geq x_{3k+2} - x_{3k-1}$	$k = 1,\ldots 4$	[48,51]
	$6 \geq x_{3k+3} - x_{3k}$	$k = 1,\ldots 4$	[52,55]

$x_1 - 15 \geq -7$ 56

$x_2 - 50 \geq -7$ 57

$x_3 - 10 \geq -7$ 58

$6 \geq x_1 - 15$ 59

$7 \geq x_2 - 50$ 60

$6 \geq x_3 - 10$ 61

$x_1 + x_2 + x_3 \geq 60$ 62

$x_4 + x_5 + x_6 \geq 50$ 63

$x_7 + x_8 + x_9 \geq 70$ 64

$x_{10} + x_{11} + x_{12} \geq 85$ 65

$x_{13} + x_{14} + x_{15} \geq 100$ 66

The starting point is (20., 55., 15., 20., 60., 20., 20., 60., 20., 20.,60., 20., 20., 60., 20.) and the solution occurs at (8.,49., 3., 1., 56., 0., 1., 63., 6., 3., 70., 12., 5., 77., 18.).

This is a representation of the problem of scheduling three generators to meet the demand for power over a period of time. The variable x_{3k+1} denotes the output from the i-th generator at time t(k).

Constraints 2 to 6 are lower limits on the power available from the first generator at t(k) k= 0,...4.

Constraints 7 to 11 are lower limits on the power available from the second generator at t(k) k= 0,...4.

Constraints 12 to 16 are lower limits on the power available from the third generator at t(k) k= 0,...4.

Constraints 17 to 21 are upper limits on the power available from the first generator at t(k) k= 0,...4.

Constraints 22 to 26 are upper limits on the power available from the second generator at t(k) k= 0,...4.

Constraints 27 to 31 are upper limits on the power available from the third generator at t(k) k= 0,...4.

Constraints 32 to 35 are lower bounds on the amount by which the output of the first generator can change from t(k - 1) to t(k) , k= 1,...4.

Constraints 36 to 39 are lower bounds on the amount by which the output of the second generator can change from t(k - 1) to t(k), k= 1,...4.

Constraints 40 to 43 are lower bounds on the amount by which the output of the third generator can change from t(k - 1) to t(k) , k= 1,...4.

Constraints 44 to 47 are upper bounds on the amount by which the output of the first generator can change from t(k - 1) to t(k) , k= 1,...4.

Constraints 48 to 51 are upper limits on the amount by which the output of the second generator can change from t(k - 1) to t(k) , k= 1,...4.

Constraints 52 to 55 are upper limits on the amount by which the output of the third generator can change from t(k - 1) to t(k) , k= 1,...4.

Constraints 56 to 61 are respectively correspondent to the blocks [32,35] , [36,39] , [40,43] , [44,47] , [48,51] , [52,55] and to k = 0 if the outputs from the generators at t(-1) were respectively 15., 50., 10..

Constraints 62 to 66 are the conditions that at each time t(k), k= 0,...4, the power generated must at least satisfy the demand.

4) The objective function is a quadratic non-strictly convex function, two of the constraints are quadratic non-strictly convex and

the other two are linear inequalities.

This is a theoretical problem given in [13].

minimize $\varphi(x) = (x_1 - x_4)^2 + (x_2 - x_5)^2 + (x_3 - x_6)^2$

$$\varphi_1(x) = x_1^2 + x_2^2 + x_3^2 \leq 5$$

$$\varphi_2(x) = (x_4 - 3)^2 + x_5^2 \leq 1$$

$$\varphi_3(x) = x_6 \leq 8$$

$$\varphi_4(x) = x_6 \geq 4$$

4) REMARKS ON TEST PROBLEMS

Test problems 1 and 2 involve non-convex functions. We can observe that in order to apply the functional linear programming algorithm, it is not necessary that all the involved functions are convex, but it is sufficient that the intersection of all constraints (subset of R^{n+1} if the objective function, or functions as in test 1 , is not linear) is a convex set.

Thus for instance the constraints can also be pseudo-convex.

Test problem number 3 is very interesting. I have solved it in the formulation given in [10] to obtain an execution time comparable with the ones referred in [10].

However we can observe that the particular features of the functional linear programming method are such that if the same problem is formulated in another way, the performance of the algorithm is greatly improved. In fact usually in this kind of problem the solution is required at different times (for instance every day) by changing only the constraints related to the initial output from each generator (namely 56 to 61). So it is a good suggestion, in order to have a rather good initial "basis" for F.L.P., to run the problem (only once) without the constraints 56 to 61, and to store the solution of it to be used as an initial basis. In this case the execution time decreases by a factor 1/8.

Unfortunately I have not run OPRQP or any augmented lagrangian method on the same problem from this better starting point, but I believe that if I had done so, the result should not have been as good as for F.L.P., because of their different theoretical features

5) REMARKS ON OTHER STRATEGIES FOR CONSTRAINED OPTIMISATION

We can observe that the latest results in the field of constrained optimisation [10] clearly indicate that Augmented Lagrange Multipliers techniques and Recursive Quadratic Programming are very efficient, especially if the performances are compared with the preexistent methods as Penalty methods [4] .

It seems that there is also a reason to use, in different kinds of problems, algorithms of these two different categories. In fact the class of functionsin which their performances are competitive is very small, while in many sorts of problems, depending on the number of constraints, on the dimension of the problem and on the cheapness of in term of time of a function evaluation, one of the two categories of algorithms is better than the other. Another great advantage of these two classes of techniques is that they are multi-purpose algorithms which can easily handle both convex problems, and more general problems .

So why this interest in Functional linear programming, which can be used only for convex problems ? The first reason is that it appears to solve problems with non-differentiable objective functions, as is the case of test problem number 1 . The second reason is that its theoretical features are really very different from each of the other algorithms, especially in three respects

- At any iteration the gradient of only 1 function is required, i.e. the evaluation of the gradient of the objective function, or of the gradient of one constraint.
- The vector of lagrange multipliers is reset at each iteration by a numerically very efficient strategy (by the inserting rule

of the revised simplex).

- It is completely independent of any kind of convex scaling of the objective function and of each constraint.

It is possible that for these reasons, at least in a restricted class of problems, Functional Linear Programming method has to be preferred both to Augmented Lagrangian techniques and to Recursive Quadratic Programming algorithms.

6) NUMERICAL RESULTS

In the following tables the performances of six different algorithms are presented.

The algorithm 1 is the Functional linear Programming method.
The algorithm 2 is the OPRQP method (a strategy based on recursive quadratic programming implemented at the Numerical Optimisation Centre of the Hatfield Polytechnic) [10]
The algorithm 3 is an augmented lagrangian method.
The algorithm 4 is the SUMT method (a method implemented at N.O.C. based on barrier functions). [4]
The algorithm 5 is the exterior SUMT method (a method based on penalty functions implemented at N.O.C.) [4]
The algorithm 6 is Murray's method [4]

The results for methods different from the first are taken from [10] as far as test problem 3 is concerned, and from [13] as far as test problem number 4 is concerned, and from [4] for test 2.

The execution time is the time of execution of a single precision fortran routine solving the problems with the prescribed accuracy of 10^{-5} on a DEC SYSTEM 10.

In table number one the execution time are given. In tables number two the number of elementary function evaluations is given. In table number two every evaluation of a single constraint or of the objective function increases by one of the number of function eva-valuation which every gradient evaluation of a single constraint or of the objective function increases by two times the dimension of the problem the number of function evaluations.

TABLE NUMBER 1 (EXECUTION TIME)

NUMBER OF THE ALGORITHM / TEST FUNCTION NUMBER	1	2	3	4	5	6
1	30.3					
2	0.92	1.9		14.1	15.6	3.1
3	15.7	40.7	43.7			
4	1.39	1.72				

TABLE NUMBER 2 (NUMBER OF FUNCTION EVALUATIONS)

NUMBER OF THE ALGORITHM / TEST FUNCTION	1	2	3	4	5	6
1	921					
2	513	2,640		4,620	36,300	76,890
3	1,436	58,695	188,240			
4	1,093	1,055				

The fact that some times and function evaluation numbers are not given is related to the fact that I did not run those algorithms on those problems.

Table number one is the more significative and seems to indicate that when F.L.P. algorithm is applicable, (for instance when the functions involved are convex) it seems competitive with the known algorithms and in these particular cases even better than them. It seems also the the algorithm works very well in high-dimensional problems.

Table number two is not related to table number one because table number one indicates the execution times when analitic derivatives were given.

Perhaps it penalizes too many methods different from F.L.P. because usually the evaluation of the gradient of a linear constraint is not expensive at all in terms of execution time, but anyway clearly indicates that if a large dimensional , with an high number of constraints, problem is run with numerical derivatives the time elapsed in function and gradient evaluations for F.L.P. is much less than the time elapsed by other algorithms.

7) APPENDIX

If φ is a quadratic nonlinear convex function, if s is a point such that $\nabla\varphi(s) = 0$, if y is a point, if μ is a scalar such that $\varphi(s)<\mu<\varphi(y)$ and if u is such that $\varphi(u) = \mu$ and $(y - s) = \lambda(u - s)$, then

$$\nabla\varphi(u)^T(y - u) \geq \nabla\varphi(x)^T(y - x) \ , \ \forall x : \varphi(x) = \mu$$

Proof

Let A be the Hessian matrix of the function

Suppose without loss of generality $\varphi(s) = 0$

In this case Taylor expansion from point imlies that

$$\varphi(x) = \frac{1}{2}(x - s)^T A (x - s) \quad , \forall x \in R^n$$

$$\nabla\varphi(x) = A(x - s) \quad , \forall x \in R^n$$

We will now prove that $\varphi(x) = \mu = \varphi(u) \Rightarrow \nabla\varphi(u)^T(y - u) - \nabla\varphi(x)^T(y-x) \geq 0$.

In fact

$$\begin{aligned}
&\nabla\varphi(u)^T(y-u)-\nabla\varphi(x)^T(y-x) = \\
&\quad = \nabla\varphi(u)^T(y-s)-\nabla\varphi(u)^T(u-s)-\nabla\varphi(x)^T(y-s)+\nabla\varphi(x)^T(x-s) = \\
&\quad = \nabla\varphi(u)^T(y-s)-\nabla\varphi(x)^T(y-s)-(u-s)^TA(u-s)+(x-s)^TA(x-s) = \\
&\quad = \nabla\varphi(u)^T(y-s)-\nabla\varphi(x)^T(y-s)-2\mu+2\mu = \\
&\quad = \nabla\varphi(u)^T(y-s)-\nabla\varphi(x)^T(y-s) = \\
&\quad = [\nabla\varphi(u)^T(u-s)-\nabla\varphi(x)^T(u-s)]\lambda = \\
&\quad = \lambda\,[2\mu-\nabla\varphi(x)^T(u-s)] \geq 0
\end{aligned}$$

We have transformed line 2 into line 3 because $\mu = \varphi(u) = \frac{1}{2}(u-s)^TA(u-s) = \varphi(x) = \frac{1}{2}(x-s)^TA(x-s)$.

We have transformed line 4 into line 5 because by assumption $(y-s) = \lambda(u-s)$.

Line 6 is greater than or equal to 0 because $2\mu = \nabla\varphi(x)^T(x-s) \geq \nabla\varphi(x)^T(u-s)$, being $\nabla\varphi(x)^T(x-u) \geq 0$ because $\varphi(u) \leq \varphi(x)$.

Q.E.D.

8) AKNOWLEDGMENTS

All the work leading to the implementation of this convex programming code has been undertaken on the CII 10070 computer of the Centro di Calcolo della Università di Genova, under the theoretical supervision of Prof. G. Treccani, being supported by the Gruppo Nazionale per l'Informatica Matematica del Consiglio Nazionale delle Ricerche.

Many suggestions were given to the author by Dr. J.J.Mc Keown,while he was a visiting professor at the Università di Genova.

Suggestions on when it is better to avoid or to admit a degenerate basis were given to the author by D.E.Sideri.

The numerical implementation of test problems two and four was performed at the Numerical Optimisation Centre of the Hatfield Polytechnic while the author visited the N.O.C. in the framework of a progetto sponsored by the Consiglio Nazionale delle Ricerche.

REFERENCES

1. Danzig G.B. (1963). Linear programming and extension.1st edition. Princeton University Press, Princeton, New Jersey, U.S.A.
2. Scarf (1967). The approximation of fixed point of a continuous mapping, In SIAM J. Appl. Math. 15 (1967) pp 1328 - 1343.
3. Colville A.R. (1968). A comparative study on nonlinear programming codes, IBM N.Y. Sci. Centre Rept. 320-2949 June 1968, p 31.
4. Biggs M.C. (1972). A new method of constrained minimization using recursive equality quadratic programming, Numerical Optimisation Centre Technical Report No. 24, The Hatfield Polytechnic. In Methods for Nonlinear Optimisation, Ed. Lootsma, Academic Press.
5. McKeown J.J. (1975). A quasi linear programming algorithm for optimising fibre reinforced structures of fixed stiffness. Comp. Methods in App. Science & Eng. 6 123-154.
6. Biggs M.C. (1976). An approach to optimal scheduling of an electric power system. In Optimisation in Action, Ed. Dixon, Academic Press.
7. Gomulka J. (1975). The reduction lemma and duality theorems for Functional Linear Programming (private communication).
8. Gomulka J., McKeown J.J. & Treccani G. (1975). Functional Linear Programming, Numerical Optimisation Centre Technical Report No. 70, The Hatfield Polytechnic.
9. McKeown J.J. (1976). A deflection-variable technique for the optimisation of fibre reinforced composite structures. Ph.D. Thesis, Imperial College, Dept. of Aerostructures.
10. Biggs M.C. (1976). A numerical comparison between two approaches to the nonlinear programming problem. Numerical Optimisation Centre Technical Report No. 77, The Hatfield Polytechnic.
11. McKeown J.J. (1977). An introduction to functional linear programming. In Towards Global Optimisation 2, Eds. L.C.W. Dixon & G.P. Szegö, North Holland Publishers.
12. Resta G., Sideri E. & Treccani G. (1977). Functional Linear Programming. In Towards Global Optimisation 2, Eds. L.C.W. Dixon & G.P. Szegö, North Holland Publishers.
13. Bertocchi M., Cavalli E. & Spedicato E. (1977). Algoritmi a lagrangiane per la minimizzazione di funzioni nonlineari con vincoli nonlineari. Quaderni I.A.C.

AUTHOR INDEX